POWER QUALITY

Edited by **Andreas Eberhard**

INTECHOPEN.COM

Power Quality
Edited by Andreas Eberhard

CBS, Edition 2016

Published by InTech
Janeza Trdine 9, 51000 Rijeka, Croatia

Publishing Process Manager Ana Nikolic

Technical Editor Teodora Smiljanic

Cover Designer InTech Design Team

Additional hard copies can be obtained from orders@intechopen.com

Power Quality, Edited by Andreas Eberhard
p. cm.
ISBN 978-953-307-180-0

INTECH

open science | open minds

Contents

Preface

Electrical power is becoming one of the most dominant factors in our society. Power generation, transmission, distribution and usage are undergoing significant changes that will affect the electrical quality and performance needs of our 21st century industry. One major aspect of electrical power is its quality and stability – or so called Power Quality.

The view on Power Quality did change over the past few years. It seems that Power Quality is becoming a more important term in the academic world dealing with electrical power, and it is becoming more visible in all areas of commerce and industry, because of the ever increasing industry automation using sensitive electrical equipment on one hand and due to the dramatic change of our global electrical infrastructure on the other.

For the past century, grid stability was maintained with a limited amount of major generators that have a large amount of rotational inertia. And the rate of change of phase angle is slow. Unfortunately, this does not work anymore with renewable energy sources adding their share to the grid like wind turbines or PV modules. Although the basic idea to use renewable energies is great and will be our path into the next century, it comes with a curse for the power grid as power flow stability will suffer.

It is not only the source side that is about to change. We have also seen significant changes on the load side as well. Industry is using machines and electrical products such as AC drives or PLCs that are sensitive to the slightest change of power quality, and we at home use more and more electrical products with switching power supplies or starting to plug in our electric cars to charge batteries. In addition, many of us have begun installing our own distributed generation systems on our rooftops using the latest solar panels. So we did look for a way to address this severe impact on our distribution network. To match supply and demand, we are about to create a new, intelligent and self-healing electric power infrastructure. The Smart Grid. The basic idea is to maintain the necessary balance between generators and loads on a grid. In other words, to make sure we have a good grid balance at all times. But the key question that you should ask yourself is: Does it also improve Power Quality? Probably not!

Further on, the way how Power Quality is measured is going to be changed. Traditionally, each country had its own Power Quality standards and defined its own power quality instrument requirements. But more and more international harmonization efforts can be seen. Such as IEC 61000-4-30, which is an excellent standard that ensures that all compliant power quality instruments, regardless of manufacturer, will produce the same results when connected to the same signal. This helps reduce the cost and size of measurement instruments so that they can also be used in volume applications and even directly embedded into sensitive loads. But work still has to be done. We still use Power Quality standards that have been written decades ago and don't match today's technology any more, such as flicker standards that use parameters that have been defined by the behavior of 60-watt incandescent light bulbs, which are becoming extinct.

Almost all experts are in agreement - although we will see an improvement in metering and control of the power flow, Power Quality will suffer. This book will give an overview of how power quality might impact our lives today and tomorrow, introduce new ways to monitor power quality and inform us about interesting possibilities to mitigate power quality problems.

Regardless of any enhancements of the power grid, "Power Quality is just compatibility" like my good old friend and teacher Alex McEachern used to say.
Power Quality will always remain an economic compromise between supply and load. The power available on the grid must be sufficiently clean for the loads to operate correctly, and the loads must be sufficiently strong to tolerate normal disturbances on the grid.

Andreas Eberhard
Power Standards Lab,
USA

Part 1

Power Quality Today and Tomorrow

1

Consequences of Poor Power Quality – An Overview

Sharmistha Bhattacharyya and Sjef Cobben
Technical University of Eindhoven
The Netherlands

1. Introduction

Modern customers use large number of sensitive devices comprising of power electronics that are quite sensitive to power quality (PQ) disturbances in the supply network. From worldwide customer surveys, it is found that complaints on PQ related disturbances (for example: harmonics, voltage dips, flicker, etc.) are increasing every year. In Europe, the quality of electricity that is provided by a grid operator has to comply with reference parameters set in the European standard EN 50160 and other specific standards or the national grid codes. In contrast, it was observed that the customer's polluting loads often interact adversely with the network components and distort the network's voltage. When the supply voltage is distorted, the customer's device draws non-sinusoidal current from the network that might be different than the sinusoidal voltage condition. This can cause many technical problems (such as extra heating, misoperation, early aging of the devices etc.) to the customer's devices at his installation. The non-sinusoidal current also causes extra losses and other problems to various network components (as example: cables and transformers). Moreover, poor PQ often has large financial consequences to the affected customers (mainly to the industries with process plants). In extreme cases, poor PQ of the electric supply can cause financial losses to the network operators and the equipment manufacturers too. All these factors led to the discussion about the responsibility sharing of PQ problems in the network. In this chapter the impacts of poor PQ will be analyzed from the perspectives of the customers, the network operators and the equipment manufacturers.

2. PQ related complaints in different countries

Every year the network operators in different countries around the world receive many complaints about PQ problems from different groups of customers. A customer complains when the operation of devices at his installation is interrupted leading to techno-economic inconveniences. It is observed that almost 70% of the PQ disturbances are originated at the customer's premises while 30% are in the network side [Emanuel & McNeil, 1997]. The Electric Power Research Institute (EPRI) conducted a five year (1990-1995) monitoring program for distribution power quality (DPQ-I) among 24 utilities throughout the United States of America. Another program DPQ-II was conducted in 2001-2002. These study results [Melhorn et al., 2005], [McNulty et al., 2002] concluded that voltage sags (dips) and swells, transient over-voltages (due to capacitor switching), harmonics and grounding related problems are the most common PQ complaints among the American customer as presented in Fig. 1.

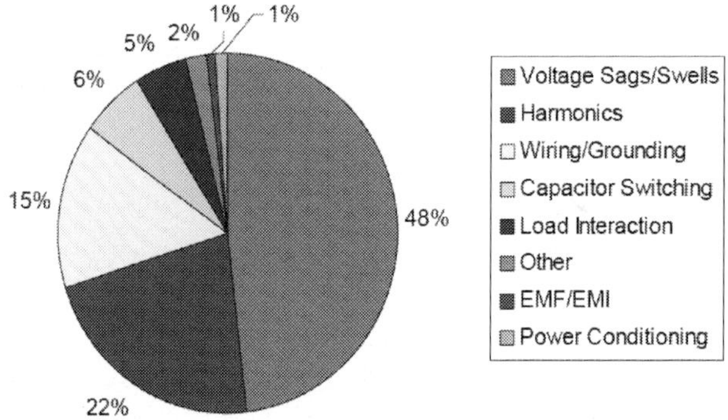

Fig. 1. PQ problems experienced by the American customers

In 2001, the European Copper Institute has done a PQ survey covering 1,400 sites in 8 countries of Europe. It is found that harmonic distortions, power supply reliability, voltage dips and electromagnetic compatibility are the most important issues for the countries of the European Union (EU) [Keulenaer, 2003]. Another PQ campaign was conducted by the Leonardo Power Quality Initiative (LPQI) among various customers in the EU-25 countries in 2004. It was concluded that on average the absolute share of impacts of power quality and reliability related problems are due to voltage dips (23.6%), short interruptions (18.8%), long interruptions (12.5%), harmonics (5.4%), transients and surges (29%) and other PQ related problems (10.7%) [Manson & Targosz, 2008]. In the United Kingdom, the customers mainly complain because of the disputed accounts and the supply standard related to the restoration time after fault interruptions. Some complaints are also about the supply quality issues such as voltage dips, harmonics and flicker [Wharmby, 1998]. In South Africa, voltage dips and transients have been identified as major PQ problem. This is because of the fact that a large part of the electricity infrastructure consists of overhead lines [Johnson & Coney, 1997].

3. Technical impacts of poor PQ

Now-a-days the customers use large number of devices at their installations that consist of power electronics. The residential customers use different domestic appliances such as televisions (TV), video cassette recorders (VCR), microwave ovens, personal computers (PC), heating-ventilation-air conditioning equipments (HVAC), dishwashers, dryers etc. The business and office equipments include workstations, PCs, copiers, printers, lighting etc. On the other hand, the industrial customers use programmable logic controllers (PLC), automation and data processors, variable speed drives (VSD), soft starters, inverters, computerized numerical control (CNC) tools and so on. Presently, many customers use compact fluorescent lamps (CFL) for lighting their installations. Many of these devices are quite sensitive to PQ disturbances. Case studies and surveys in different countries around the world have been done to estimate the impacts of poor PQ to the customers. However, until now, only few cases are surveyed to analyze the technical and non-technical inconveniences of poor PQ to the network operators. Nevertheless, a theoretical estimation of technical losses on different network components because of various PQ disturbances can be done to get an indication of possible impacts of poor PQ in the network.

3.1 For customers

From various surveys, it was generally noticed that industries are vulnerable to long and short interruptions (that are considered as 'reliability issues' in the power system analysis). Voltage dip is the main PQ problem for the semiconductor and continuous manufacturing industries, and also to the hotels and telecom sectors. Harmonic problems are perceived mainly by the commercial organizations and service sectors such as banks, retail, telecom etc. Another PQ problem that draws high attention is the presence of transients and surges at the customer's installation. In 2001, the Leonardo Power Quality Initiative (LPQI) surveyed in eight countries of the European Union (EU) [Keulenaer, 2003] and declared that the customers report a complaint to the network operators when they suffer one of the inconveniences as shown in Table 1 at their sites due to poor PQ of the electric supply.

Perceived inconvenience	*Affected devices*	*Reported PQ problem*
Computer lock-ups and data loss	IT equipments (that are sensitive to change in voltage signal)	Presence of earth leakage current causing small voltage drops in earth conductors
Loss of synchronization in processing equipment	Sensitive measurements of process control equipment	Severe harmonic distortion creating additional zero-crossings within a cycle of the sine wave.
Computer, electronics equipments damage	Electronic devices like computer, DVD player etc.	Lightning or a switching surge
Lights flicker, blink or dimming	Flickering, blinking or dimming of lighting devices, and other visual screens	Fast voltage changes leading to visible light flicker
Malfunctioning of motors and process devices. Extra heating, decreased operational efficiency and premature aging of the equipments	Motors and process devices	Presence of voltage and current harmonics in the power supply
Nuisance tripping of protective devices	Relays, circuit breakers and contactors	Distorted voltage waveform because of voltage dip
Noise interference to telecommunication lines	Telecommunication system	Electrical noise causing interference signals

Table 1. Customer's reported complaints in EU-8 as per LPQI survey

In 2008, another report was published by the LPQI in which PQ survey was conducted among the customers of the EU-25 countries [Manson & Targosz, 2008]. It was reported that loss of synchronization of processing equipment is an acute problem in the industries (mainly for the continuous manufacturing process plants). Lock ups of computers and switching equipment tripping are the second largest problem for industries. For the service and transport sectors, circuit breaker tripping and data loss have been identified as the main problems caused by poor PQ. It was noticed that main sources of PQ disturbances in the industries are the motor driven system and static converters. In contrast, PQ problems in the service sectors are mainly originated from various electronic equipments. Fig. 2 illustrates the LPQI survey results that indicate the frequency of different PQ consequences to the industries and the service and transport sectors as a percentage of cases analyzed.

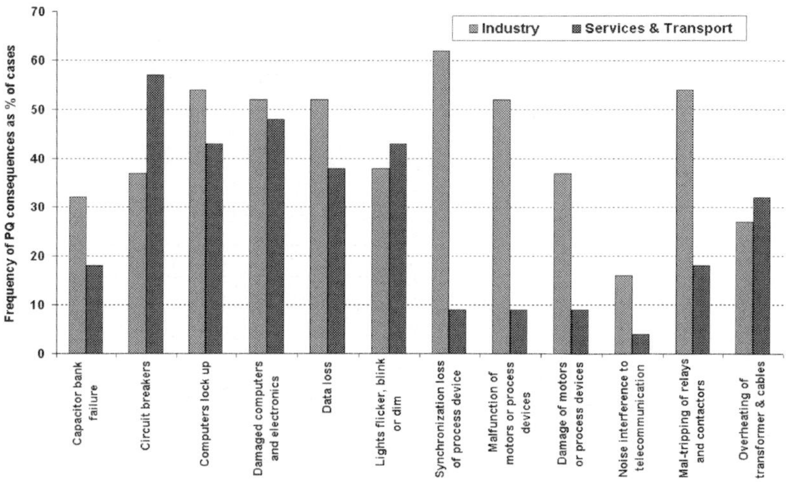

Fig. 2. Consequences of poor PQ as experienced by the customers

Fig. 3 shows the survey results [Manson & Targosz, 2008] of the devices that mostly get affected by one of the PQ problems in different installations in the EU-25 countries. It shows that electronic equipments are the most vulnerable to PQ disturbances both in the industries as well as in the service and transport sectors.

In 2000, the EPRI and CEIDS consortium conducted a PQ survey [Lineweber & McNulty, 2001] among the industrial customers in the USA. It was declared that the most affected devices in the industries because of poor PQ are computers and micro processor based devices (43%), variable speed drives (13%), lighting equipments (8%), motors (5%), relays (1%) and other equipments (30%). The 4th benchmarking report [CEER, 2008] of the European Regulators also gave an overview of PQ related costs in different countries of the world.

3.2 For network operators

In the last decades, customers across the globe have become more aware of PQ related disturbances at their installations. Due to large amount of PQ emissions also from the customers' sides, it is difficult for the network operator to maintain high voltage quality at a customer's point of connection. Moreover, a guaranteed high quality of voltage supply requires large investments in the network. Among various PQ problems, mainly harmonics in the network often interact adversely with the network components and cause inconveniences to the network operators. The operations of power electronic devices produce harmonic currents that lead to additional harmonic power flow and increase network's total apparent power demand while decrease true power factor of the network. Large harmonic current can also cause overloading and extra power losses in the network components. In extreme cases, it can lead to high thermal stresses and early ageing of the network devices. Imposing penalties to the harmonic producing customers is not presently feasible because of the lack of proper measuring devices. Harmonic currents when combined with high grid impedance increases voltage distortions in the network and in extreme situation can shift zero-crossing points of the supply voltage waveform. This increases noise and electromagnetic interference in the network. Transformers, cables and power-factor correction (PFC) capacitors are the network components that mainly get affected by PQ disturbances and are discussed briefly in the following sub-sections.

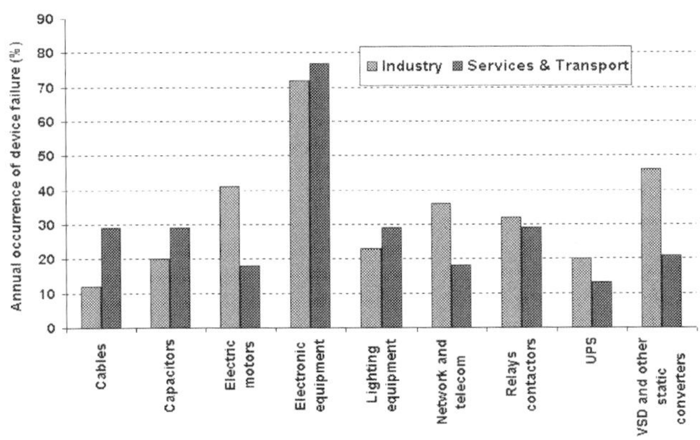

Fig. 3. Equipments affected by PQ problems in different sectors

3.2.1 Effects on transformers

Presence of harmonic current increases the core losses, copper losses, and stray-flux losses in a transformer. These losses consist of 'no load losses' and 'load losses'. No load loss is affected mainly by voltage harmonics, although the increase of this loss with harmonics is small. It consists of two components: hysteresis loss (due to non-linearity of the transformers) and eddy current loss (varies in proportion to the square of frequency).

The load losses of a transformer vary with the square of load current and increase sharply at high harmonic frequencies. They consist of three components:

- Resistive losses in the winding conductors and leads
- Eddy current losses in the winding conductors
- Eddy current losses in the tanks and structural steelwork

Eddy current losses are of large concern when harmonic current is present in the network. These losses increase approximately with the square of frequency. Total eddy current losses are normally about 10% of the losses at full load. Equation (1) gives total load losses (P_T) of a transformer when harmonics are present in the network [Hulshorst & Groeman, 2002].

$$P_T = P_{CU} \times \left(\frac{I_L}{I_1}\right)^2 + P_{WE1} \times \left(\sum_1^n \left(\frac{I_n}{I_1}\right)^2 \cdot n^2\right) + (P_{CE1} + P_{SE1}) \times \left(\sum_1^n \left(\frac{I_n}{I_1}\right)^2 \cdot n^{0.8}\right) \tag{1}$$

Where,

P_{CU} = total copper loss
P_{WE} = eddy current losses at 50Hz (full load)
P_{CE1} = additional eddy current losses at 50Hz (full load)
P_{SE1} = stray losses in construction parts at 50Hz (full load)
I_n = rms current (per unit) at harmonic 'n'
I_L = total rms value of the load current (per unit)
I_1 = fundamental component of load current (per unit) at 50Hz frequency
n = harmonic number

Other concern is the presence of 'triple-n' harmonics. In a network, mainly the LV nonlinear loads produce harmonics. With a MV/LV transformer of Δ/Y configuration, 'triple-n' currents circulate in the closed delta winding. Only the 'non triple-n' harmonics pass to the upstream network. When supplying non-linear loads, transformers are vulnerable to overheating. To minimize the risk of premature failure of transformers, they can either be de-rated or use as 'K-rated' transformer which are designed to operate with low losses at harmonic frequencies. Increased loading can cause overstressing of transformer and the chance of its premature failure. This effect is usually expressed in terms of 'loss of lifetime'. The hot-spot temperature is used for evaluation of a relative value for the rate of thermal ageing as shown in Fig. 4. It is taken as unity for a hot-spot temperature of 98°C with the assumption of an ambient temperature of 20°C and hot-spot temperature rise of 78°C. Equation (2) shows the calculation of relative ageing rate (V) as a function of hot-spot temperature θ_h [Najdenkoski et al., 2007].

$$V = \frac{\text{Ageing rate at } \theta_h °C}{\text{Ageing rate at } 98°C} = 2^{(\theta_h - 98)/6} \tag{2}$$

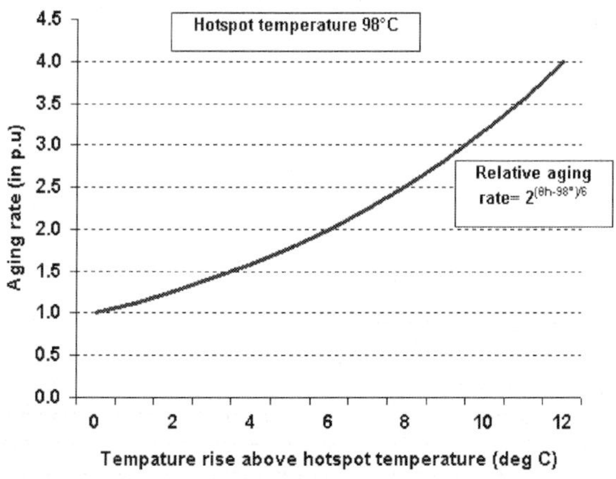

Fig. 4. Aging of a transformer with increased temperature above hotspot

3.2.2 Effects on cables

Harmonic currents have two main effects on cables:

- Additional 'ohmic losses' (I^2R losses) in the line and neutral conductors of a cable because of increased rms value of current due to harmonics. This causes increased operating temperatures in a cable.
- Harmonic currents along with the grid impedances cause harmonic voltages across various parts of the network. This harmonic voltage increases the dielectric stresses on the cables and can shorten their useful lifetime.

Resistance of a cable is determined by its DC value plus skin and proximity effect. The eddy current, which is generated due to the relative motion of the electromagnetic field and circulating current in a conductor, is the root cause of skin effect. The current tends to flow on the outer surface of a conductor. It increases the effective resistance of the conductor and eddy current losses, mainly at high frequencies. The 'proximity effect' is because of the mutual inductances of parallel conductors and the changing magnetic field of the nearby conductors. Both the skin effect and the proximity effect are dependent on the power system frequency, conductor size, the resistivity and the permeability of the material. The presence of harmonics in the cables influences conductor's resistance and further increases its operating temperature. This can eventually cause early aging of the cables. Heat generated in a cable consisting of 'm' conductors and harmonic component 'n' is given by equation (3) [Fuchs et al., 1986].

$$Q(m) = \sum m \cdot I_n^{\,2} \cdot r_{ac}(n) \tag{3}$$

Where,

$Q(m)$	= heat generated in a cable per unit length
m	= number of conductors in the cable
$r_{ac}(n)$	= conductor resistance for n^{th} harmonic per unit length
I_n	= effective or rms value of nth harmonic current

Thermal degradation of an electric device is mainly caused by temperature rise beyond the rated value. When the operating temperature deviates from the rated temperature, the life expectancy of a cable is changed and can be calculated by equation (4) [Fuchs et al., 1986].

$$\rho = \rho_{rat} \cdot e^{-\left(\frac{E}{K}\right)\cdot\frac{\Delta\theta}{\theta_{rat}(\theta_{rat}+\Delta\theta)}} \tag{4}$$

Where,

ρ	= lifetime referred to $\theta = \theta_{rat} + \Delta\theta$
ρ_{rat}	= lifetime referred to $\theta = \theta_{rat}$
$\Delta\theta$	= temperature rise in relation to θ_{rat} in Celsius
θ_{rat}	= cable rated temperature in Kelvin
K	= Boltzmann constant
E	= material's activation energy

Laboratory measurement is done to identify cable's temperature rise for different loadings conditions with linear and nonlinear loads respectively. In the test, a 4x2.5 mm^2 XVB-F2 type cable with copper conductor, PVC insulation and extruded polyethylene shield is used. This is a common type installation cable in Belgium. Fixed values of set-up currents (5A, 10A, 15A and 20A) are applied to the liner as well as the nonlinear loads. Incandescent lamps are used to represent linear load and PCs for nonlinear loads (that have total harmonic current distortion of 40% approximately). Fig. 5 (a) shows a comparison of the cable's Cu conductor temperature rise with linear and nonlinear loads for different set up currents [Desmet et al., 2004]. The ambient temperature was recorded 21°C at the test site. It was found that with increasing set up currents, the neutral conductor's current (because of 'triple n' harmonics) increases sharply for nonlinear loads as shown in Fig. 5 (b). This increases total heat content of the cable and raises its temperature. In the test, it was noticed that the temperature of Cu conductor of the cable exceeded its permissible temperature limit for high nonlinear loads (while set up currents are always maintained below the limit of maximum cable current capacity). This can cause life time reduction of the cable as described in equation (4).

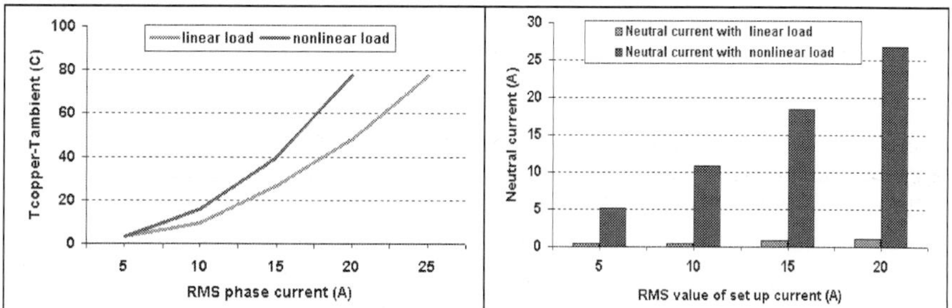

(a) Relative temperature rise of a cable's conductor with linear and nonlinear loads

(b) Increase of neutral currents with increasing fundamental current loadings

Fig. 5. Accelerated temperature rise in a cable with nonlinear loading

3.2.3 Effects on PFC capacitors

Power-factor correction (PFC) capacitors are provided to draw currents with a leading phase angle to offset lagging currents drawn by the inductive loads such as an induction motor. In the presence of a non-linear load, the impedance of a PFC capacitor reduces as the frequency increases, while the source impedance is generally inductive which increases with the frequency. The presence of voltage harmonics in the power system increases the dielectric losses in the capacitors at high operating temperature and reduces the reliability. In extreme situation, harmonics in the network can cause reduction of operational life time of a PFC capacitor. The dielectric loss in a capacitor is calculated by equation (5) [Fuchs et al., 1986].

$$P_{loss} = \sum_{n=1}^{n=40} C(\tan\delta)\cdot \omega_n . V_n^2 \tag{5}$$

Where,

$\tan\delta$ = $R/(1/\omega C)$ is the loss factor

ω_n = $2\pi f_n$ = harmonic frequency

V = rms voltage of nth harmonic

In the electricity networks, PFC capacitors are used to improve power factor of the network. However, with the capacitor and the stray inductance of the network components, a parallel resonant circuit can be formed. This causes very large (often localized) harmonic voltages and currents to flow, often leading to the catastrophic failure of the capacitor system. To reduce the chance of resonances in the network, tuned PFC capacitors can be used to filter harmonic components.

3.3 For equipment manufacturers

When an equipment manufacturer introduces a device in the market, he guarantees it to perform satisfactorily at sinusoidal supply voltage condition (as per the standard requirements). However, its optimum performance is not guaranteed when the supply voltage is distorted. Laboratory experiments show that devices produce higher amount of harmonic currents when the supply voltage is distorted. Table 2 compares total harmonic current distortions of some household devices under the sinusoidal and distorted supply voltage condition (with 5% total harmonic voltage distortion (THD_V) in the network) [Bhattacharyya et al., 2010].

Device	Total current harmonic distortion (THD_I) with respect to the total rms current drawn by the device	
	under clean voltage condition	under polluted voltage condition (THD_V=6%)
TV	48%	55%
Personal computer (PC)	87%	89%
Refrigerator	10%	18%
Compact fluorescent lamp (CFL)	72%	79%

Table 2. THD_I of devices under clean and polluted network voltage conditions

It might also happen that when specific types of devices are connected in large quantity at a certain part of the network, some orders of harmonic currents increase significantly resulting in large distortion of the network voltage. This can result in abnormal operation of other network devices. The device manufacturer, however, can not be blamed directly for such a situation as he is not responsible for his devices' operations under a distorted supply voltage condition.

A relatively new concept is developed to measure harmonic fingerprint of a device to estimate its harmonic current emission behaviour under various polluted voltage conditions. A harmonic fingerprint is a database that contains a large set of harmonic current measurements for a device at different conditions of the supply voltage. A device is tested separately for a clean sinusoidal voltage condition. Subsequently, the supply voltage is polluted with various order harmonics (from 2nd harmonic up to 25th harmonic order), having an amplitude variation of 1% to 10% (with a step of 1%) and also a phase shift of 0° to 360° (with a step of 30°) and respective harmonic currents are measured. It is observed that with a polluted supply voltage, harmonic current emission of a device can change significantly. Reference [Cobben et al., 2007] gives more information on harmonic fingerprint development method. From a harmonic fingerprint plot, it is often possible to identify internal characteristic of a device. In Fig. 6 harmonic current emissions are plotted for a CFL when the 25th harmonic voltage pollution is increased from 1% to 10% of the fundamental voltage for phase shifts of 0°, -30°, +30°, -180°, and +150° respectively [Bhattacharyya et al., 2010]. It shows that the harmonic current emission of a CFL increases almost 10 times when the 25th harmonic voltage pollution (with 0° phase shift) at the supply terminal is changed from 1% to 10% of its magnitude.

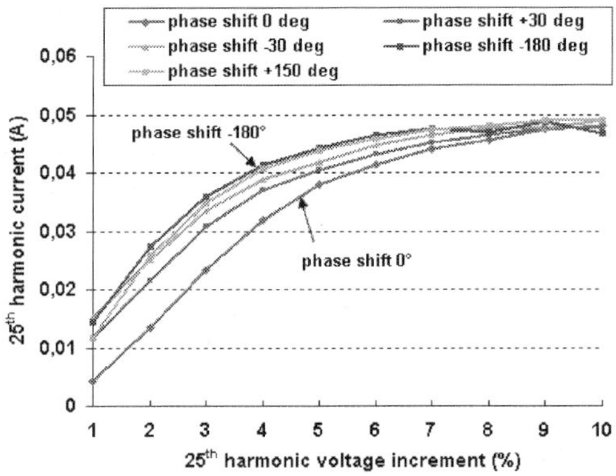

Fig. 6. Harmonic currents of a CFL for 25th harmonic voltage pollutions

It was also found that mutual interactions occur among various harmonic-producing devices in the network. Furthermore, a distorted supply voltage (as background pollution) influences harmonic emission behaviour of the loads. Fig. 7 shows examples of a PC and a CFL when they are tested with clean sinusoidal voltage and with two polluted voltage conditions of THD_V as 3.2% (marked as 'average b.g pollution') and 6% (named as 'high b.g pollution') respectively. Harmonic current emission of a device can change significantly with the background supply voltage distortions.

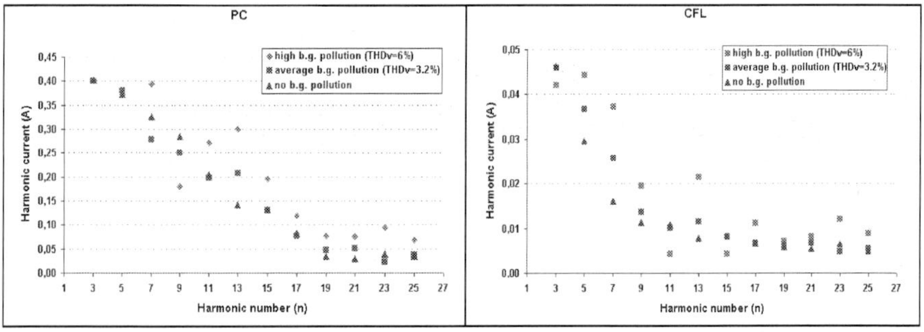

Fig. 7. Emissions of devices under background pollution conditions

4. Financial impacts of poor PQ

Power quality disturbances can have significant financial consequences to different customers and the network operators. It is quite hard to estimate correct financial losses of poor PQ as many uncertainties are involved. Therefore field surveys, interviews and case studies are carried out to get an indication of the costs of poor PQ. From literatures, many analyses are found on PQ costs for various types of customers. In contrast, very limited information is available on PQ cost for the network operators. As the cost evaluation of poor PQ is a complicated issue, the CIRED/CIGRE 'Joint Working Group'- JWG C4.107 was formed to develop a systematic approach for estimating various costs related to PQ problems. This group proposed methodologies to determine PQ costs for the customers as well as the network operators.

4.1 For customers

From worldwide customer surveys on the electric supply, it is found that voltage dip is one of the PQ problems that causes large inconveniences and has significant financial impacts to various industrial process equipments. The actual financial losses are customer specific and depend mainly on customer category, type and nature of activities interrupted and the customer size. Also, financial losses are event specific and different severity could incur different losses to various customers. It is noticed from different surveys that short interruptions and voltage dips are the major contributors to financial losses in terms of PQ related costs. The European Power Quality survey report declared that PQ problems cause a financial loss of more than 150 billion Euros per year in the EU-25 countries [Targosz & Manson, 2007]. The survey was done over two years period during 2003-2004 among 62 companies from different industries and service sectors. It was found that 90% of the total financial losses are accounted to the industries. Fig. 8 shows the percentage shares of total financial losses on various PQ aspects in the EU-25 countries. It shows that 56% of total financial loss in EU-25 is a result of voltage dips and interruptions, while 28% of the costs are due to transients and surges. Other financial losses (16%) are because of harmonics, flicker, earthing and EMC related problems.

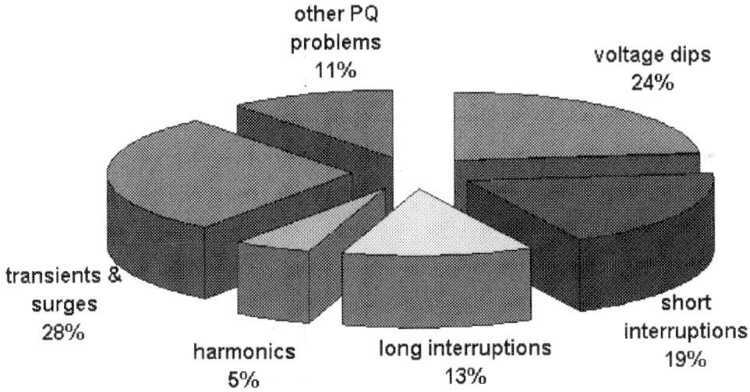

Fig. 8. Percentage share of PQ and interruption costs EU-25 countries

As per the proposal of the CIRED/CIGRE JWG 4.107 group [Targosz & Manson, 2007], two distinct methods of measuring the economic impact of poor PQ have been identified.

- The first method is a direct method, which is an analytical approach to consider the probabilities and impacts of the events. This method leads to a precise answer, but mostly it is difficult to obtain correct input values.
- The second method is an indirect method, which considers historical data for analysis and the customer's willingness to pay for solving PQ problems.

Total cost of a PQ disturbance for a production company consists of expenditures in various accounts as follows:

- Staff cost - this is the cost because of personnel rendered unproductive for disrupted work flow.

- Work in progress - this category includes the costs of raw material involved in production which is inevitably lost, labour costs involved in the production, extra labour needed to make up lost production etc.
- Equipment malfunctioning - if a device is affected, the consequences can be slow down of the production process, extra 'idle' time.
- Equipment damage - if an equipment is affected, consequences can be complete damage of the device, shortening of its life time, extra maintenance, need of stand-by equipment etc.
- Other costs - the costs paid for penalties due to non-delivery or late delivery, environmental fines, costs of personal injury (if any), increased insurance rate etc.
- Specific costs - this category includes extra energy bill due to harmonic pollutions produced by non-linear devices, fines for generating harmonic pollution in the network (if applicable). Reduction of personal working efficiency; related health problem due to flicker can also be included in this cost category.
- Savings - there are some savings in the production too. It includes saving from the unused materials, saving from the unpaid wages, savings on energy bill etc.

In a typical continuous manufacturing sector large financial losses are incurred by the lost work-in-progress (WIP) which is (most of the cases) about one third of total PQ costs. Also, the slowing down of processes and labour costs are quite significant in this sector. In other sectors, the situation is not very clear with the labour cost and equipment related costs. In the public services like hotels and retail sectors, PQ impact is measured as slowing down their business activities, in terms of revenue lost. In the industries the losses are mainly because of voltage dips, interruptions and transient surges. Fig. 9 shows the distribution of PQ costs in various accounts in industries and service sectors, as estimated in the LPQI survey for the EU-25 countries [Manson & Targosz, 2008].

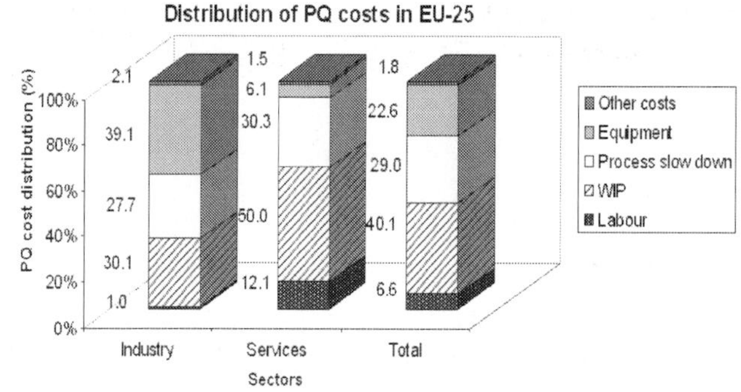

Fig. 9. PQ cost distributions per sector in EU-25

PQ cost estimation survey was also performed by the EPRI and CEIDS consortium for the American industries in 2000. It was estimated that the US economy loses annually 119 billion dollars to 188 billion dollars due to voltage dips, short interruptions and other PQ problems [Lineweber & McNulty, 2001]. Digital economy and continuous manufacturing industries are found to be the most affected sectors. It can be remarked here that PQ cost data, obtained from different surveys, is quite difficult to compare as the references of representations in different surveys often vary. Hence, proper evaluation method of the analysis is required for correct interpretation of the cost data. Another report [McNulty et al., 2002] estimated the costs of momentary and 1 hour outages for various sectors in the USA. Similar type of survey was also conducted by UMIST, UK in 1992 [Kariuki & Allan, 1996] to estimate costs of outage to different customer groups. Table 3 compares both the findings of these surveys. It shows that outage costs in different sectors in UK and US vary significantly, except for the industrial customers suffering momentary outages.

Sector	Survey done for UK customers Costs per outage per customer (€/event/customer)[1]		Survey done for US customers Costs per outage per customer (€/event/customer)[2]	
	momentary	1 hour	momentary	1 hour
Residential	-	0.84	1.63	2.02
Commercial	13.8	127.2	454	664
Industrial	1440	5160	1420	2375

Note: Various original cost data are converted to equivalent Euros for better comparison. The conversion rates are taken as: 1 £ = 1.20 €[1] and 1 $=0.75 €[2].

Table 3. Costs of outage as experienced by different customer groups

It is quite difficult to make a general conclusion on financial losses in different industries as the PQ cost and the cost of outage due to interruption depend largely on the customer's installation characteristic and the devices involved. Among the industries, there can be a wide range of variety in device usages and their sensitivity to PQ problems. The same is also applicable for the commercial sectors.

4.1.1 Estimation of financial losses caused by process failure

When a disturbance occurs in an industrial process plant because of a voltage dip event, it can cause appreciable financial losses for the plant owner. The financial losses for an industrial customer can be determined as shown in equation (6) [Milanovic & Jenkins, 2003].

$$CL = MV - CM - CE + AC \tag{6}$$

Where,

CL = combined financial losses

MV = market value of goods that could be manufactured during the time of process outage. Also, called "opportunity lost: sales and profit forfeited".

CM = cost of material which could be used up in a production process

CE = cost of energy that could be used up in a production process

AC = additional costs connected with a loss of supply

Alternatively, total financial losses because of a PQ event can also be expresses by equation (7). Both equations (6) and (7) are equivalent and any of it can be applied for calculating the financial losses due to voltage dips.

$$CL = EE + (RL - VE)\cdot(r + s) + FC \tag{7}$$

Where,

EE = extra expenses incurred because of the failure (€/per failure)

RL = revenue lost per hour of plant downtime (€/per hour)

VE = variable expenses saved per hour of plant downtime (€/per hour)

r = repair or replacement time after a failure (hours)
s = plant start-up time after a failure (hours)
FC = any fixed costs (€/per failure)

Financial losses due to a voltage dip are very much influenced by the customer's load compositions and process layout at the plant site. Also, different base references are used in various case studies to represent voltage dip financial data (for example: sometimes the reference of representation is per event cost, sometimes per kVA or total installed capacity of the plant, and sometimes per hour cost or total cost in a year etc.). Fig. 10 provides a range of financial losses due to voltage dip event in different industries that can be used for macro-level planning purpose [Andersson & Nilsson, 2002].

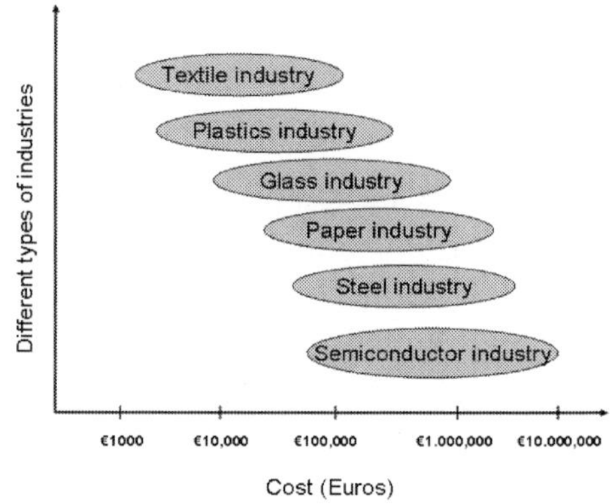

Fig. 10. Indicative financial losses in different sectors due to voltage dips

4.1.2 Estimation of financial losses caused by harmonics

Harmonic voltages and harmonic currents both can cause failure or abnormal operation of a customer's devices and can have financial consequences to the customer. Generally, harmonics in the network cause three types of problems:

- Additional energy losses (in the customer's transformers, connection cables, motors, neutral conductors etc.)
- Premature aging of a device
- Abnormal operation or misoperation of a device

Two methods can be applied to estimate costs related to harmonics: deterministic and probabilistic. The deterministic method is applied when all information related to the calculation (such as knowledge of network's operating conditions, various devices' operating conditions and absorbed power levels, the variation rate of the electrical energy unit cost and discount rate etc.) are available without any uncertainty. The present worth of operating costs of all components (represented as '$(Dw)_{y,\,pw}$') in a considered installation for a period of N_T can be estimated as shown in equation (8) for a harmonic rich environment [Caramia et al., 2009].

$$Dw = \sum_{y=1}^{N_T} (Dw)_{y,pw} = \sum_{y=1}^{N_T} \frac{(Dw)_y}{(1+\alpha)^{y-1}} \tag{8}$$

$(Dw)_y$ is the sum of extra operating costs for all components (due to energy losses) at the customer's installation in all the time intervals for a specific year 'y' under consideration. 'α' is the present worth discount rate and N_T is the period of years under consideration. Reference [Key & Lai, 1996] gives estimation of harmonic related losses in office building. In this case, about 60 kW electronic loads (mainly computers) are connected that operates 12 hours per day for 365 days in a year. It was found from the analysis that those offices were paying extra energy bills of 2100 dollars each year (which is approximately 8% of the total energy bill) because of 21,9 MWh additional energy losses annually due to harmonics at the customer's installation. It was also observed that the station transformer became overloaded when supplying to only 50% of its capacity equivalent non-linear loads.

The premature aging caused by harmonic pollution involves incremental investment costs (Da_k) for the kth device during the observation period. This is shown by equation (9) [Caramia et al., 2009], where $(C_{k,ns})_{pw}$ and $(C_{k,s})_{pw}$ are the present worth values of total investment costs for buying the kth device during its life in non-sinusoidal and sinusoidal operating conditions respectively.

$$(Da_k)_{pw} = (C_{k,ns})_{pw} - (C_{k,s})_{pw} \tag{9}$$

The evaluation of costs of misoperation is the most complex. It is often difficult to determine if the degradation of a device's performance is only due to harmonics or other PQ disturbances or due to other types of overloading. To estimate cost of a device's abnormal operation, it is required to get information of that device's characteristic under harmonic conditions, the activity for which the device is used and the relative importance of it in that process activity. Reference [Carpinelli et al., 1996] describes an investigation on a wide range of devices used in the commercial and industrial sectors and concluded that estimating the cost of abnormal operation requires extensive information on a device's behaviour in the presence of harmonics, the activity in which the device is used and the economic values of all items contributing to lower productivity.

Most of the cases it is difficult to gather all information related to costs as harmonics have mainly long term impacts and have relatively less visual immediate effects. Therefore, probabilistic method of cost estimation can also be applied when some of the calculating parameters are uncertain or not known correctly. Equation (10) shows a general equation to estimate the expected economical value of present worth of harmonics '$(D)_{pw}$' for a device in the probabilistic analysis [Caramia et al., 2009]. It includes a probability density function 'E' which describes its statistical feature of failure probability of the device, '$(Dw)_{pw}$' represents present worth of additional energy losses and '$(Da)_{pw}$' is the present worth of incremental investment due to premature aging of the device.

$$E(D)_{pw} = E(Dw)_{pw} + E(Da)_{pw} \tag{10}$$

4.1.3 Consequences of voltage flicker

Voltage flicker is an annoying problem for the customers. Most of the times, it does not have high financial impact. However, it causes inconveniences to the people when frequent flickering (of light and computer screens) occurs at their work-places or homes. From field studies it was found that voltage flicker can cause severe headache, epilepsy and other vision related illness to the customers. So, the affected people have to go for medical supervisions that might involve appreciable expenses. It was estimated from the LPQI survey [Manson & Targosz, 2008] that the cost consequences due to flicker related problems can be up to 10% percent of an organization's employment costs.

4.2 PQ costs for the network operators

Poor PQ has impacts to the network components (such as cables, transformers, capacitor banks etc.) that might suffer excessive heating, overloading, reduced energy efficiency, undesired tripping and early aging. Not many case studies are done yet to estimate the financial impacts of poor PQ for network components. However, it is clear that the network operators experience additional losses in the networks because of harmonic currents, mostly originated by various customers' devices. Also, the costs of unwanted tripping of protective devices or control equipments in the network can be significant as those can lead to unplanned supply interruption. The cost of reduced equipment's lifetime due to early aging is also very high, especially for expensive network devices. A transformer is expected to have a lifetime of at least 30-40 years. It is possible that it has to be replaced in 10 years earlier due to its early aging caused by increased harmonic pollutions in the network. Most of the time, the effects of harmonics are hidden and not immediately visible.

Light flicker is another PQ problem that has drawn high attention even though it has lesser financial impacts than harmonics and voltage dip problems. It can cause bad reputation of the network operator as a service provider in the electricity business. Also, when a customer complaints to the network operator about flicker problem, then an inspection engineer has to be sent to the site to supervise the problem to take necessary action. All these cause extra cost for the network operator.

In an existing network, PQ performance can be improved by rearranging and reinforcing the network. Also, regular maintenance strategy has to be adopted to enhance the lifetime of the network components and reduce failure rate. All these require appreciable investments. Implementing a mitigation device is another method to increase PQ performance level in the network at a desired level. The decision on adapting one of those strategies can be done after carefully analyzing PQ problem experienced by the network operators and customers and their relative financial losses in comparison to the investment required. Furthermore, it is also possible to tighten the PQ (or EMC) standards for the equipments so that they become less sensitive to disturbances. This requires a regulatory change and also can increase the equipment's manufacturing cost.

4.2.1 Harmonics in the network

At present, no published case study shows that the network components are failing because of poor PQ. However, it is often found that extra losses are occurring in the network components because of additional harmonic currents in the network. In the future electricity infrastructure, with the increased usage of many power electronics devices, harmonics can become a problem in the networks. Regular PQ monitoring can probably indicate actual PQ situation in the network. Various costs of harmonics are categorised as the operating cost (e.g. increased power losses), the aging cost (e.g. reduced lifetime cost) and the cost due to equipment's maloperation. All those costs can be calculated by the same method as described in section 4.1.2. When large harmonic currents flow in a network, the network operators often can notice its impacts immediately as the network components get overloaded. In extreme situation, it can cause tripping of the protection device. Reference [Papathanassiou et al., 2007] estimates an increase of only around 0.15%-0.20% of the total losses in the network components (lines and transformers) because of existing level of harmonics in the Greek networks.

4.2.2 Consequences of voltage dips

A voltage dip event can disrupt the operation of sensitive devices that might lead to partial or complete interruption of a customer's power supply. The effects of voltage dips mainly depend on the type of customer, the usage of the power supply and the electricity demand of the installation. Fig. 11 shows the EPRI-PEAC survey [McNulty et al., 2002] results of the US industrial customers (mainly continuous manufacturing process industries and digital economy sectors). It shows that different durations of outages at the customer's installations can have varying impacts to them. For a 1 sec of power failure (or very deep voltage dip), around 56% of the total surveyed industrial customers suffer process interruptions of 1.1-30 min at their installations, while 27% customers suffer a sudden outage at their sites for a duration of 1 minute or less. Similarly, when the supply power failure is for 3 minutes, it was noticed that around 82% of the surveyed customers would face a process interruption in the range of 1.1-30 minutes and 15% of them experience process outages for 30 minutes to 2 hours. In this survey, it was estimated that for an industrial customer the average process outage time after 1 second power supply failure is 21 minutes. However, it is only an indicative value. The real process outage time can vary largely among different industries depending on their type of operations and sizes.

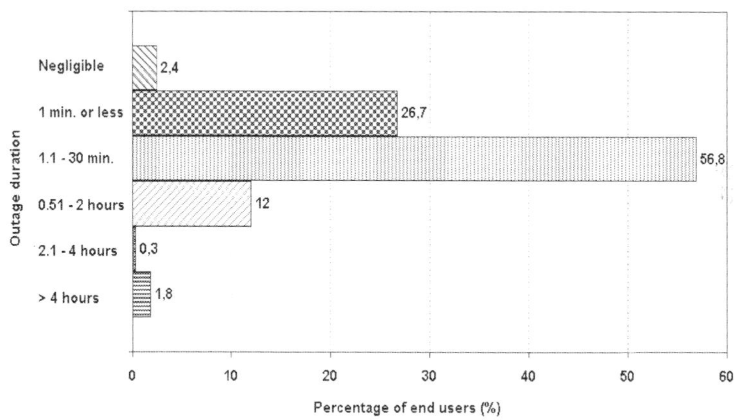

Fig. 11. Outage durations in different industries for 1 sec power failure

Generally, the industrial customers demand large quantities of electricity. Therefore, when a voltage dip event in the network disrupts the power supply of many industries in a specific part of the network, it can have significant financial impacts to the concerned electricity service providers (due to loss of 'kWh units' of electricity tariff).

5. Other impacts of poor PQ

Another aspect that has grown interest in the electricity service sector is the 'customer satisfaction' index. It depends mainly on the mutual relationships between a customer and network operator; and indicates the commercial quality of the electricity distribution service. Commercial quality generally relates to the individual agreement between network operator and customer. However, only some of these relations can be measured and regulated through standards or other legal instruments. When more customers are not satisfied with the PQ of the supply, the network operator may lose the trust of the customers. In the extreme situation, the national regulators may take action against the network operator to solve the problem. Presently, in many countries the network operators are obliged to verify PQ complaints of individual customers. They should provide a voltage at a customer's terminal that has to fulfil the applicable standard requirements. However, a customer's responsibility regarding various PQ requirements at the point of connection is not yet well defined. Therefore, it is a challenge for the network operator to maintain both the technical and service quality to satisfy the customers' needs.

When a customer buys a product from the market, the 'brand name' plays an important role in decision making of his purchase. A device when is sold in the market, the device manufacturer guarantees its performance as per the relevant product standards (of IEC or other internationally recognized standard) under clean voltage condition. However, in reality, the devices operate in an electricity environment with distorted supply voltage. Thus, the connected devices generate different harmonic currents than that at sinusoidal voltage condition. In certain situations, some specific order harmonic current can exceed the limits of the standards. Also, it produces extra losses and may operate abnormally that leads to the decrement of its lifetime. All these can bring doubts in the customer's mind about the quality of the device and indirectly on the device manufacturers supplied product's quality. Thus, the 'commercial quality' of the equipment manufacturer might get affected too.

6. Responsibility sharing among various parties

Many PQ problems in the network are contributed by customer's nonlinear devices and most of those devices are susceptible to failures under polluted supply voltage condition. In addition, some PQ problems such as voltage dips, transients are mainly originated in the network side. Improving a network to reduce PQ problem costs huge amount of investment for a network operator; while poor PQ also accounts large financial losses to the customers. Therefore, following parameters have to be considered for PQ cost analysis in the networks:

- Network's mitigation cost (such as changes in network infrastructure, placing a PQ mitigation device etc.)
- Compensation for the extra losses in network components
- Extra costs to handle customer's complaints (effort in finding the problem, network intervention for modification and improvement)
- Customer's willingness to pay extra money (and tariff) to minimize PQ disturbances at their installations
- Evaluation of cost of extra-quality by introducing individual 'PQ contract' scheme
- Costs aspects that concern manufacturers in designing equipment for improving PQ (emission and immunity of equipment)
- Total market size (that means the number of customers involved) for a specific PQ solution under consideration

To implement PQ mitigation, it is required to evaluate different alternatives to improve PQ performance in the network. Therefore, a systematic approach is followed to first find out the responsibility of each of the involved parties in the network as shown in Fig. 12.

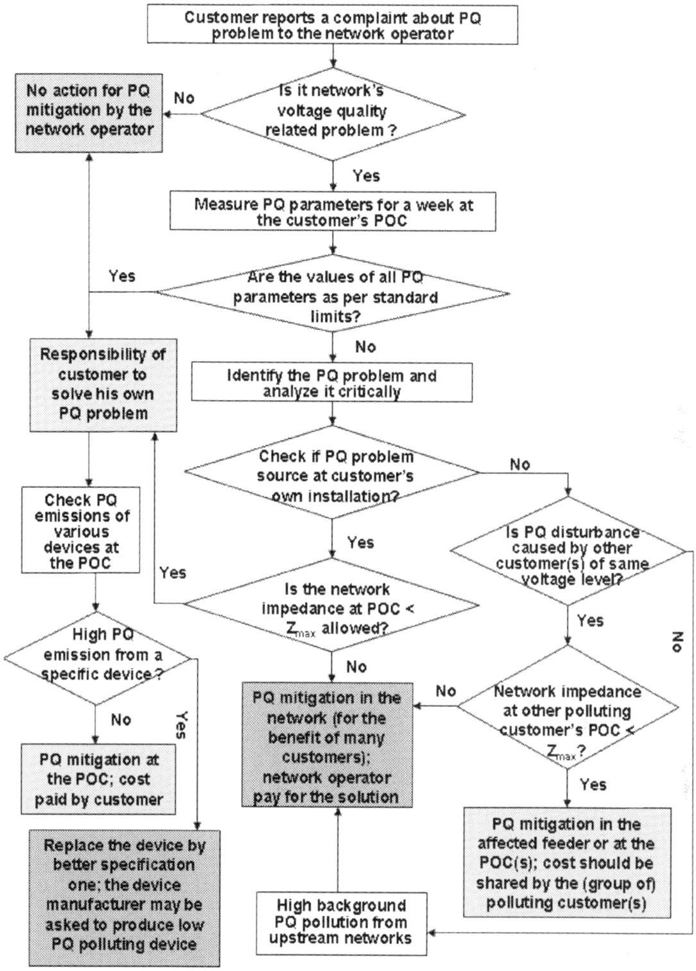

Fig. 12. Decision making flow-chart on PQ solutions

One of the main influencing factors on the decision making is the number of complaints that a network operator gets from the customers because of poor PQ of the supply. First, the source of PQ problem in the network is to be detected. It is also important to know about the network impedance at the POC where customer is facing PQ related inconveniences. After analysing the problem, network operator may ask the customer either to install a mitigation device at his terminal or to pay for the solution, if the problem is caused by the customer's device. On contrary, if the network impedance at POC is found higher than the maximum (fundamental and harmonic) impedance allowed for the customer's connection type, the network operator will be responsible to solve the PQ problem. Depending on the sensitivity of the problem, the network operator has to implement mitigation method at the complaining customer's POC, or in the feeder or elsewhere in the network to fulfil the needs of a large group of customers. Thus, following situations are possible while deciding on an investment for a PQ solution:

- A mitigation method is chosen at the customer's terminal or in his installation and the investment is also done by the customer.
- Mitigation method is adopted at the customer side, but the investment is done by the network operator.
- A mitigation method can be implemented in the network (such as: network a reconfiguration, laying an extra cable, installing storage or filters, etc.) while the investment can be shared by the network operator and the customer, depending on the situation.

Finding an optimal PQ solution and defining PQ related responsibility of each of the parties connected in the network are complicated issues. The standards that are presently available give limiting values for some specific PQ parameters that are valid only for sinusoidal voltage condition. Hence, the standards also need to be adjusted to restrict PQ emissions at a customer's installation. Alternatively, various devices can be manufactured with stricter specification so that they produce less PQ pollution in the network. Moreover, all the concerned parties would work together to find out the best solution. Fig. 13 illustrates mutual responsibility sharing on PQ aspects among the network operator, the customer and the equipment manufacturer.

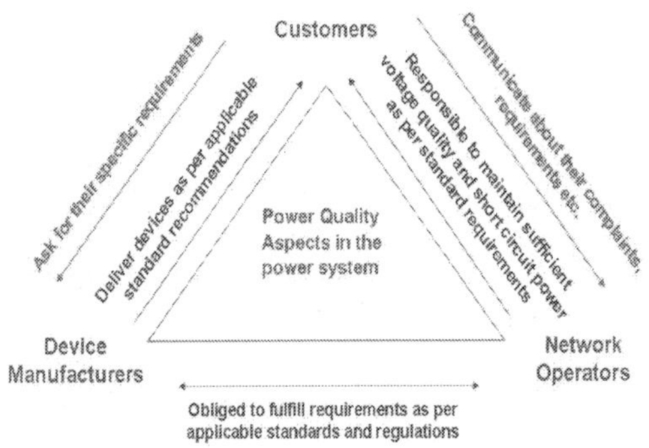

Fig. 13. Mutual responsibilities among various parties in the network

6. Conclusion

Poor PQ causes techno-economic inconveniences to different parties connected to the network. PQ problems such as voltage dips and harmonics can have significant techno-economic impacts to the customers and the network operators. It was found from the LPQI survey (2004) in the EU-25 countries that electronic equipments, electrical motors, variable speed drives and static converters are the most affected equipments in the industries. The other affected devices are cables, capacitors, lighting equipments and relay contactors. Estimating the financial losses for a customer because of poor PQ is quite complicated as it includes various direct (immediately visible costs) and indirect (long-term) costs. The methodologies for calculating the costs related to voltage dips and harmonics are described in this chapter. It was found that continuous manufacturing process plants and digital industries are the most vulnerable sectors for PQ related disturbances. It was also noticed that the network operators can have significant inconveniences too because of poor PQ in the network. Network components suffer extra losses, reduced operational efficiency, abnormal tripping and premature failure because of harmonics in the network. Various reference case studies are mentioned where poor PQ related financial losses for the customers and network operators have occurred. It was found that light flicker is technically not a big problem for the network. However, when the customers suffer from flicker related disturbances and complain frequently to the network operators, it can have some financial consequences. The 'commercial quality' aspect of the electricity is also briefly discussed that emphasizes on the customer's satisfaction about the received electricity.

The presence of polluting devices in the network often distorts the supply voltage. Under a polluted grid condition, a customer's devices behave differently than the sinusoidal voltage condition. The present standards give requirements on PQ parameters only for clean sinusoidal voltage condition. In future, those standards need to be modified to make them more appropriate for the real distorted network voltage condition. Also, the standards should specify clearly the responsibility of the customer regarding various PQ parameters at the connection point of his installation with the network. To select an optimum PQ mitigation method in the network, a detailed cost-benefit analysis is to be done for the involved parties.

7. References

Andersson, T.; & Nilsson, D. (2002). Test and evaluation of voltage dip immunity. *STRI AB and Vatenfall AB 27,* November 2002.

Bhattacharyya, S.; Cobben, S.; & Kling, W. (2010). Harmonic current pollution in a low voltage network, *Presented as a panel list paper in IEEE PES GM 2010 conference,* July 2010, Minnesota.

Caramia, P.; Carpinelli, G.; & Verde, P. (2009). Power quality indices in liberalized markets. *A John Wiley & Sons Ltd.,* ISBN: 978-470-03395-1.

Carpinelli, G.; Caramia, P.; Vito, E. Di.; Losi, A.; & Verde, P. (1996). Probabilistic Evaluation of the Economical Damage due to Harmonic Losses in Industrial Energy System. *IEEE Transactions on Power Delivery,* vol. 11, No. 2. April 1996.

CEER (2008). 4th Benchmarking report on quality of electricity supply 2008. Issued by: Council of European Regulators ASBL, Brussels, December 2008. Ref.: C08-EQS-24-04.

Cobben, S.; Kling, W.; & Myrzik, J. (2007). The making and purpose of harmonic fingerprints. *Proceeding of 19th International Conference on Electricity Distribution (CIRED 2007).* Vienna.

Desmet, J.; Putman, D.; D'hulster, F.; Belmans, R. (2003). Thermal analysis of the influence of nonlinear, unbalanced and asymmetric loads on current conducting capacity of LV-cables. *Proceeding of IEEE Bologna PowerTech conference, Bologna, Italy,* June 2003.

Emanuel, A. E.; McNeil, J. A. (1997). Electric power quality, *Annual Rev. Energy Environ,* Worcester Polytechnic Institute, Massachusetts, USA. Ref no. 1997.22:263-303.

Fuchs, E.F.; Roesler, D.J.; & and Kovacs, K.P. (1986). Aging of electrical appliances due to harmonics of the power system's voltage. *IEEE Transactions on Power Delivery,* Vol. PWRD-1, no. 3, July 1986.

Hulshorst, W.T.J.; Groeman, J.F. (2002). Energy saving in industrial distribution transformers. *KEMA report,* reference 40130017-TDC 02-24226A.

Johnson, P.; Coney, R. (1997). Introducing national quality of supply (power quality) standards the South African experience. *10th International Conference on Electromagnetic Compatibility, Conference,* publication no. 445 0 IEE 1997, September 1997.

Kariuki, K.K.; Allan, R.N. (1996). Evaluation of reliability worth and value of lost load. *IEE proceedings: General Transmission and Distribution,* vol. 143.

Keulenaer, H. De. (2003). The hidden cost of poor power quality, *Leonardo Energy,* European Copper Institute, October 2003. Available: www.leonaro-energy.org

Key, T.S.; & Lai, J.S. (1996). Costs and benefits of harmonic current reduction for switch-mode power supplies in a commercial office building. *IEEE Transactions on Industry Applications,* vol. 32, no. 5.

Lineweber, D. ; McNulty, S.R. (2001). The cost of power disturbances to industrial & digital economy companies, *EPRI IntelliGrid Initiative (A Primen report from EPRI & CEIDS),* Ref no. 1006274, June 2001. Available : www.epri.com/ceids

Targosz, R.; Manson, J. (2007). PAN European LPQI power quality survey, *Proceedings of 19th International Conference on Electricity Distribution (CIRED 2007),* May 2007, Vienna.

Manson, J.; Targosz, R. (2008). European power quality survey report *Leonardo Energy,* November 2008, www.leonardo-energy.org.

McNulty, S.; Primen; & Howe, B. (2002). Power quality problems and renewable energy solutions. *Submitted to Massachusetts Renewable Energy Trust,* September 2002.

Melhorn, C.J.; Maitra, A.; Sunderman, W.; Waclawiak, M.; Sundaram, A. (2005). Distribution system power quality assessment phase-II: voltage sag and interruption analysis. *Copyright IEEE,* paper no. PCIC- 2005-13.

Milanovic, J.V.; Jenkins, N. (2003). Power quality cost for manufacturing industries. *Presented on EDF workshop in Paris,* Paris.

Najdenkoski, K.; Rafajlovski, G.; & Dimcev, V. (2007). Thermal aging of distribution transformers according to IEEE and IEC standards. *Power Engineering Society General Meeting, 2007,* IEEE, ISBN: 1-4244-1296-X.

Papathanassiou, S.; Kasmas, N.; Drossos, N.; Stavropoulos, D. (2007). A practical evaluation of distribution network losses due to harmonics. *Proceeding of 19th International Conference on Electricity Distribution (CIRED 2007),* Vienna, June 2007.

Wharmby, B. (1998). Power quality in the privatised UK electricity industry. *Proceedings of Industry Applications Conference,* at 33rd IAS annual meeting, volume: 3, pp: 2288-2291, 1998.

2

The Impact of Power Quality on the Economy of Electricity Markets

Hector Arango, Benedito Donizeti Bonatto,
José Policarpo Gonçalves de Abreu and Carlos Márcio Vieira Tahan
Itajubá Federal University (UNIFEI) and University of São Paulo (USP)
Brazil

1. Introduction

1.1 Basic concepts

The main goal of this chapter is the economic valuation and prioritization of those investments which are intended to improve the power quality of the system. While most of the research work in the field of power quality was aroused by the scientific challenges to be overcome, it is fair to recognize that the practical importance of power quality hinges on the economic impact of such phenomena.

Thus, economic analysis of power quality is of the outmost significance. Such analysis, according to present day corporate doctrine, implies the assessment of the changes in the market value which are produced when an ideal scenario - where electric energy is supplied without any imperfection - is disturbed by a series of distortional events, thus leading to a loss of quality. In brief, we want to relate imperfection level with market value.

In a more practical line of thought, we wish to answer specific questions like this: how many equivalent dollars of market value are lost when the severity index of some type of imperfection increases - say - by one unit? Or - opposingly - how many dollars are won because of a decrease in severity?

Moreover, given an intervention in the markets which diminishes the index, we seek to estimate the monetary benefit of such quality improvement. As we also know the capital needed to accomplish such intervention we can introduce in the market model and the change in value added may be subsequently calculated. Then, the relation between that change and the capital invested can be compared against a hurdle rate which functions as a yardstick for the merit of the intervention. This can be used also in order to prioritize capital investments on power quality (this rather straight way of investment prioritization, becomes more complicated when the uncertainties inlaid in the quality issues are introduced).

1.2 The quality market model outline

Figure 1 depicts the electricity market model without considerations on quality issues. Essentially, such market is based upon the capacity of the consumers to draw value from electrical energy through an usage system (the "EE usage" rectangle on upper right). This system is fed across a supply network (the rectangle below the first one). These two systems are the physical assets from which the creation of users value (S) results. The payment of users is the firm revenue (R). So the difference (U-R) can be seen as a surplus (S) obtained from the acquisition of energy. Subtracted from (R) are the expenses need to operate the physical assets, taxes and the remuneration of capital (financial assets), the remaining value (V) goes to the investors, as seen at the left, (S) and (V) sum up to produce the public value (W) added to the society. Maximization of (W) is the ultimate goal of regulation mechanisms (Arango et al., 2008a).

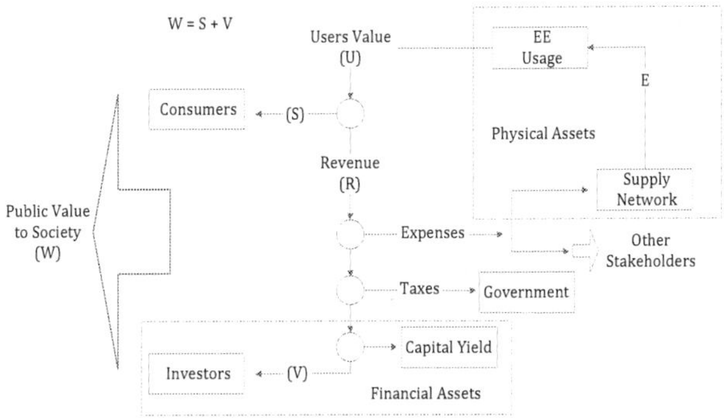

Fig. 1. Diagram of the monetary streams (someones are virtuals) taking place in an electricity market.

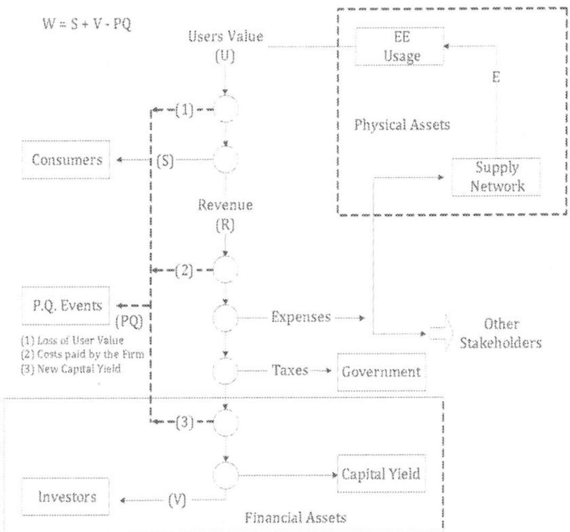

Fig. 2. Insertion of Power Quality events in the Electricity Market Model.

Figure 2 shows how the economic impact of quality events can be introduced in the previous market model. Due to the heterogeneous nature of such events they influence the customers, the firm or both. There is also an additional impact due to the capital invested in measures aimed to power quality improvement. Anyway, each of these influences can be duly taken on account and the net impact of quality on consumers' surplus and the

companies' economic value added can be calculated. (Those influences are located in Fig. 2 as circles (1), (2) and (3)).
In this way, this new market model allows a comprehensive treatment of power quality (P.Q.) either from the point of view of the utilities, customers or society.

2. The consumers representation

We shall begin with a brief explanation of the consumers attitude in regard to the electrical energy (E.E.).

2.1 Basic assumptions

a. From any quantity (E) of Electrical Energy the consumer gets a value of use expressed in monetary terms, as a concave value-function U(E), also known as a Von Neumann-Morgenstern utility curve (Von Neumann & Morgenstern, 1944). The simplest function of such kind is a quadratic one, expressible as:

$$U(E) = a.E - \tfrac{1}{2}\, b.(E)^2 \tag{1}$$

where (a) implies the Eagerness to use (E), while (b) portraits the Satiation due to the quantity already spent.

b. The energy is sold at a fixed Tariff (T). Then, the purchase of a quantity (E) will demand a Payment P(E) given as:

$$P(E) = T.E \tag{2}$$

c. The difference between Value and Payment - whenever positive - represents a Surplus S(E) which the user draws from the transaction. Based on expressions for the equalities (1) and (2) results the following expression for the Surplus.

$$S(E) = (a-T)E - \tfrac{1}{2}\, b.(E)^2 \tag{3}$$

2.2 The purchasing of energy

Under the hypothesis above we postulate that the fundamental mechanism governing the purchase of E.E. can be verbalized as follows (Kreps, 1990; Jehle & Reny, 2000):

"The consumer buys such quantities (E*) of energy that maximize his (her) surplus."

Then, (E*) must satisfy the well-known first-order condition:

$$\frac{dS}{dE}\Big|_{E=E^*} = (a - T) - bE^* = 0$$

Thus, for any a>T, it results the optimal quantity purchased by the customer:

$$E^* = (a-T)/b \tag{4}$$

This important equation relates consumption with price in the electricity market, in terms of the customer preference and the practiced tariff. The reverse equation is also very often useful:

$$T = a - bE \tag{5}$$

From (2), (3) we shall have the optimized surplus:

$$S^* = (b/2)E^2 \quad (6)$$

and the corresponding payment:

$$P^* = aE - bE^2 \quad (7)$$

which is also the Revenue (R) of the firm.

2.3 The diagram of economic flow of consumption

The relationships above may be held together in a Diagram of Economic Flow as shown in Figure 3:

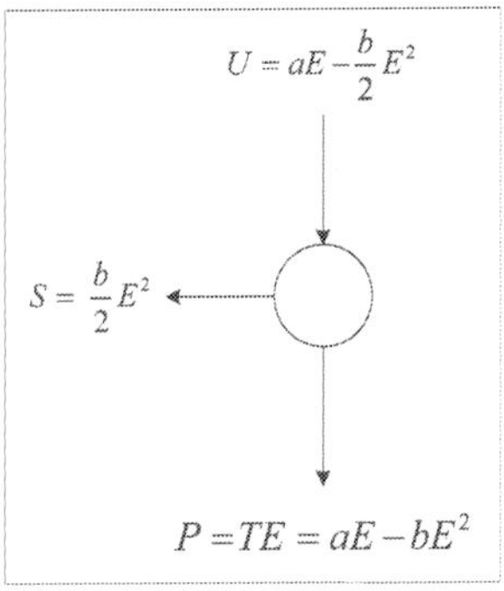

Fig. 3. Diagram of economic flow of a customer or cluster of customers.

For a = 4300 US$/MWh, b = 500 US$/MWh^2 and E = 8 TWh. The diagram is depicted in Figure 4:

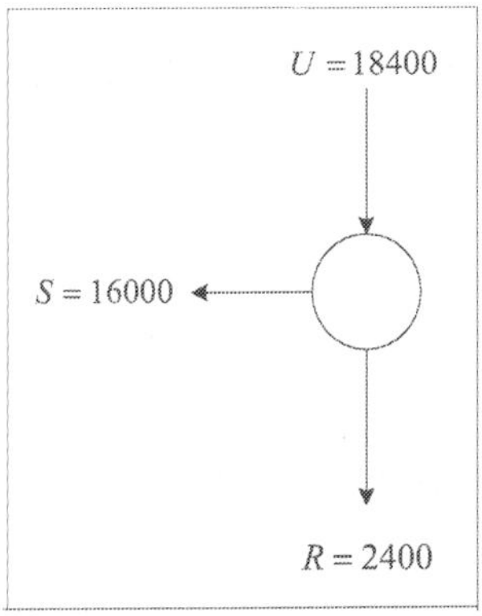

Fig. 4. Diagram of economic flow with reasonable values for the monetary streams occurring in real world.

Then, T = a - bE = 4300 - 4000 = 300 US$/MWh, as it should be.

2.4 Practical considerations

The customers parameters (a,b) are not readily known in practice. Nevertheless, they can be related to information that is far more available. First, we are familiar with the purchased annual energy (E) and the firm revenue (R), thus we are also aware of T = R/E. Second, there is a basic economic parameter describing the consumers attitude with regard to the energy which was being thoroughly studied and measured: it is the Elasticity Consumption-Price (ε), defined as:

$$\varepsilon = -(T/E)(dE/dT) \quad (8)$$

Recalling the fundamental equations (4) and (5) it can be devised a formula linking (ε) with (a) as:

$$a = (1+ \varepsilon^{-1})T \quad (9)$$

On the other hand, we always have the formula (5) which is reiterated here.

$$a - bE = T$$

For example, let be E = 8 TWh, T = 300 US$/MWh, ε = 0,075=7,5%
Then, from (9) we shall have:

$$a = (1 + 1/0{,}075)300 = 4300 \text{ US\$/MWh}$$

and using (5) it results:

$$b = (a - T)/E = (4300 - 300)/8 = 500 \text{ US\$/MWh}^2$$

Note that:

$$R = (4300 \times 8) - (500 \times 8^2) = 2400 \text{ MUS\$} = 8 \text{ TWh} \times 300 \text{ US\$/MWh}$$

as it should be.

Observation:

ε varies with the point of operation of the customer. The value used in the calculus must be the elasticity computed for the same (T) obtained above. When ε was obtained for a different tariff T′, i.e.,

$$\varepsilon' = -(T'/E)(dE/dt) = T'/bE = T'/(a - T') \quad (10)$$

We must retrieve ε as:

$$a = (1+ \varepsilon^{-1})T = (1+(\varepsilon')^{-1})T' \quad (11)$$

Then:

$$(1+ \varepsilon^{-1}) = (1+(1+ (\varepsilon')^{-1})T/T$$

which results in:

$$\varepsilon' = 1/((1-\varepsilon^{-1})(T/T') - 1)$$

Then

$$\varepsilon = 1/(1+(\varepsilon')^{-1})(T'/T) - 1 \quad (12)$$

For example, let be T′ = 150 US$/MWh. Then, we shall have:

$$\varepsilon' = 1/27,66666667$$

or:

$$\varepsilon' = 0,036144578$$

giving, as expected,

$$\varepsilon = 1/(28,66666667 \times (150/300)^{-1}) = 0,075$$

For example, suppose the elasticity is known to be ε' = 0,036144578 for T′ = 150 US$/MWh. Then, we shall have ε = 0,075 for T = 300 US$/MWh.

3. The firm representation

3.1 A few words on capital and rates of return

We start with a brief description of capital. Total capital (B) is composed by equity (A) - contributed by the Shareholders - and Debt (D) - due to the Bondholders.
It is customary to define a Leverage Factor (δ) such that ($D = \delta B$) to express the level in which the firm is indebted. Moreover:

1. An interest must be paid over debt at a convened debt-rate (r_D).
2. A remuneration should remain available over equity at an expected rate of return (r_A).
3. Interest is not subjected to taxation. Remuneration, instead, is fully taxable on a tax rate (t).
4. r_D and r_A combine to produce the weighted-average cost of capital (WACC) where:

$$WACC = r_w = (1-\delta)r_A + \delta(1-t)r_D \quad (13)$$

where t is the tax rate.

3.2 Financial statements

We shall represent the firms by a statement of their performance over a given period or financial exercise.
A statement begins always with the total Revenue (R), from which is deducted the Cost, thus leading to the so called Result.
In order to make clear our choice of Statement, two types of such instruments are to be considered:

1. Accounting statement
   Is the traditionally used and published (according to rules of USGAAP - United States Generally Accepted Accounting Practices), which Cost includes expenses and interest on debt (D). Its Result is called Net Profit (L).
2. Economic value statement
   Cost includes not only expenses and interest, but the yield on the total capital (B). Its Result is called Economic Value Added (V) (Martin & Petty, 2000).

In the following, we shall use the Economic Value Statement type, recognizing that it represents better the modern principles of corporate finance. Nevertheless, in order to highlight the distinctive features, some examples will be worked on both statements.

3.3 Modeling the firm

The supplier is moulded as a system (Jensen & Meckling, 1976) whose input is the revenue (R). The various components of the economic cost (C) are then subtracted. As (C) includes de yield of all the capital (debt and equity) the remaining quantity (output or result) stands for the value (V) created above the expectations of investors. The firm surplus (V) is known as Economic Value Added or Economic Profit.

One of the distinctive features of our representation is the proposal of a cost structure, that is to say, an explicit dependence among the various costs with the basic variables: supplied energy (E) and capital investment (B).

The proposed structure, i.e. the supplier model, is presented as a diagram of economic flow in Figure 5.

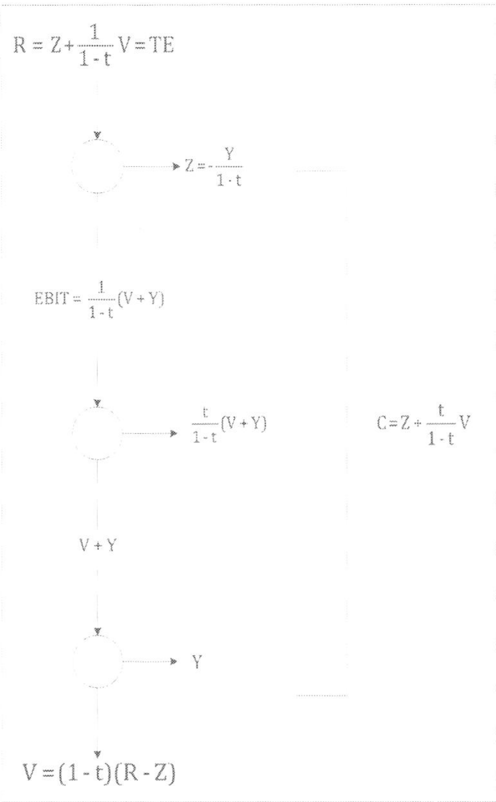

Fig. 5. Diagram of the Economic Value Statement of a distribution company (DISCO).

where the symbols mean:

R = Receipt; T = Tariff; C = Economic Cost; Z = Total Cost at Equilibrium; dB = Depreciation of Physical Assets; Y = $r_w B$ = Yield of Capital (equity plus debt); V= Economic Value Added or economic profit; EBIT = Earnings before interest and taxes

Note that the cost functions C(E, B) are depending also from certain parameters (e, p, d, w) portraying the operational and corporative efficiency of the firm on one hand and its technological standards on the other.
Note that, a distinctive feature, our market model brings a cost structure, that is to say, a functional dependence among each component of cost and the fundamental variables, which are quantities of supply and capital investment. In our case the variables are the energy purchased (E) and the investments in scale (B) and quality (Q).
As cost components we have the functions:
eE - Operational expenses, energy bought from generation companies and other charges.
$p(E^2/B)$ - Cost of energy lost in the electric network.
dB - Depreciation of physical assets.
r_wB - Reward to investors

3.4 The firm with optimal allotment of capital

The investment B has an optimal value B* for which Z attains its minimal value Z*. B* is obtained by zeroing the derivative of Z(E, B) with respect to B:

$$\frac{\partial Z}{\partial B} = -\frac{pE^2}{B^2} + k = 0 \tag{13}$$

Then

$$B^* = \left(\frac{p}{k}\right)^{\frac{1}{2}} E \tag{14}$$

and

$$Z^* = \left(e + 2(pk)^{\frac{1}{2}}\right) E = cE \tag{15}$$

Note that the average price remains constant.

3.5 The firm in economic equilibrium

Another condition of interest implies V=0. Then we shall have:

$$R = C = Z \tag{16}$$

In this case, the supplier firm is said to be in Economic Equilibrium.
When the firm is simultaneously with optimal investment and an economic equilibrium we shall have:

$$R = C = Z = cE \tag{17}$$

where

$$c = e + 2(pk)^{1/2} \tag{18}$$

As will be explained later, a regulated firm should belong to this particular class.

4. Peculiarities of electricity markets

4.1 The company as a natural monopoly

It is important to note that the distribution company (DISCO) owns the electric network which allows the physical delivery of E.E. to the customers. Any other firm wishing to compete in the business should construct another network replicating the geographical layout of the original one. It is obvious that the entry of such competitor will decrease the overall market efficiency (Kupfer & Hasenclever, 2002).
In such conditions, the disco has the market power required to impose whatever tariff which it has in mind. Clearly, the disco goal is to maximize its economic value added (V). Thus, it can be easily concluded that such tariff ought to satisfy:

$$\frac{dR}{dE} = \frac{dZ}{dE} \tag{19}$$

Assuming that the company optimizes its investments, the equality (19) will lead to:

$$a - ZbE = c \tag{20}$$

Then,

$$E_M = (a\text{-}c)/2b \tag{21}$$

and the "monopolistic" tariff will be:

$$T_M = (a+c)/2 \tag{22}$$

Substituting (21) in the formula of (V), given in Figure 3, we shall have:

$$V_M = (1\text{-}t)(R\text{-}Z)_M = (1\text{-}t)\ ((a\text{-}c) - bE_M)\ E_M = (1\text{-}t)((a\text{-}c)^2/4b) \tag{23}$$

4.2 The regulated firm

It is widely accepted that monopolistic tariffs impose on the customers an unfair weight. Being the electricity supply a public service, it should be regulated in order to attain maximum generation of social welfare (Beesleyand & Littlechild, 1989).
As social welfare is a debatable and elusive notion, regulators usually adopt, as a reasonable proxy, the quantity (Friedman, 2002):

$$W = S + V \tag{24}$$

i.e., the sum of the consumers' surplus and the value created by the utility, (which was called Market Output). Nevertheless, the unconstrained maximization of W leads to nonsensical results implying the firm's collapse. The constrained solution requires that the firm be operated in economic equilibrium. Then, for optimum investment, the regulated company is compelled to operate satisfying:

$$aE - bE^2 = cE \tag{25}$$

Thus, leading to:

$$E_R = (a\text{-}c)/b \tag{26}$$

and then

$$T_R = c$$

4.3 Example of a regulated operation of a firm

In order to enlighten the discussion on monopolist and regulated firms, Figure 6 shows the same distribution company supplying the same customers when operating free (i.e., using T_M) and when regulated (T_R).

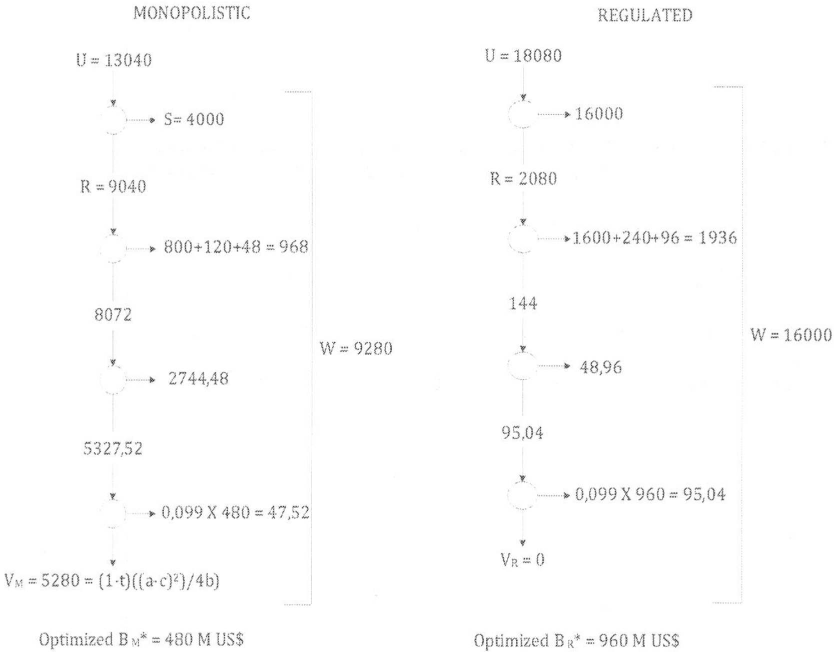

(a) Monopolistic (b) Regulated

Fig. 6. Monopolistic and regulated diagram of economic flow of a given firm.

Let be an utility and its costumers for the following parameters.

Firm:

e = 200 US$/MWh
$p = 3600$ US$^{2}/MWh^2
d = 0,10
r_w = 0,099
c = 260 US$/MWh

Customers:

a = 4260 US$/MWh

$b = 500$ US$/MWh^2
Note: the parameters already given correspond to a stylized albeit of a real medium-size utility supplying a cluster of customers, typical of a developing country.
In such case we shall have:

Monopolistic

$T_M = 2260$ US$/MWh
$E_M = 4$ TWh

Regulated

$T_R = 260$ US$/MWh
$E_R = 8$ TWh
leading to the diagram of economic flow of Figure 6:

4.4 The hurdle rate

We shall assume that every distribution utility seeks to maximize its performance, expressed as the Result of its exercise Statement. But a question arises: what type of statement is to be adopted?
As was said, there are two basic types, whose results are net profit and economic value added, respectively.
On the other hand, consider a project applicable to the company. We want to know if this project is convenient.
We shall define the hurdle rate (h) as the one for which the project does not change the Result produced by the company.
The hurdle rate can be obtained for both types of statement according to the incremental analysis on Figure 7.

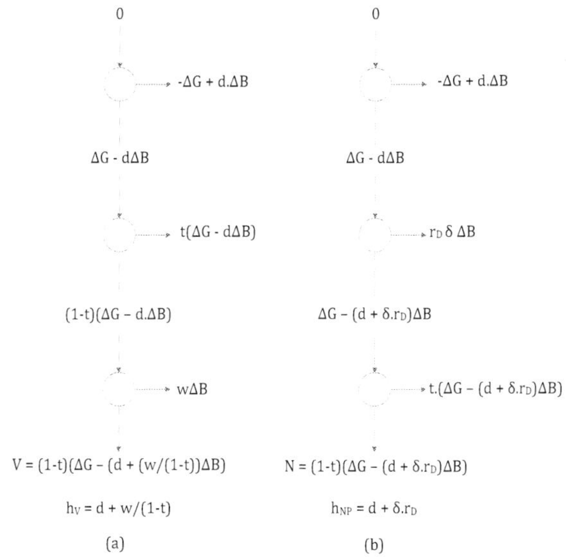

Fig. 7. Hurdle rate for economic value added and net-profit statements.

We can see that the hurdle rate depends heavily on the type of statement, that is to say, on the Result to be maximized. Substituting r_w in Figure 7 (a) for its value in (13) we shall have:

$$h_V = d + \frac{1-\delta}{1-t} r_A + \delta r_D = h_{NP} + \frac{(1-\delta) r_A}{1-t} \tag{27}$$

That is to say, the hurdle rate required to comply with economic value added is greater than the one for net profit by excess equal to:

$$\frac{1-\delta}{1-t} r_A \tag{28}$$

Example:
Let be: d=0,05, r_A=0,1595, r_D=0,075 and δ=0,55. Then, r_w=0,099 and:

$$h_V = 0{,}05 + \frac{0{,}099}{0{,}66} = 0{,}20 \qquad h_{NP} = 0{,}05 + 0{,}55x0{,}075 = 0{,}09125$$

$$h_V = h_{NP} + 0{,}10875 = \frac{1-\delta}{1-t}\gamma_A = 0{,}10875$$

For the sake of comparison, Figure 8 depicts the accounting statements for the monopolistic and regulated operation of the same firm under the hypothesis of leverage factor δ = 0,55, r_A = 0,1595 and r_D = 0,75.

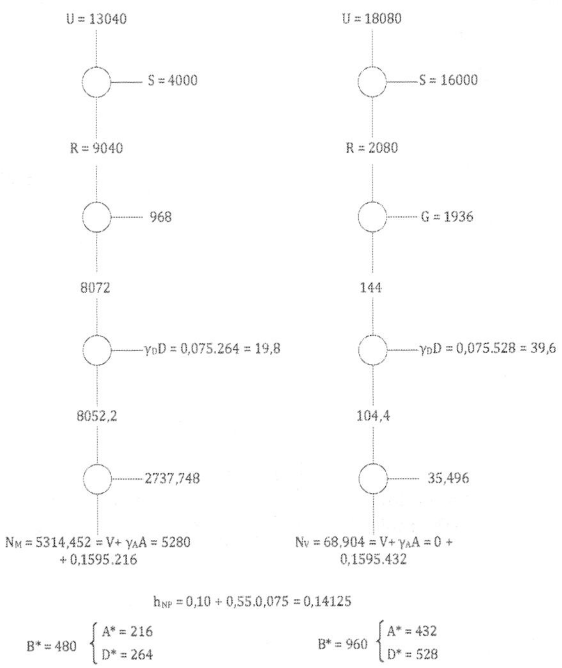

Fig. 8. Net profit statements for the data.

Note that L - V is always equal to the equity yield $r_A.A$ as appears at the bottom of both diagrams of economic flows.

4.5 The electricity Market Model (M.M.)

The (M.M.) is obtained by simply joining the two representations already made, as shown in simplified form in Figure 9.

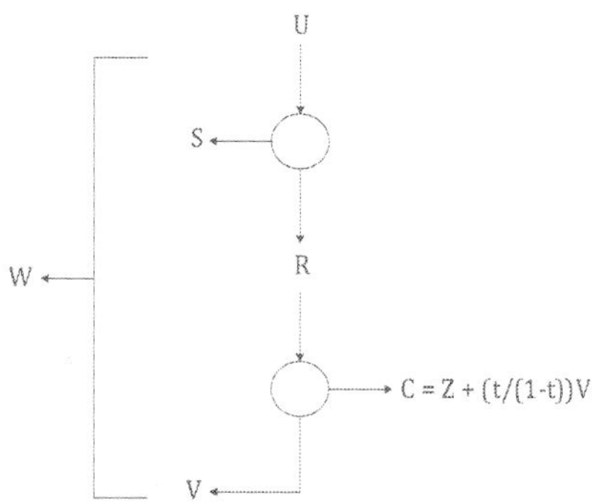

Fig. 9. The basic Electricity Market Model.

Now the total diagram can be interpreted as a single system whose input is the value of use of the electric energy and whose output is the sum of the consumers' surplus and the company economic value added.

5. The Insertion of quality in the electricity Market Model

5.1 Basic considerations

The approach to power quality is built on the notion of quality event, that is to say, any occurrence leading to some imperfection of the product (electric energy) or service (electric supply). Quality events are of the most different nature and consequences. In order to assess their effects, (especially for regulatory purposes) each type of event is associated to some numeric index of severity.

As an example, take the important class of events which cause interruptions on the electrical supply. This type of imperfections is customarily described by the average (or expected) time during which customers are de-energized (DTD) and the average frequency of interruption (DFD). These are technical indexes which affect both consumers and suppliers.

The functional dependence between such indexes and their economic impacts may be highly complex and heterogeneous. Thus, in order to get a unified and simple picture we shall assume that quality can be introduced as an additional cost (C_Q) (Arango et al., 2008b, Arango et al., 2010), which must be paid by consumers and producers in various proportions.

5.2 The quality cost

We shall call (C_Q) quality cost. The structure of (C_Q) will be assumed as:

$$C_Q = qE^2/Q \tag{29}$$

In this formula Q represents the capital investment which is devoted mainly to enhance power quality. Please note that Q adds to the capital (B) already considered, which is chiefly invested on the network size. In this line, Q and B shall be usually referred as "quality" and "scale" investment, respectively.
Of course, more realistic assumptions for the economical impact of quality may be done, but the approach above allows obtaining analytic answers leading to a better and faster understanding of such impact. Otherwise, such analysis gives satisfactory results in a broad class of practical problems regarding power quality.
The electricity market model including quality issues is depicted in Figure 10.

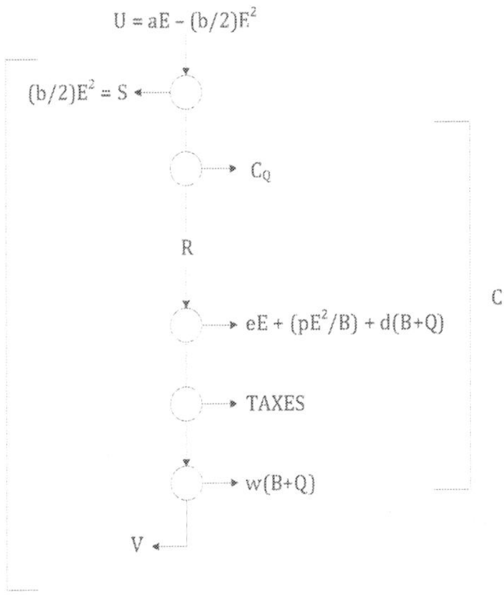

Fig. 10. The electricity market model considering Power Quality.

5.3 The purchasing of polluted energy

The consumption of imperfect energy is governed by the same principle of surplus maximization introduced in section for perfect energy. In the present case we shall have:

$$S = U - R - C_Q = (a\text{-}T)E - (b/2 + q/Q)E^2 \tag{30}$$

Then

$$\left.\frac{dS}{dE}\right|_{E=E^*} = a - T - \left(b + \frac{2q}{Q}\right)E = 0 \tag{31}$$

Thus

$$E^* = \frac{a - T}{b + \frac{2q}{Q}} \tag{32}$$

and

$$S^* = \frac{1}{2}(b + 2q/Q)E^2 \tag{33}$$

5.4 Optimizing the investment on quality

Note that investment on quality Q can be optimized in the same way that the investment on network scale B. Then, we shall have:

$$Q^* = \left(\frac{q}{k}\right)^{\frac{1}{2}} E \tag{34}$$

and

$$C_Q^* = (qk)^{\frac{1}{2}} E \tag{35}$$

leading to an optimized cost

$$\tilde{c} = e + (pk)^{\frac{1}{2}} + (qk)^{\frac{1}{2}} = c + (qk)^{\frac{1}{2}} \tag{36}$$

where c comes from (18).

5.5 The regulation of Power Quality

Power quality regulation adjusts to the following rationale:

The basic regulated variable is price (i.e., the electricity tariff). Moreover, recall that the main goal of regulation is the maximization of social welfare which is attained when the supplier company operates at equilibrium, that is to say, with zero economic profit. Additionally, the agency requires that the company have "efficient costs and prudent investments". Such criteria are met trough a benchmark procedure, searching for the parameters that should be adequate for a virtual or reference company adapted to the real conditions of supply (such as area to be served, number of customers, network technology, etc.) while investment of capital in quality must be optimal in the sense discussed in 4.3.

In such conditions, the agency obtains the average tariff by dividing the economic cost by the energy to be sold.

The crucial point of this process is the calculus of the quality cost. Observe that the parameter q comes from the reference company and Q* is the optimized capital. We assume that the corresponding quality cost C_Q is univocally related to the severity index. The index value which corresponds to the (C_Q*) of the reference company is obtained as the practical target to be pursued by the regulated firm. In other words, the agency includes in the total cost allowed to the company the C_Q* associated with the target index, expecting that the customers receive exactly such level of quality.

If a particular customer is committed to a lesser quality level, the agency will require that the firm pay to him (her) a monetary compensation (usually as a discount in the next electricity bill) equal to the difference between the actual and the target cost of quality. This implies an effective tariff for which the consumer pays only for the quality level that he or she is actually receiving. The process described above is presented in Figure 11.

The company parameters in the case outlined are:
e = 200 US$/MWh p = 3600 US$²/MWh² q = 1600 US$²/MWh² d = 0,10 w = 0,099
Assuming E = 8 TWh the customer parameters come to be:
a= 4320 US$/MWh b = 500 US$/MWh²
and the optimal investments result in
B* = 960 MUS$ Q* = 640 MUS$
This leads to the flow diagram of Figure 11. In this example the investment in quality is one half of the optimum level.

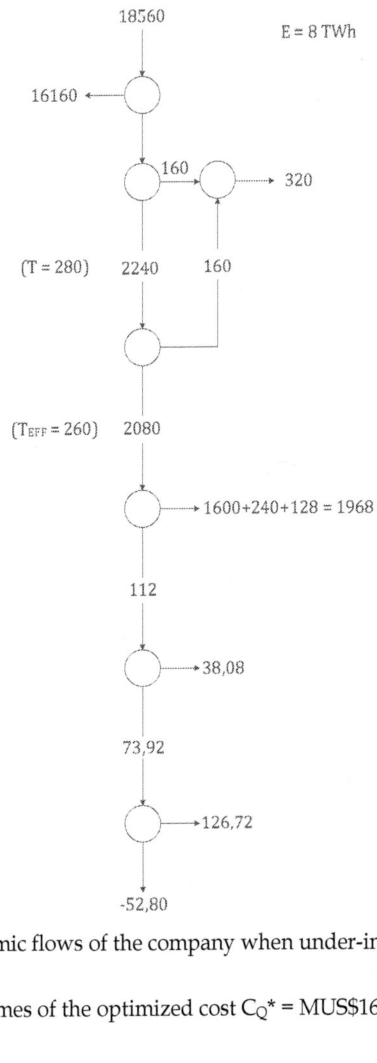

Fig. 11. Diagram of economic flows of the company when under-invested, with compensation.

Then, the cost C_Q is two times of the optimized cost C_Q^* = MUS$160.

But the company is required to return to the customer the difference MUS$320 - MUS$160 = MUS$160 as a consequence, the consumer pays as effective tariff equal to 260 US$/MWh, which corresponds to the degradated energy which he (she) receives.
Note that the company in such conditions has a negative V. Thus, compensation also acts as a disincentive for underinvestment and a stimulus for optimal use of capital.

6. Conclusions

Along the chapter we have laid the foundations of an Electricity Market Model, named TAROT - Tariff Optimization, which:

a. allows for the consumers representation, featuring the surplus that they draw for the purchasing of energy;
b. considers the basic goal of the regulatory frame, namely the maximization of the social value generated by the market transactions;
c. favors the insertion of the quality levels of supply expressed as quality costs and quality improvements as capital investment;
d. unifies the regulatory frame - initially restricted to a cap-price of the energy disregarding its imperfections - optimizing both the scale and the quality portions of total invested capital.

Other applications can be devised on the same lines of thought. Perhaps the main advantage of this approach lies in this integrative potential which allows treating problems of very diverse nature and characteristics under a unified insight.

7. References

Arango, H.; Abreu, J. P. G.; Bonatto, B. D.; Tahan, C. M. V.; Kagan, N. & Gouvea, M. R. (2008). A Model for Electricity Markets: The impact of Regulation on Value, *Proceedings of the International Conference on the European Electricity Market (EEM 2008)*, Lisbon - Portugal, May 28-30, 2008.

Arango, H.; Abreu, J. P. G.; Bonatto, B. D.; Tahan, C. M. V.; Kagan, N. & Gouvea, M. R. (2008). Modelling the Influence of Power Quality on the Creation of Market Value, *Proceedings of the 13th International Conference on the Quality of Power (ICHQP 2008)*, Wollongong, New South Wales, Australia, Sept. 28 -Oct. 1st, 2008, IEEE/PES, USA.

Arango, H.; Abreu, J. P. G.; Bonatto, B. D.; Tahan, C. M. V.; Kagan, N. & Gouvea, M. R. (2008). The Influence of Quality on the Creation of Economic Value in Electricity Markets, *Proceedings of the 14th International Conference on the Quality of Power (ICHQP 2010)*, Bergamo, Italy, Sept. 26-29, 2010, IEEE/PES, USA.

Beesleyand, M. E. & Littlechild, S. C. (1989). The Regulation of Privatized Monopolies in the U.K.. *RAND Journal of Economics*, Vol. 20, No.2, Autumn 1989, pp. 454-472.

Friedman,L. (2002). *The Microeconomics of Public Policy Analysis*, Princeton University Press.

Jehle, G. A. & Reny, J. P. (2000). *Advanced Microeconomic Theory*, 2nd ed., Addison Wesley - Longman.

Jensen, M. C. & Meckling, W. H. (1976). Theory of the firm: Managerial Behaviour, Agency Costs and Ownership Structure, *Journal of Financial Economics*, Vol. 3, 1976, pp. 305-360.

Kreps, D. (1990). *A Course in Microeconomic Theory*, Ed. Harvester & Wheatshaf.

Kupfer, D. & Hasenclever, L. (2002). *Economia Industrial*, Ed. Campus-Elsevier.

Martin, J. D. & Petty, J. W. (2000). *Value Based Management*, Harvard Business School Press, Boston, 2000.

Von Neumann, J. & Morgenstern, O. (1944). *Theory of Games and Economic Behavior*, V. Press, Princeton, USA.

Part 2

Power Quality and Applications

3

Power Quality in Public Lighting Installations

Peter Janiga and Dionyz Gasparovsky
Slovak University of Technology in Bratislava
Slovakia

1. Introduction

Power quality in public lighting networks is yet not sufficiently studied today. New conditions and requirements arise as a consequence of rapid application of new technologies like LED lighting, lighting control and monitoring, adaptive lighting systems etc. Optimization of reliable and efficient service of public lighting systems can be only achieved if behaviour of individual elements within the system are well described.

For erection, operation and re-construction of public lighting networks it is important to have knowledge on behaviour of the network in different conditions. Because currents primarily depend on the supply voltage, it is essential to analyze electrical parameters not only under standard conditions but also under operational conditions. These comprise e.g. distorted or regulated voltage supply.

Non-linearity of electrical parameters of luminaires seems to have significant influence on power quality. In luminaires with conventional ballast, non-linearity is generated by discharge lamp as well as choke. The older type of lamp or choke (in aged existing systems), the worse characteristics. Luminaires with electronic ballasts do not provide this effect and are helpful to maintain "clear" networks. However, in the field of public lighting, unlike in interior lighting with fluorescent lamps for instance, electronic control gears still do not reach any comparative popularity. LED luminaires of diverse quality now intrude the market though technology level, as agreed by experts on many forums, do not compete with available traditional lighting approaches, is known for unsolved problems and needs a series of further developments. From the electrical point-of-view, control gears for LED lamps are similar to electronic ballasts. Supply voltage provided by switch-type source is rectified first. Semiconductor components of rectifier act as non-linear elements. Though these control gears have to be equipped with filter of harmonics, many LED luminaires still lack such circuits.

For assessment and investigation of power quality characteristics it is indispensable to understand the behaviour of individual elements and then the behaviour of network as a whole. Distribution network impacts the public lighting network (via transformer) and vice versa. To describe the nature of this two-way influence in details is very complicated, thus certain simplifications in order to speed up calculations are applied. For example, nominal values of harmonic voltages or currents can be used instead of distorted (non-harmonic) characteristics. Or for accurate calculation of steady-state network with discharge lamps measured values can be used instead of complicated description of discharge parameters.

2. Elements of public lighting networks

2.1 Luminaires

Luminaires with discharge lamps (fig. 1 on the left) are the major luminaire types in public lighting systems. LED luminaires (fig. 1 on the right) are installed still rarely and in most of cases only as experimental or demo projects. Discharge lamp as a non-linear element is responsible for deformation of current. Properties of light sources at nominal voltage supply are very similar in production of harmonic frequencies. Fig. 2 depicts V-A characteristics of most commonly used lamps in public lighting - compact fluorescent lamps PL-L and high-pressure sodium lamps. Lamp power has only minimum affect to the shape of V-A characteristic. But if the supply voltage is distorted (from ideal sinusoidal curve), deformation of characteristic is more than evident; this is a consequence of changes in gas ionization inside the lamp's burner.

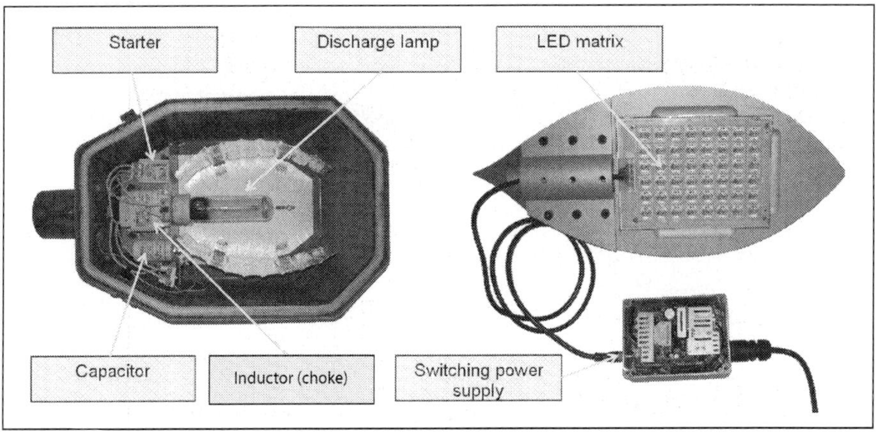

Fig. 1. Luminaire with discharge lamp and LED luminaire

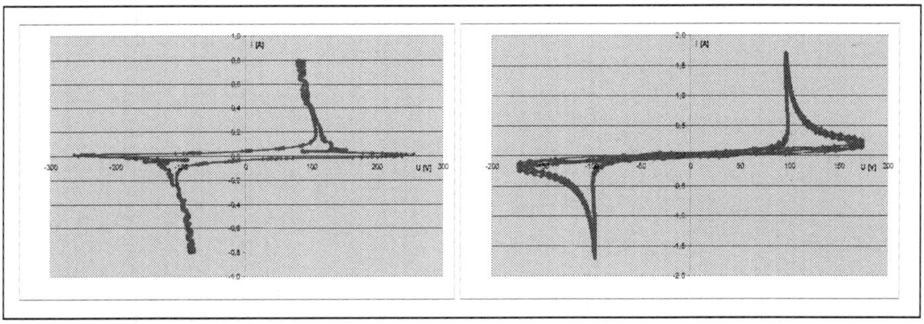

Fig. 2. V-A characteristics of compact fluorescent lamp PL-L (left) and high-pressure sodium lamp (right)

Lamp's ambient temperature is another factor having a significant influence on gas ionization. Low-pressure discharge lamps (including popular PL-L compact fluorescent lamps) are particularly sensitive to low temperatures, therefore, installation of luminaires with this type of lamps should be carefully deliberated. There exist solutions with "thermal cap" on the CFL's tube cold point (tip of the lamp) to supress this effect.

Order of harmonic	HPS lamp 70 W	HPS lamp 100 W	CF lamp 36 W	MH lamp 150 W	MH lamp 400 W
0	0,0 %	0,0 %	0,0 %	0,7 %	0,1 %
1	100,0 %	100,0 %	100,0 %	100,0 %	100,0 %
2	0,3 %	0,3 %	0,6 %	0,4 %	0,3 %
3	18,0 %	18,0 %	21,6 %	17,3 %	13,3 %
4	0,3 %	0,2 %	0,4 %	0,2 %	0,2 %
5	4,5 %	3,6 %	2,2 %	3,1 %	2,7 %
6	0,2 %	0,2 %	0,5 %	0,2 %	0,2 %
7	3,6 %	3,3 %	3,2 %	2,6 %	1,3 %
8	0,2 %	0,1 %	0,4 %	0,2 %	0,2 %
9	1,7 %	1,3 %	1,3 %	0,7 %	0,2 %

Table 1. Harmonic content of a public lighting luminaire with inductive ballast supplied by ideal sinusoidal voltage

Key: HPS = High Pressure Sodium, CF = Compact Fluorescent, MH = Metal Halide

Current flowing through a discharge lamp depends (besides other factors) on the age of lamp. This effect is namely sharp for metal halide lamps. Approaching the end of lamp's life, rectifying effect may occur. Therefore, control gear should be equipped with overload protection circuits.

Table 1 shows that luminaires with magnetic ballasts generate mainly the third harmonic of the current. Values in table 1 have been acquired by measurements of new, duly aged lamps, and it is clear that since the begin of operation the direct-current component is already present. Compensation of power factor in luminaires with conventional ballast is fixed to a particular value and during lifetime of the lamp the system is variably undercompensated or overcompensated.

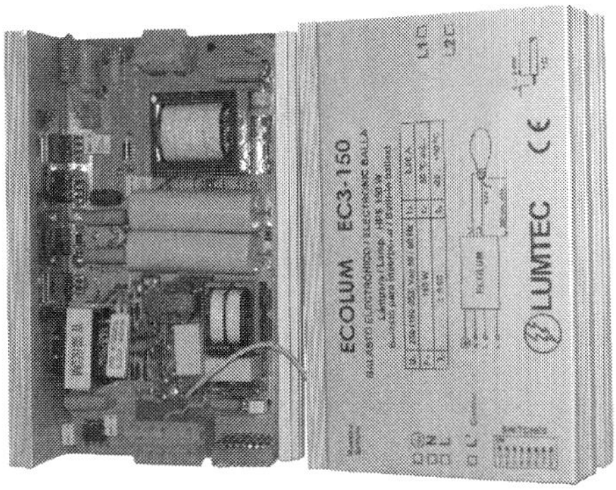

Fig. 3. Electronic ballast

Situation is different for luminaires with electronic ballasts (Fig. 3) and for LED based luminaries (Fig. 1 on the right), where voltage is rectified prior to further modification of supply parameters. Here a different spectrum of harmonics can be observed (see table 2). Harmonic content depends on properties of harmonic filter, if there is any. For high-quality filters the content of harmonics can be decreased down to minimum values. In comparison to luminaires with inductive ballasts, harmonic content is not so much influenced by quality of power supply.

Order of harmonic	LED luminaire without filter of harmonics	LED luminaire with filter of harmonics	HPS lamp 150 W (electronic ballast with filter of harmonics)
0	0,0 %	9,2 %	0,0 %
1	100,0 %	100,0 %	100,0 %
2	0,1 %	1,4 %	0,4 %
3	21,6 %	63,0 %	8,9 %
4	0,2 %	1,0 %	0,3 %
5	6,2 %	29,3 %	2,0 %
6	0,2 %	0,8 %	0,1 %
7	3,6 %	18,2 %	1,4 %
8	0,2 %	0,8 %	0,1 %
9	1,8 %	18,7 %	1,1 %
10	0,1 %	0,9 %	0,0 %
11	1,4 %	11,8 %	0,8 %

Table 2. Harmonic content of a public lighting luminaire with electronic ballast

In electronic ballasts internal wiring has to fulfil requirements to avoidance of interference with surrounding devices that are operated on higher frequencies. In EU, emissions shall be in accordance with the norm EN 55 015.

Electronic ballasts (EB) used in public lighting in comparison to similar devices used in interior lighting can have circuits for regulation of output power according to preset switching profile (ON-OFF-DIM cycles with reduced power in dimming mode). Electronic ballasts cannot be connected to networks with central voltage regulator because decreased supply voltage for ballast may malfunction the lighting operation. Light dimming in EB is possible thanks to controlled high-frequency oscilator.

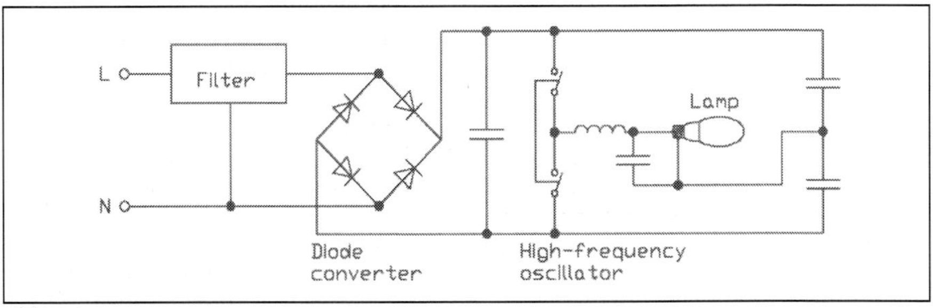

Fig. 4. Circuit diagram of an electronic ballast

Input filter (see fig. 4) limits the efect of current deformation and protects against random high-frequency pulses that may appear in supply network (e.g. as a result of switching processes). Right to the rectifier bridge there is a coupling capacitor for smoothing of rectified voltage. Next stage of signal processing consist of two transistors used to generate a high-frequency signal for maximized power. The frequency must be above the range of audible sound, i.e. 20 kHz (upper limit lies about 100 kHz). Choke is wired in series with lamp. Choke is responsible for limitation of current through lamp. Capacitor connected parallel to the lamp is used for creation of the ingition voltage.

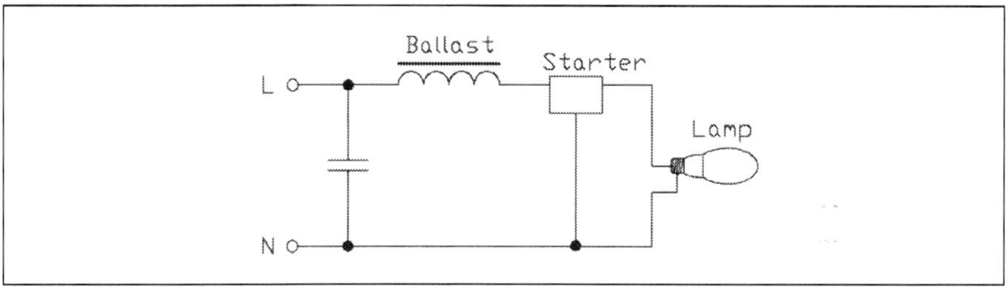

Fig. 5. Circuit diagram of a magnetic ballast

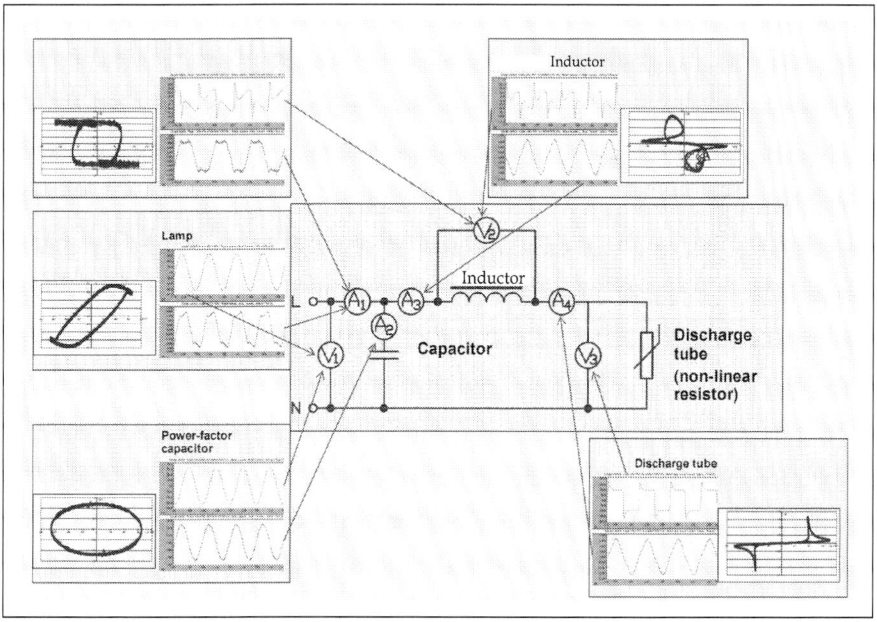

Fig. 6. Behaviour of voltage and current in a discharge lamp

Magnetic (conventional) ballasts (energy efficiency index EEI = B1, B2, C, D according to CELMA) consist of choke, capacitor and starter. Thermal losses are prevailing in this type of ballasts. Losses are originated in windings of the choke by flowing currents and also by hysteresis of the core. Level of losses depends on mechanical construction of the ballast and wiring of its windings.

Simple choke is a dominant ballast type for high-pressure discharge lamps. It has single winding on a core, due its construction it is simple and cheap. Choke is connected in series with lamp (Fig. 5). Provides only limited regulation possibilities by input voltage. Ignition current is very high, thus, the whole circuit must be constructed to withstand such currents. During operation of magnetic ballast, non-harmonic currents flow as a result of non-linearity of the lamp. Fig. 6 shows V-A characteristics of individual components of a luminaire in a steady-state condition. Besides non-linear V-A characteristic of lamp a hysteresis of choke is also depicted. This hysteresis also contributes to the resulting deformated current flowing through a luminaire. Compensating capacitor shifts the current phase in order to get better power factor. But only phase of the first harmonc is affected, higher-order harmonic currents have a phase shift. If this type of ballast has a 20 % content of harmonics (related to the first harmonic), this part of current remains with a phase shift and thus energy losses are increased.

Luminaires with conventional ballasts are usually equipped with compensating capacitor which is used to increase the PF factor and to compensate the reactive power. Values of active and reactive power vary during stabilization of discharge (fig. 7). Shortly after switch-on of luminaire the total current is overcompensated, because current flowing through the choke and lamp are smaller than nominal current.

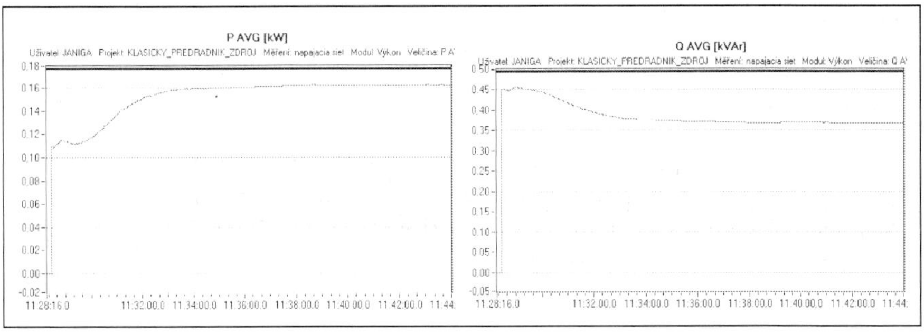

Fig. 7. Active and reactive power during start-up of the luminaire with inductive ballast

Measurements of capacitors from outdated roadlighting luminaires (over 30 years of operation) showed that the drop is very small, practically neglectable. Although these capacitors are thermally stressed, measured capacity value was still within the tolerable range.

2.2 Cables

Cables realize conductive connections between network nodes. Requirements to technical parameters of cables used in public lighting systems are established in the norm IEC 60 446. These heavy-current leads can be supplied by additional control wires.

Network lines can have one of these forms:

- **Overhead conductors** – non-isolated wires on concrete masts or wooden poles
- **Buried cables** – mainly in city centres or residential areas where emphasize is given to visual "invisibility" of infrastructures

- **Overhead cables** combining the easy access for maintenance of overhead lines and compactness of cabled wires; these are used for public lighting reconstruction as replacement of bare conductors (e.g. to avoid short-circuits caused by winds) or to bypass local faults etc.

Only resistance comes to calculation as an input parameter. Reactance is much lower, therefore neglectable. Measurements in a real network with 9 luminaires confirmed low influence of reactance. Harmonic content do not vary with cable length (distance from switchboard) but only depends on degree of voltage distortion, small differences in characteristics due to the age of lamp and deviations of electrical parameters as shown on Fig. 8.

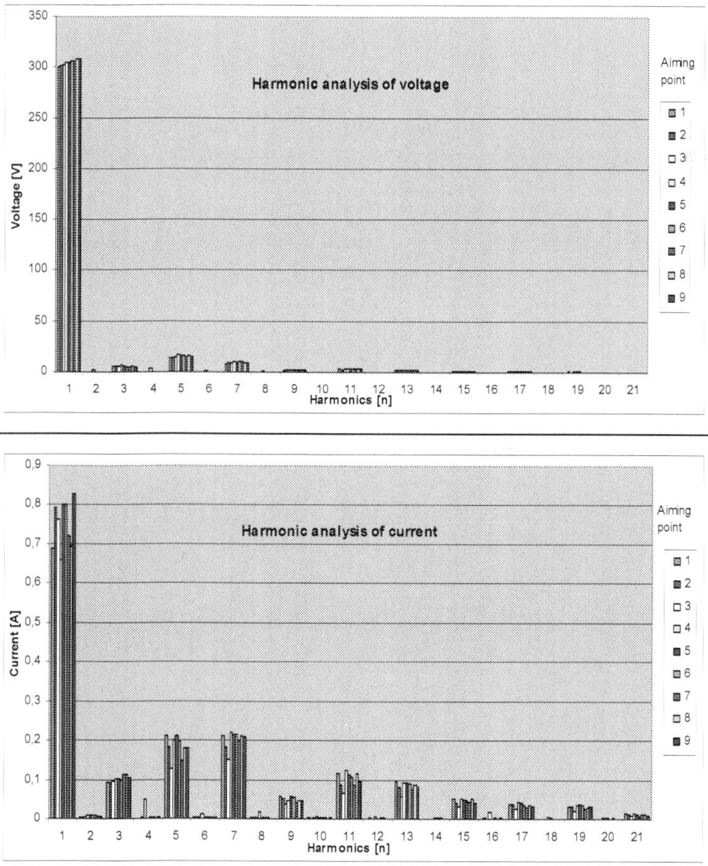

Description of the measured lighting system:
Spacing of light poles: 30 m
Network type: buried cables
Material of conductors: aluminium core, 16 mm^2 cross-section

Fig. 8. Harmonic content versus length of line

Public lighting systems are supplied by means of primary leads, terminated in a distribution box (Fig. 9) equipped with main fuse, kWh-meter, lighting control and circuit breakers on outputs where secondary network is connected. Secondary side is branched to one or more sections. Secondary network is typical for tree-type topology with linear branching, loops are constructed only in small number of cases, sometimes formed unintentionally as a result of unprofessional maintenance. Burried cables are usually connected directly to pole's terminal block and continue from pole to pole. In some countries, junction of smaller size cable by means of T-connector is preferred instead. In case of overhead lines (bare or isolated), always an auxiliary wire enters the luminaire, terminated on its terminal block.

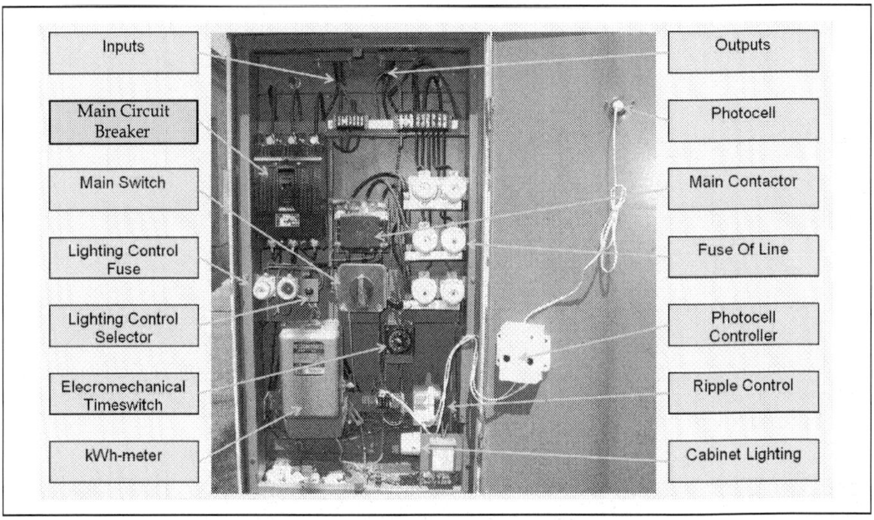

Fig. 9. Public Lighting Switchboard

Quality of network is determined by several factors like materials of cables or conductors, arrangement and fixation of lines, load of conductors. Fault rate of lines has a strong impact on operational costs because any servicing of buried cables is very time-consuming, personnel demanding and therefore expensive.

Many older systems of public lighting utilize common PEN conductor shared with distribution network, e.g. supplying residential houses. This solution is based upon efforts to save one wire years ago. Today such approaches are not allowed due to many reasons but still there are many older systems operated this way. Shortcomings are evident - overload of the PEN conductor and, what is even more important, mutual influence of two otherwise separate networks. This problem only concerns overhead lines. If re-construction of lighting systems is planned, one of the most important measure is to fully separate the public lighting network from distribution network.

3. Problems in public lighting networks related to power quality

Insufficient power quality characteristics often do not appear immediately but after certain period of time, e.g. after long-term stress of isolation due to overload of conductors. It is therefore very important to know and recognize possible difficulties, problems and their consequences.

3.1 Distortion of supply voltage

Currents flowing through public lighting networks can be calculated from nominal values, i.e. current values at ideal sinusoidal voltage.

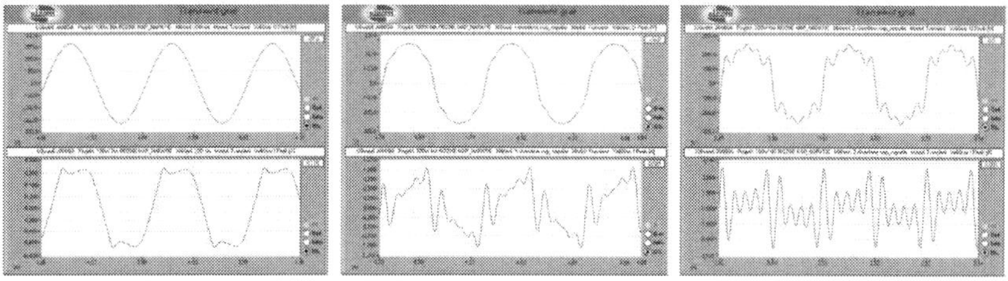

Harmonics	U [%]	I [%]
1	100,0	100,0
3		18,9
5		4,2
7		3,4
9		1,4

Harmonics	U [%]	I [%]
1	100,0	100,0
3	10,9	30,5
5	4,8	37,7
7	3,8	44,2
9	1,9	25,8

Harmonics	U [%]	I [%]
1	100,0	100,0
3	12,2	35,6
5	13,1	104,8
7	13,7	157,2
9	13,8	201,2

Fig. 10. Example of the current wave deformation depending on supply voltage deformation for a luminaire with conventional ballast

In a real operation, waveform of voltage always less or more deviates from its ideal shape. How this voltage deformation influences to the deformation of current waveform in a luminaire, is depicted on Fig. 10.

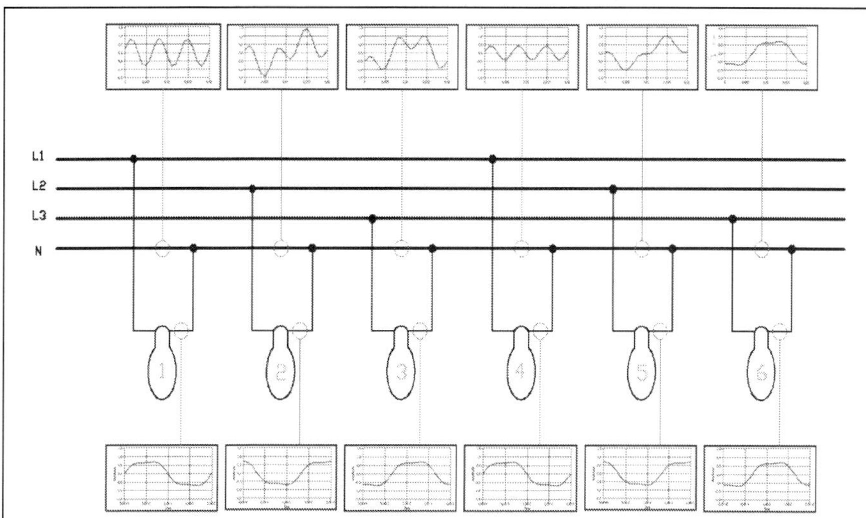

Fig. 11. Emersion of triplen harmonics in the PEN conductor in public lighting networks

Increase of voltage distortion results in much higher increase of the current deformation. In three-phase networks the content of triplen harmonics may cause significant overload of the PEN conductor, leading to degradation of its insulation and consequent cable faults. Fig. 11 shows the emersion of these currents in PEN conductor.

Voltage distortion may also be initiated by noise, defined as an undesired broadband spectrum signal superponed to the basic harmonic voltage wave. Electronic devices generate the major part of noise in networks. Luminaires with conventional ballasts are sensitive to the power quality of supply voltage and even small steps are transmitted and magnified to the current waveform. Such small variation of voltage can be caused by semiconductor regulators, supply source etc. Fig. 12 shows how small noise can impact the waveform of current in a luminaire with conventional ballast due to gradient of voltage. These small and rapid changes are transmitted via choke to the discharge lamp, influencing the gas ionization in its burner.

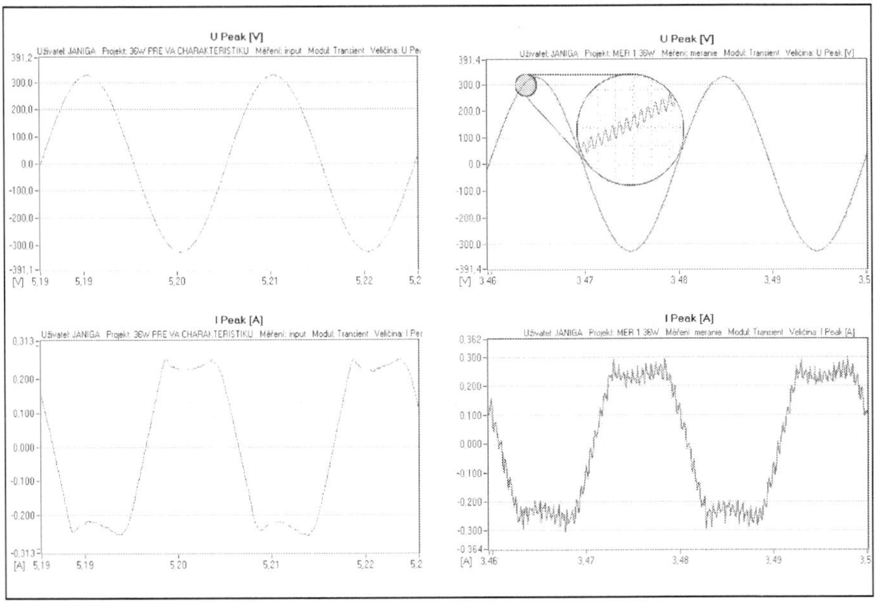

Fig. 12. Influence of the power quality of supply source to the waveform of current (supply by sinusoidal voltage and supply by voltage with small oscillations)

Operation of luminaires with electronic ballasts can help to eliminate flows of non-harmonic currents. Smoothing of the current waveform can be performed by means of filters to less or more extent, depending on particular requirements. Generally spoken it is impossible to obtain again an ideal undistorted waveform. Example is shown on Fig. 13. When current is passing through the zero point, small portion of harmonics is not filtered. In this case, resulting value of THD_I is 10 %. Electronic ballasts in comparison to inductive ballasts benefit from almost zero phase shift. Thus, the network do not transmit neither any distortion power nor any reactive power.

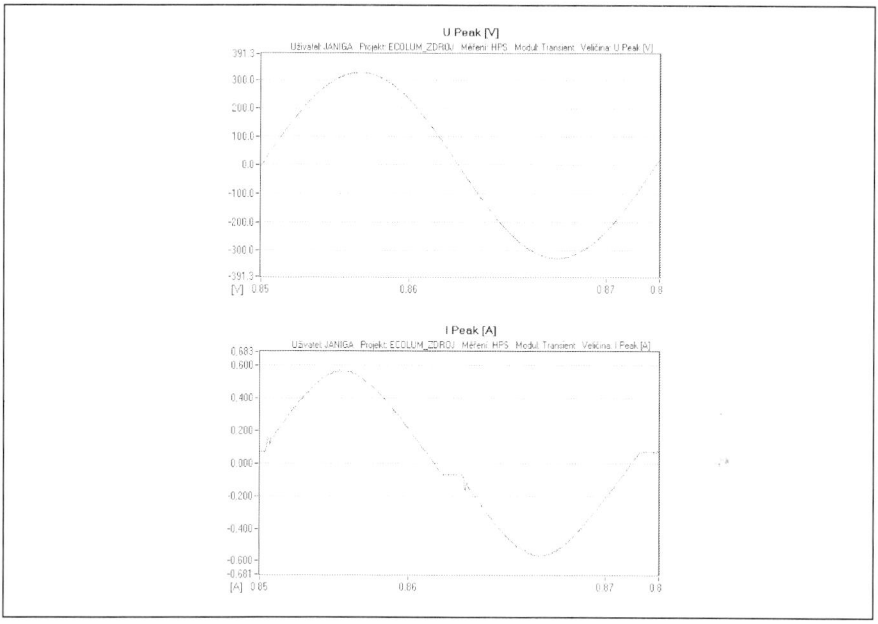

Fig. 13. Waveform of current for a luminaire with electronic ballast and filter of harmonics

3.2 Out-of-range voltage

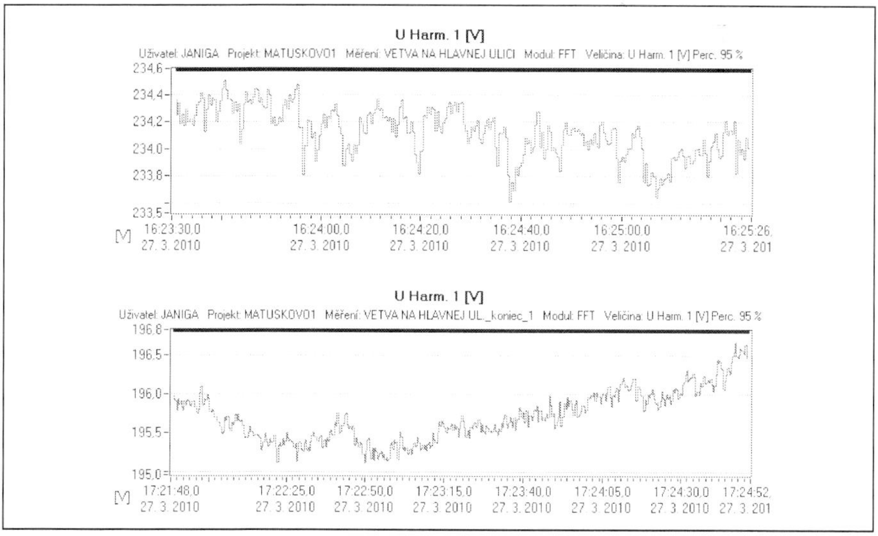

Fig. 14. Voltage at the begin (green) and end (red) of a network branch

For dimensioning of cross-section of conductors of public lighting lines it is not sufficient to take into account solely the load carrying capacity, it is also necessary to check the voltage drops. This is particularly crucial for networks with central voltage regulators. Combination of an excessive voltage drop with decrease of supply voltage may lead to situations when discharge lamps located farther from the distribution box will not be able to maintain the stability of discharge and stop to lit. This problem concerns mainly luminaires with conventional ballasts and the effect depends on rated lamp power. The higher power the longer discharge (given by distance between electrodes in lamp's burner) and the higher voltage needed to keep the discharge in a stable burning state. Electronic ballasts can compensate smaller variations of voltage but these are not suitable (or applicable) for light dimming by voltage regulation.

Fig. 14 depicts results of measurement (Fig. 15) of voltage in a network with high-pressure mercury lamps (250 W) and buried cables. Topology of the networks is linear and its length is 1,5 km. Cable cores are made of aluminium with 16 mm^2 cross-section. In this network the voltage-related problem is strengthen during start-up of the operation when higher (so called starting) currents flow through conductors. This moment the luminaires are most sensitive to the level of supply voltage. Long tube compact fluorescent lamps PL-L (widely used in public lighting systems) at lower temperatures and voltage below 195 V were not able to start-up reliably.

Fig. 15. Measuring power quality in public lighting network

3.3 Short-time voltage variation

Transition effect shown on Fig. 16 was caused by third-party electricity consumer. This situation is very dangerous for luminaires because step changes of voltage may induce extensive current pulses on choke and/or lamp. It is almost impossible to track the reasons of faults of luminaires or network because there is no continuous monitoring of voltage or current. In spite of the fact that these faults are induced to the public lighting network from outside, service costs for repair of damaged part of system have to be covered by network operator. Passive elements cannot detect such small voltage changes and are therefore helpless in protection against them. The only option is limitation of current through network.

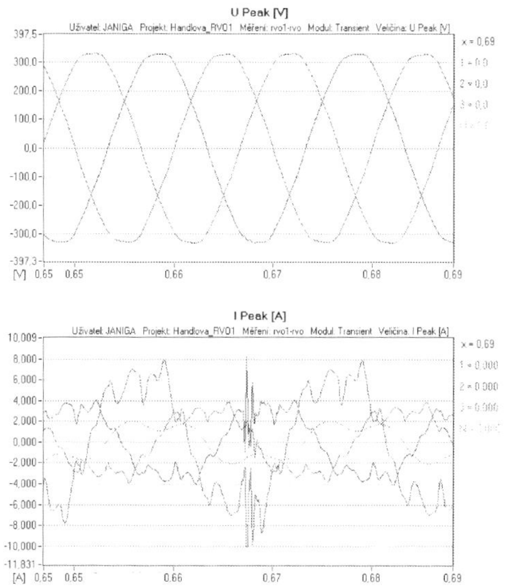

Fig. 16. Rapid voltage variation which induced a short-time transient effect

3.4 Incorrectly set voltage regulator

If a lighting regulator is installed, at the end of longer lines the voltage may drop below the range of stable operation even if recommended settings are satisfied. Regulator with smooth start-up is a particular case, voltage during operation of a regulator is shown on Fig. 17.

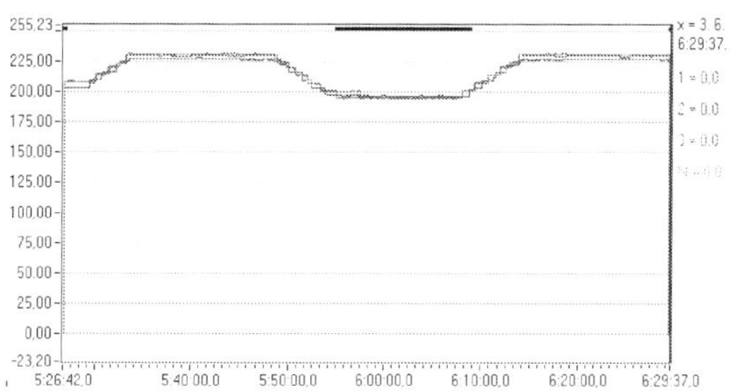

Fig. 17. Output voltage of a regulator

If the regulator starts with 210 V at the begin point of a branch (just on output of the distribution box with regulator), voltage at the end point of this line may fall below the level of ignition voltage for luminaires with metal halide lamps or compact fluorescent lamps (at low temperatures in particular).

4. Properties of network from the power quality point-of-view

4.1 Level of voltage and current waveform deformation

Internal wiring of a luminaire is shown on Fig. 18. Deformated voltage induces deformation of current on two parallel sub-circuits, one with capacitor and the other one with choke and lamp.

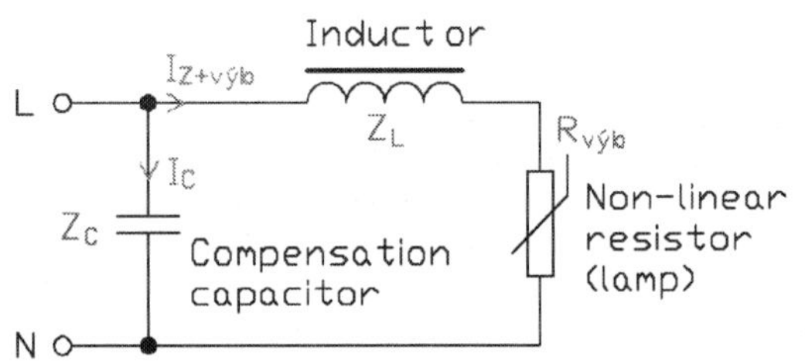

Fig. 18. Simplified arrangement of elements of a luminaire with conventional ballast

Impedance of capacitor is lower for higher-order harmonics than for the basic harmonic. If internal resistance is neglected, impedance of capacitor is given by the expression as follows:

$$Z_C = \frac{1}{j\omega C} \tag{1}$$

Description of impedance of the other sub-circuit is more complicated if supplied by distorted voltage, because behaviour of electrical parameters is linked to chemical and physical processes in filling gas and ability of gas to response to voltage variations what is not the subject of this publication.

The ballast choke has for higher-order harmonic voltages increased impedance and the lamp is typical for non-linear resistivity. There is a certain form of memory effect - same level of voltage in general do not correspond with the same level of current. If capacity and inductance of the lamp (discharge) is neglected, impedance of the choke and lamp can be calculated as follows:

$$Z_{L+vyb} = R_L + j\omega L + R_{vyb} \tag{2}$$

Situation for luminaires with electronic ballast or switch-type source is different. Behaviour is determined by V-A characteristics of rectifying diods. Resulting characteristics of the current can be determined on the basis of calculated or measured impedance as portion of voltage and impedance for individual harmonics.

But if it is not possible to determine the function describing the luminaire's impedance, one of the ways how to describe the influence of distorted voltage to problems appearing in public lighting networks - is to use measured values.

Basic harmonic	Higher order harmonics of voltage			Higher order harmonics of current [%]							
	n	[%]	[V]	1	2	3	4	5	6	7	8
220				100	0,3	16,7	0,3	4,5	0,3	3,7	0,4
220	2	2	4,2	100	5,6	16,5	0,3	4,5	0,4	3,7	0,5
220	3	5	10,5	100	0,2	13,8	0,3	4,4	0,4	3,5	0,3
220	4	1	2,1	100	0,2	16,6	7,6	4,5	0,3	3,5	0,1
220	5	6	12,6	100	0,1	17	0,2	69,6	0,3	3,7	0,2
220	6	0,5	1	100	0,2	16,7	0,2	4,4	5,8	3,7	0,3
220	7	5	10,5	100	0,2	16,8	0,1	4,4	0,3	76,9	0,3
220	8	0,4	0,8	100	0,2	16,5	0,3	4,5	0,4	3,7	8,6
220	9	1,5	3,1	100	0,2	16,8	0,1	4,5	0,2	3,3	0,3
220	10	0,4	0,8	100	0,3	16,6	0,2	4,5	0,2	3,4	0,1
220	11	3,5	7,4	100	0,2	16,7	0,1	4,5	0,2	3,3	0,2

Basic harmonic	Higher order harmonics of voltage			Higher order harmonics of current [%]							
	n	[%]	[V]	9	10	11	12	13	14	15	16
220				1,7	0,3	1,2	0,1	1,0	0,1	0,7	0,2
220	2	2	4,2	1,5	0,3	1,3	0,2	1,0	0,2	0,6	0,1
220	3	5	10,5	1,8	0,3	1,4	0,4	1,0	0,3	1,0	0,3
220	4	1	2,1	1,8	0,1	1,4	0,2	0,8	0,1	0,8	0,1
220	5	6	12,6	1,7	0,1	1,7	0,2	1,0	0,2	0,9	0,2
220	6	0,5	1	1,9	0,4	1,1	0,3	1,0	0,2	0,9	0,2
220	7	5	10,5	1,7	0,2	1,4	0,2	1,4	0,2	1,4	0,3
220	8	0,4	0,8	1,4	0,4	1,4	0,4	1,2	0,3	0,5	0,3
220	9	1,5	3,1	30,1	0,4	1,9	0,5	0,9	0,4	0,8	0,1
220	10	0,4	0,8	1,9	10,8	1,5	0,6	0,5	0,6	0,9	0,3
220	11	3,5	7,4	2	0,4	93,6	0,6	0,9	0,2	0,1	0,2

Basic harmonic	Higher order harmonics of voltage			Higher order harmonics of current [%]					
	n	[%]	[V]	17	18	19	20	21	Sum of odd triplen harmonics
220				0,6	0,2	0,6	0,2	0,6	19,7
220	2	2	4,2	0,7	0,2	0,5	0,2	0,6	19,2
220	3	5	10,5	0,5	0,2	0,8	0,3	0,4	17,0
220	4	1	2,1	0,6	0,1	0,6	0,2	0,6	19,8
220	5	6	12,6	0,2	0,3	0,6	0,2	0,4	20,0
220	6	0,5	1	0,5	0,2	0,6	0,2	0,6	20,1
220	7	5	10,5	0,5	0,3	0,8	0,2	1,1	21,0
220	8	0,4	0,8	0,8	0,2	0,7	0,3	0,5	18,9
220	9	1,5	3,1	0,7	0,3	0,5	0,5	1,2	48,9
220	10	0,4	0,8	0,8	0,4	0,5	0,2	0,7	20,1
220	11	3,5	7,4	0,7	0,2	0,7	0,2	1,6	21,3

Table 3. Influence of the voltage waveform deformation on current waveform deformation

Influence of distorted voltage can be evaluated from partial deformations. Table 3 is an illustration of measured values of distorted current when luminaire with inductive ballast is supplied by voltage distorted within the limits prescribed by the EN 50 160 standard. During performance of individual measurements the basic harmonic frequency was distorted by superposition of only one harmonic of higher order. From measurement results follow that even if the voltage satisfies requirements of the norm regarding acceptable range of distortion, deformation of the current waveform is too high. Highlighted values in table 3 point to the case when a voltage harmonic influences the same harmonic of current. Sum of odd triplen harmonics of current for 9th harmonic of voltage exceeds 33 % (separatedly highlighted in the table) and in three-phase systems may cause increased overload of neutral wire in comparison to phase wires.
It can be concluded that limit values of voltage deformation cause significant current deformation. However, due to voltage derivation in particular time periods it is not possible to sum up deformations caused by individual single harmonics.

4.2 Starting current

To describe the behaviour of current during start-up as well as operation is too complicated as its waveform depends on several varying factors. Some of them vary with the age of luminaire (and its relevant parts), some with time of start-up, some depend on supply voltage and some are influenced by temperature etc. Start-up current can be easily measured.

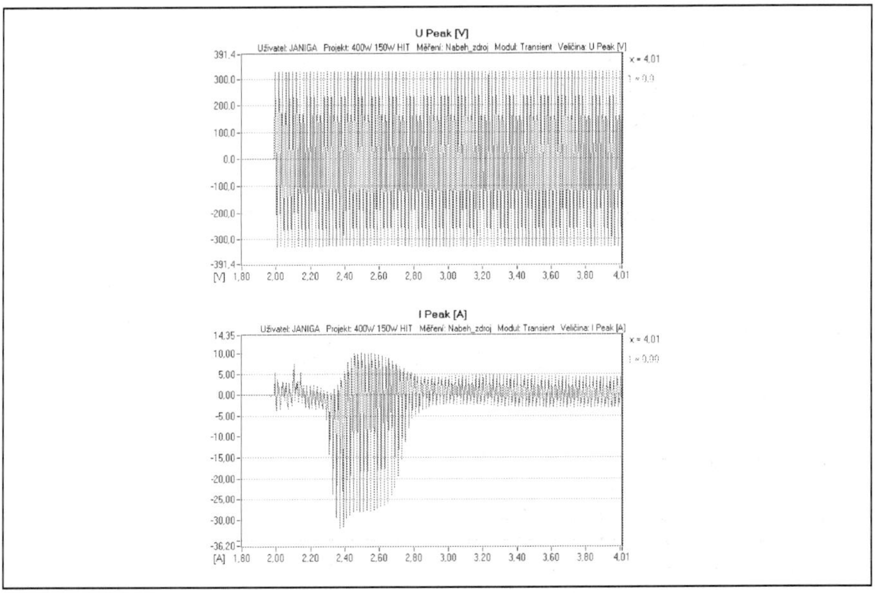

Fig. 19. Pulse start-up current in a luminaire with conventional ballast and 400 W metal halide lamp

This current can reach several tenths of ampers. Start-up duration depends on lamp type and if the start-up begins from cold state or warm lamp after e.g. mains interruption. For metal halide lamps it is particularly important to allow the lamp cool down before attempts to start again. Start-up of luminaires with conventional ballasts may invoke high pulse start-up current or high re-ignition currents after short-time mains interruption.

Pulse current may occur first of all in luminaires with inductive ballasts. In luminaires with electronic ballast or switch-type source no pulse current is emerged. Transient phenomena are invoked by switching of discharge lamp through a choke. As it can be seen on fig. 19, pulse current may reach multiples of nominal current and lasts up to several periods.

With the age of metal halide lamps the rectifying effect is continuously increasing and therefore also the direct current component of the transient process. If the voltage in network is low or if discharge lamps come close to their end of life, due to unstability of arc the repetition of pulse current appearence in short periods of time may cause serious problems in network.

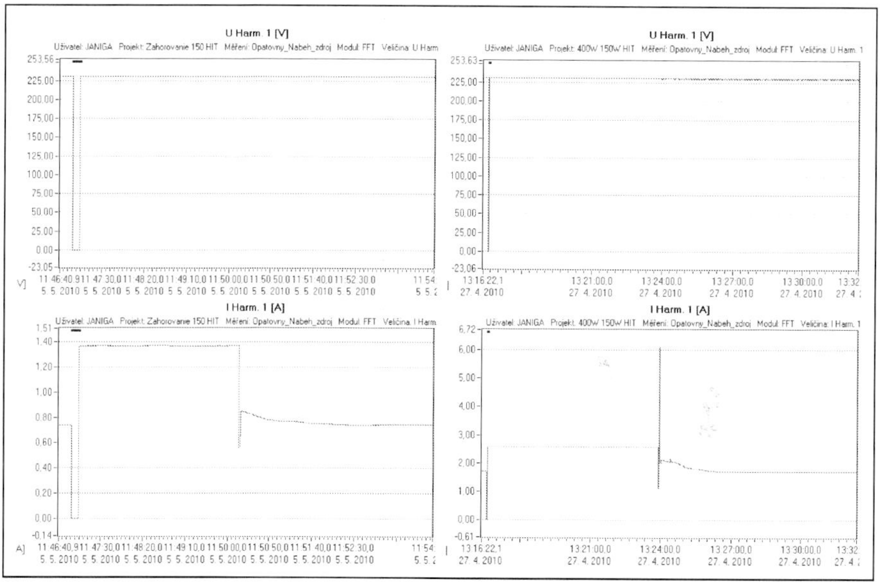

Fig. 20. Pulse re-ignition current in a luminaire with conventional ballast and 400 W metal halide lamp

Hot start-up of discharge lamps after short-time interruptions of voltage supply invoke, in general, different currents than during normal start-up. These characteristics depend on lamp type as well as choke. Re-starting current after voltage interruption is generally much higher than steady-state current in normal operation, as illustrated on fig. 20. Power factor is much worse as well because current flows only through the compensating capacitor. This situation may last as long as the gas in lamp's burner cools down. The duration depends on chemical composition of gas and may be up to 15 minutes.

Similar kind of problem may arise if a luminaire with fault lamp is continued to operate. Then its capacitor remains permanently connected to the network and contributes to the aggravation of power factor. This is a long-lasting problem in public lighting networks.

5. Model of public lighting network

5.1 Algorithm

Model of public lighting network is based on the method of loop currents. Luminaire as a non-linear consumer is substituted by a current source. Flow direction of this current source runs from positive to negative pole. Calculation of the steady state and determination of voltages in nodes of the network is processed by iterations. First step is to assume the same supply voltage on terminals of all luminaires. Current flowing through luminaire at given supply voltage is then measured and used to define the current source which will substitute a luminaire. Voltage values are determined in the next step. The iterations are then repeated until deviations are smaller than pre-defined boundary condition.

This model based upon substitution of luminaires by current sources seems to be the most suitable and practically appliable method for analyses of steady-state public lighting networks due to its simplicity because mathematical description of non-linear luminaire is overcomplicated. Assumptions of the model comprise single supply source and tree-structure of the network what is the prevailing situation in public lighting systems.

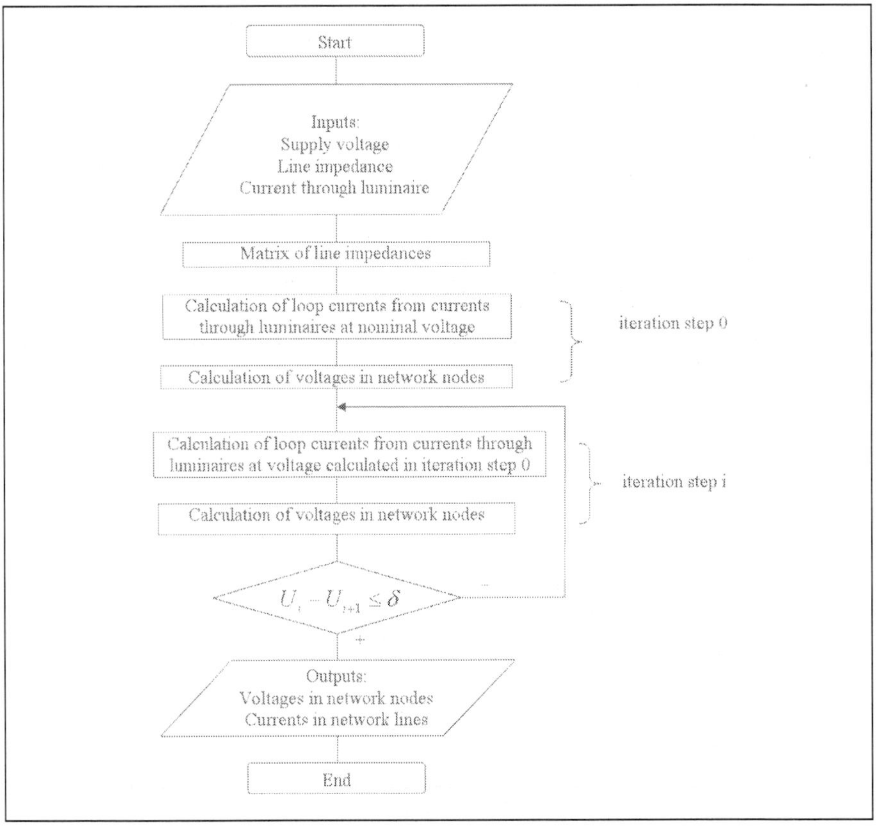

Fig. 21. Algorithm of calculation of steady-state public lighting network

5.2 Impedance of lines

In the framework of initial calculation, matrix of the line impedances $\overline{Z}_v$ is to be assembled. It is a rectangular matrix of $n \times n$ dimensions where n is the number of luminaires in network.

$$\overline{Z}_v = \begin{pmatrix} Z_{11} & Z_{11} & \dots & Z_{11} \\ Z_{21} & Z_{22} & \dots & Z_{2n} \\ \vdots & \vdots & \ddots & \vdots \\ Z_{n1} & Z_{n2} & \dots & Z_{nn} \end{pmatrix} \qquad (3)$$

The Z_{ii} element stands for impedance through which the ith current loop is closed. If the shortest current loops are chosen, what means that current loops have common current sources only, then for single-phase network the matrix $\overline{Z}_v$ is diagonal and the Z_{ij} element only appears in three-phase networks. An element apart from diagonal is a sum of impedances that are common for the ith and j[th] current loop. Fig. 22 shows an example of single-phase network and similarly fig. 23 shows example of three-phase network with impedances through which loop currents are closed as well as the manner how loop currents are determined.

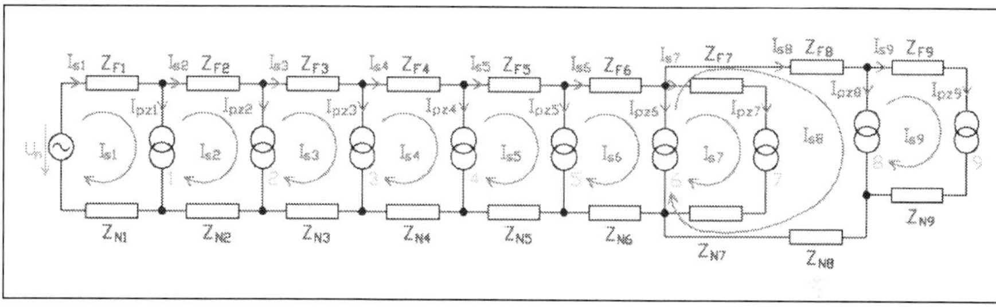

Fig. 22. Equivalent diagram of a network with current sources as substitutions of luminaires (example of a single-phase network)

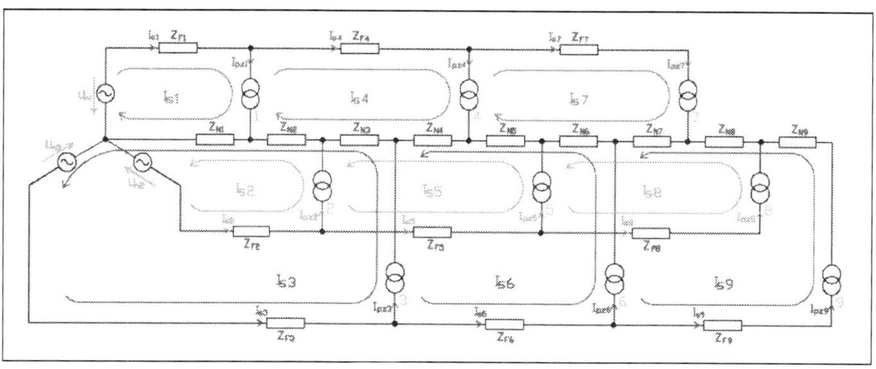

Fig. 23. Equivalent diagram of a network with current sources as substitutions of luminaries (example of a three-phase network)

Elements on the diagonal of matrix $\overline{Z}_v$ can be calculated using the following formula:

$$Z_{ii} = Z_{Fi} + \sum Z_{Ni} \tag{4}$$

where Z_{ii} - line impedance through which the ith loop is closed
Z_{Fi} - phase conductor impedance through which the ith loop is closed
Z_{Ni} - neutral conductor impedance through which the ith loop is closed
If minimum loop currents are defined then in three-phase networks the elements $\overline{Z}_v$ apart from diagonal only consist of the neutral conductor impedance common for several loops.

$$Z_{ij} = \sum Z_{Nij} \tag{5}$$

where Z_{Nii} - impedance of neutral conductor common for i[th] and j[th] loop

5.3 Loop currents

Direction of current flows in network and assumed direction of current loops used for calculations are indicated on fig. 22 and 23. The loop current method does not allow to pass more than one loop current through a current source due to linear dependant equations describing the network. In the public lighting network model, however, the currents are known and the system of equations has a single solution. In zero step of iteration it is assumed that luminaires are supplied by equal level of voltage. Currents are consequently measured at the specified voltage. If currents flowing through luminaires are known, it is possible to calculate currents in lines in the zero step. Currents in phase conductors are in fact loop currents entering the calculation of voltages and the calculation is processed from the end of tree structure where only current from a single luminaire flows.

$$I_{si} = I_{pzi} \tag{6}$$

where I_{si} - current in i[th] loop
I_{pzi} - current flowing through i[th] luminaire.
Remaining loop currents are calculated as sum of currents flowing through those luminaires that are supplied via line of the given loop.

$$I_{si} = \sum I_{pz} \tag{7}$$

5.3 Network node voltages

Voltages related to a loops can be calculated by multiplication of the matrix of line impedances with the matrix of loop currents.

$$\overline{U}_s = \overline{Z}_v . \overline{I}_s \tag{8}$$

Rewriting the equation above into matrix form we obtain:

$$\begin{pmatrix} U_{s1} \\ U_{s2} \\ \vdots \\ U_{sn} \end{pmatrix} = \begin{pmatrix} Z_{11} & Z_{12} & \cdots & Z_{1n} \\ Z_{21} & Z_{22} & \cdots & Z_{2n} \\ \vdots & \vdots & \ddots & \vdots \\ Z_{n1} & Z_{n2} & \cdots & Z_{nn} \end{pmatrix} . \begin{pmatrix} I_{s1} \\ I_{s2} \\ \vdots \\ I_{sn} \end{pmatrix} \tag{9}$$

Network node voltages can be then calculated from the matrix of voltages related to loops. The matrix element U_s represents a voltage difference on i[th] luminaire and voltage in node from which the ith luminaire is supplied.

$$U_{si} = U_{napi} - U_i \tag{10}$$

5.4 Model outputs

The model provides these main outputs:

- **Voltage on teminal blocks of luminaires**: these values allow to investigate if operation of the network do not induce voltages that may damage luminaires.
- **Currents flowing through the network**: these values allow to determine the backward impact of public lighting network on the supply network. It also allows to determine the current in neutral conductor which is often overloaded in three-phase networks.

Voltage and current values can be then used for calculation of parameters defining the distortion of voltage and power flows in the network.

Because calculations are based on iterations, it is suitable to use software tools (see fig. 24).

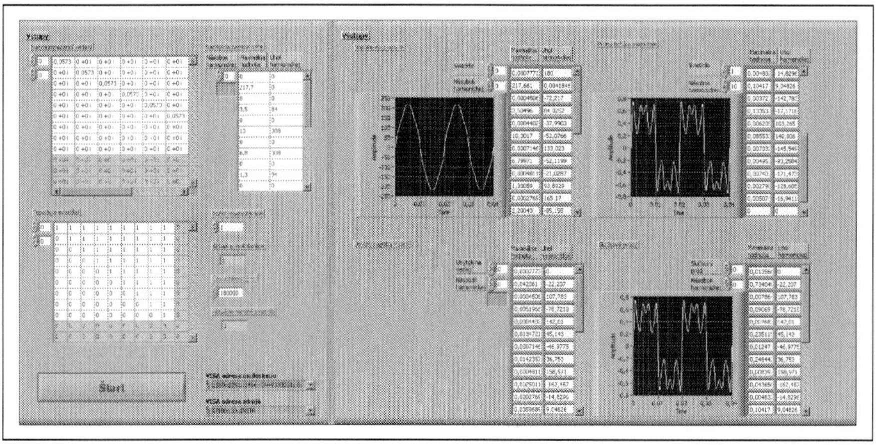

Fig. 24. Software for calculation non-harmonic voltages and currents in public lighting network

6. References

[1] EN 50 160: Voltage characteristics of electricity supplied by public distribution networks

[2] EN 60 555 - 2: Power supply harmonic emissions

[3] IEEE 519-92: Recommended practice for monitoring electric power quality

[4] Acha, E., Madrigal, M. (2001). Power systems harmonics: Computer Modelling and Analysis, John Wiley & Sons, Ltd, New York

[5] Arrillaga, J., Watson, N. R., Chen S. (2000). Power system quality assessment, John Wiley & Sons, Ltd, Chichester

[6] Baggini, A. (2008). Hadbook of Power Quality, John Wiley & Sons, Ltd, Chichester

[7] Belan, A., Eleschova, Z., Smola, M. (2005): Resonance Overvoltages in Electric Power Networks. Proceeding of International Scientific Conference „2005 IEEE St. Petersburg PowerTech". St. Petersburg, Russia 2005. ISBN 5-93208-034-0, pp. 465
[8] Bollen, M. (1999). Understanding Power Quality Problems: Voltage Sags and Interruptions, Wiley-IEEE Press
[9] Chapman, D. (2001). Harmonics - causes and effects, Leonardo Power Quality Application Guide, Part 3.1
[10] Dugan, R.; McGranaghan, M. F.; Beaty, H. W. (1996). Electrical Power Systems Quality, McGraw-Hill, New York
[11] Eleschova, Z., Belan, A., Mucha, M. (2006): Synchronous Generator with a Salient Pole Rotor - Source of Harmonic Distortion. WSEAS Transactions on Power Systems, ISSN 1790-5060. Issue 7, Volume 1., pp. 1196-1201
[12] Onaygil, S., Guler, O., Erkin E., Saygin, H. (2005): Energy Efficient Lighting by Compact Fluorescent Lamps (CFLs) at the Residences, Leon'05, CIE Midterm Meeting and International Lighting Congress, 12-21 May, Leon, Spain
[13] Sankaran, C. (2002). Power Quality, CRC press
[14] Sokansky, K., Novak, T., Blaha, Z. (2009): Assessment of parameters of lights for public lighting. 10th International Scientific Conference "Electric Power Engineering", May 12-14, 2009 Kouty nad Desnou, Czech Republic. pp. 194-196
[15] Sokansky, K., Novak, T. (2008): Energy savings in public lighting. Przeglad Elektrotechniczny, Volume 84, Issue 8, pp. 72-74
[16] Szathmary, P. (2003). Power quality, PRO s.r.o., Banska Bystrica
[17] Gasparovsky, D.; Herdova, B. (2006) Public lighting reconstruction, Typhoon s.r.o and Energy Centre Bratislava, Bratislava

4

Power Quality Impact of High Capacity End-Users

N. Golovanov[1], G. C. Lazaroiu[2], *Student Member, IEEE*, M. Roscia[3] and D. Zaninelli, *Senior Member, IEEE*

[1]*Department of Electrical Engineering, University Politehnica of Bucharest, Splaiul Independentei 313, Bucharest,*

[2]*Department of Electrical Engineering, University Politehnica of Bucharest, Splaiul Independentei 313, Bucharest,*

[3]*Department of Progettazione e Tecnologie, Università di Bergamo,*

[4]*Department of Electrical Engineering, Politecnico di Milano, piazza Leonardo da Vinci 32, Milano 20133,*

[1,2]*Romania*

[3,4]*Italy*

1. Introduction

The random load variation, as in case of the arc furnaces, determines random voltage variations at the supply bus-bars., the can affect the power quality delivered to other customers, supplied by the same bus-bars [1-3]. The voltage flicker level, sensed by the human eye as variations of the light sources flux is determined using the flickermeter, which considers the irritability sensation of the human eye when a threshold level of the supply voltage is exceeded.

To quantify the amplitude of such disturbance, reference is made to the instantaneous flicker sensation $S(t)$. The value of $S(t)$ is measured by means of an instrument, the so-called flickermeter. The setting-up of the flickermeter model is done in Simulink, the simulation tool, in the MATLAB environment. The main advantage of this software product is that it makes it possible to support an object programming procedure, with a circuit representation obtained by means of graphic blocks associated with the different power system components.

The flickermeter model is a conventional analog one, in agreement with the models proposed in current standards that of the necessity to comparative analyses between innovative equipment and conventional architectures, whose disturbances were measured using analog instruments [4].

The present paper deals with the case study of an existing ac arc furnace facility. The 110kV monitoring campaign was conducted using the ION 7600 equipment. The main power quality indices regarding harmonics and flicker levels are reported in the paper with reference to IEEE standards.

2. AC Furnace

The ac furnace, from the system operator point of view, is a highly polluting load: the real and imaginary power variations necessary for the network causes severe harmonics and voltage flicker such that the system operator must establish limits for the disturbances generated by the load. Independently from the type and amplitude of the admissible disturbances, the critical point in analyzing the electrical perturbation of an ac arc furnace is constituted by the high voltage bus, where the network operator evaluates the conformity with the contractual parameters. These parameters are stipulated by standards [5].

Fig. 1 shows the layout of an ac arc furnace supply system. As it can be seen, the intermediate voltage level is of high importance due to the fact that here the filtering processes occur.

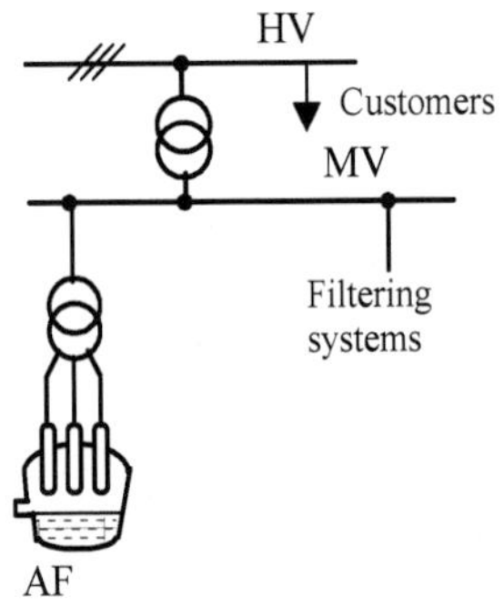

Fig. 1. AC arc furnace connection layout

The power quality evaluation considering the voltage flicker impact is evaluated in accordance with [3]:

- considered data correspond to a week observation period;
- every 10 minutes consecutives values of the short-term voltage index are analyzed;
- the values obtained when the bus voltage is outside the range V±15% or when voltage sags with depth higher or equal to 15% of the rated voltage are considered not valid;
- the index P_{lt} is computed considering 12 valid and consecutives values of the P_{st};
- the number of valid values, N, of P_{lt} measure is determined;
- the number, N_1, when P_{lt} exceeds unity is determined;
- the condition $N_1/N \leq 0{,}05$ is checked.

3. UIE Flickermeter

Since, from the point of view of analyzing the signal, the flicker is a low-frequency modulation of the network voltage at 50 Hz, the purpose of the flickermeter is to separate the carrier from the modulating wave, weight the effects of the latter based on human sensitivity to the disturbance, and return the instantaneous flicker sensation signal.

In the history of electronics, many kinds of architecture have been used to create such instruments. Some of them are based on analog filtering of the voltage at the common point of coupling in order to extract the instantaneous disturbance level and apply it to a predefined weighting curve (shown in Fig. 2) for weighting its incidence and ascertaining its severity.

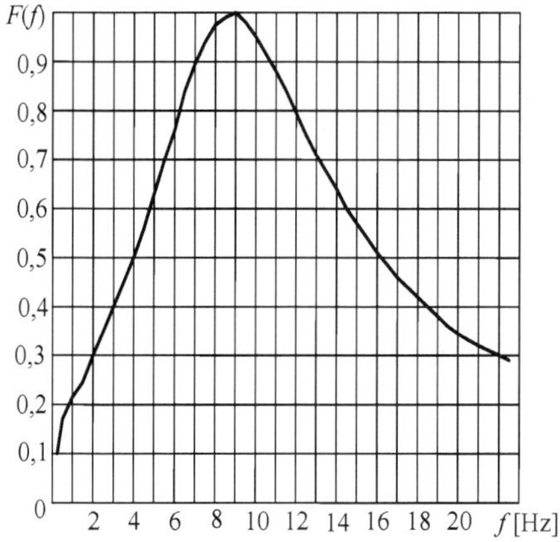

Fig. 2. Weighting curve proposed by UIE (International Union for Electro-thermal Applications)

Fig. 3 shows the layout of the UIE flickermeter, a standardized instrument for measuring the flicker obtained by simulation and by stochastic analysis of the response of the lamp-eye-brain chain to voltage fluctuations.

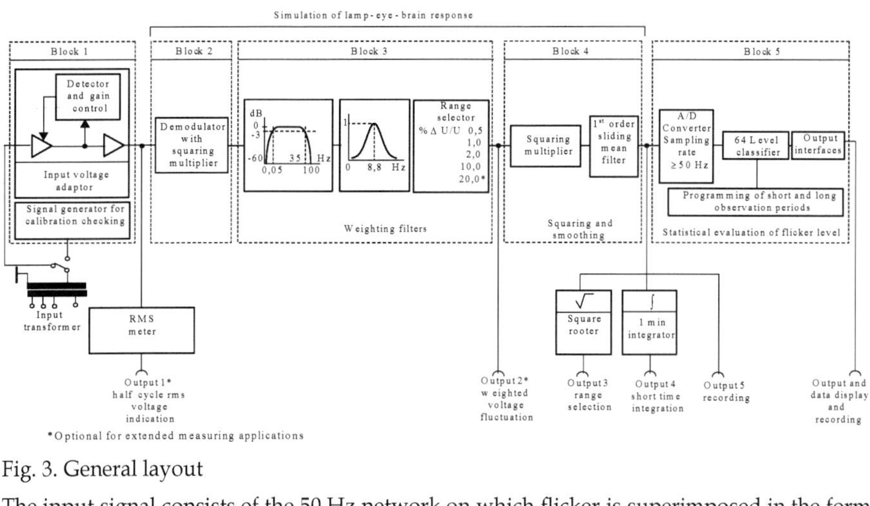

Fig. 3. General layout

The input signal consists of the 50 Hz network on which flicker is superimposed in the form of amplitude modulation and enters Block 1, the so-called "normalization block", after which a transformer reduces its voltage level to values compatible with the electrical specifications of the electronic components located downstream.

Block 1 contains a signal generator for checking the flickermeter setting in the field and a circuit for normalizing, at the internal reference level, the rms value of the input voltage at network frequency.

In this way, flicker measurements are made independent of the actual input voltage, whose variations (ΔV) are expressed as a relative value of the rated voltage $\Delta V/V_n$. This function is obtained by acting on the gain of an amplifying stage by means of a suitable control signal obtained for comparison between the effective output voltage value and a reference signal.
The subsequent blocks (from 2 to 4) react to the signal by simulating the behavior of the lamp-eye-brain chain.
Block 2 is a quadratic demodulator for separating the modulating fluctuation (ΔV) from the carrier wave (network frequency). Block 3 consists of two filters in cascade and by a scale selector of the sensitivity that may precede or follow the selection filter circuit. The first filter eliminates from the output voltage of the quadratic demodulator the dc component and that of the dual frequency in respect of network frequency. The second filter, centered at 8.8 Hz, imposes the form of the response in flickermeter frequency on the modulating fluctuation.
The "weighting filter" block simulates the response in frequency on the sinusoidal fluctuations in the voltage of the chain consisting of a lamp filled with inert gas with a spiral filament (60 W-230 V), followed by the human eye. The response function is based on the perceptibility threshold ascertained, for each frequency, out of 50% of the persons taking part in the experiment.
The structure of the flickermeter may be represented through the transfer functions contained in the block diagram in Fig. 4.
This condenses all the signal manipulations required for obtaining in exit the trend of the instantaneous flicker disturbance in accordance with the information supplied by the current standard [4]. In addition, this diagram considers the steps necessary for setting up efficient electronic implementation.

- Block A has the dual function of expressing the instant input voltage as a p.u. value and that of simulating the behavior of the quadratic demodulator;
- Block B has the job of cutting off the dc voltage component, eliminating the signal of the carrier wave and the high-frequency fluctuations found at the exit of the quadratic demodulator;
- Block C simulates the response in the frequency of the human eye to the voltage fluctuations of an incandescent lamp supplied by a variable sinusoidal voltage;
- Block D squares the signal exiting from Block C by simulating the non-linear perception of the flicker in the eye-brain chain;
- Block E is a low-pass filter that performs the smoothing action.

In Fig. 5 we find implementation of the functionalities of Block A, in Fig. 6 those of Block B, in Fig. 7 those of the Butterworth filter, in Fig. 8 those of Block C, in Fig. 9 those of Block D, and in Fig. 10 those of Block E.

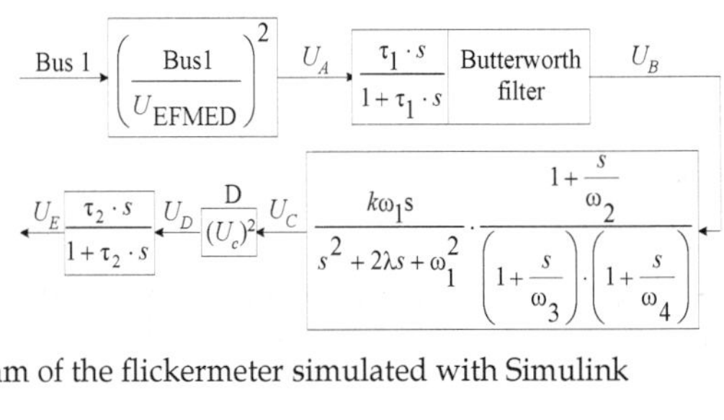

Fig. 4. Block diagram of the flickermeter simulated with Simulink

Once we have obtained the trend in respect of the instantaneous flicker sensation, we have, from the latter, to calculate the indices that make it possible to evaluate the severity of the disturbance injected into the network [4].

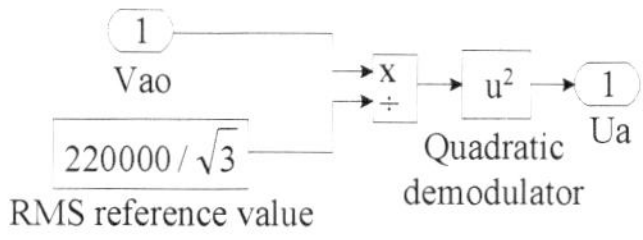

Fig. 5. Diagram of Block A

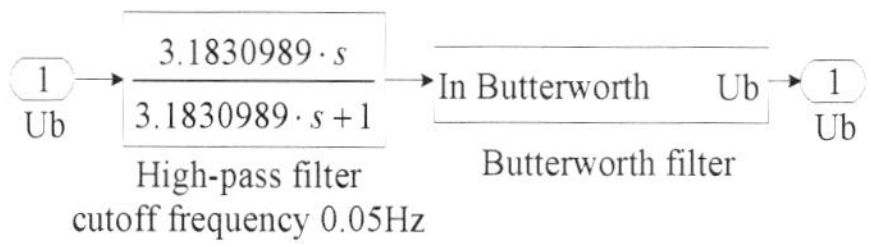

Fig. 6. High-pass filter

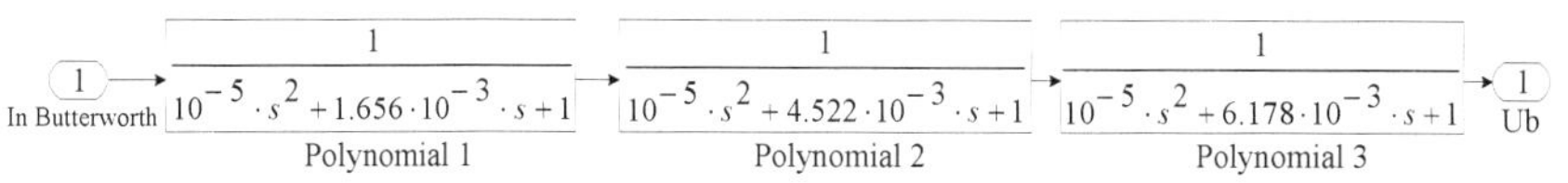

Fig. 7. Butterworth Filter Diagram.

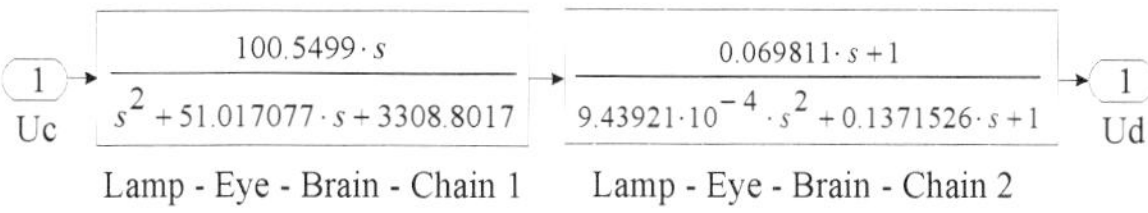

Fig. 8. Diagram of Block C

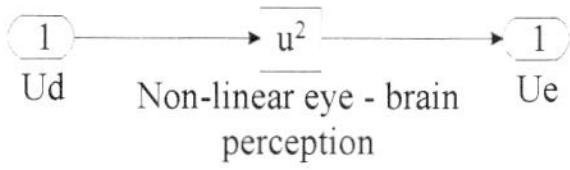

Fig. 9. Diagram of Block D.

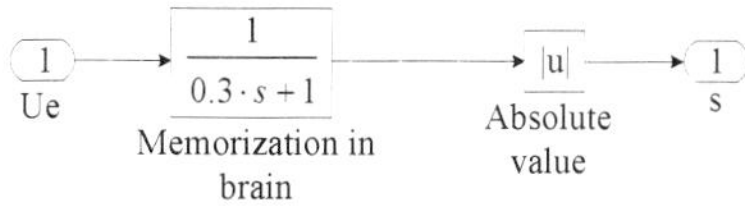

Fig. 10. Diagram of Block E

Fig. 11 shows the voltage waveform at the Block 1 output for a sinusoidal waveform of the rms voltage (Fig. 11(a)) and respectively for a square waveform (Fig. 11(b)). Outputs of the Block 2 for the two considered waveforms are shown in Fig. 11(c) and respectively Fig. 11(d).

The waveforms of the voltages at the output of the Block 3 are shown in Fig. 11(e) and Fig. 11(f). The outputs of the Block 4 are representative for the instantaneous flicker sensation and must be reported to the irritability curve of the human eye. These waveforms are illustrated in Fig. 11(g) and Fig. 11(h), [6].

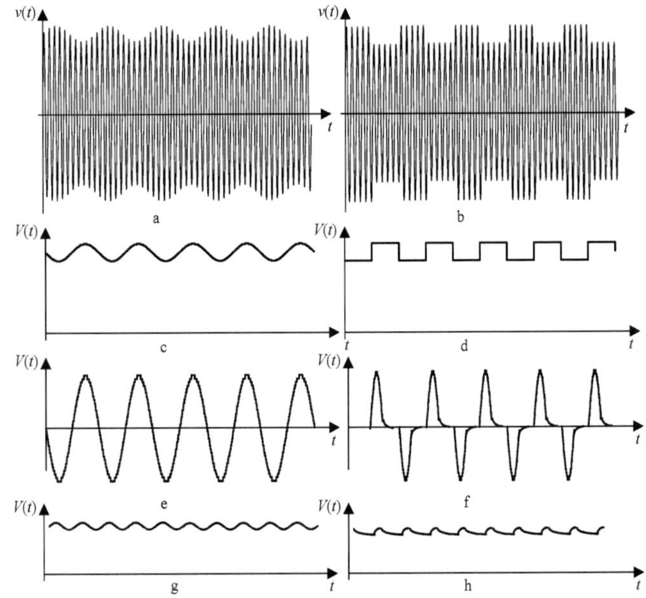

Fig. 11. Waveforms of the voltages in different points of the of the flickermeter scheme, for sinusoidal and square inputs

4. Case of application

The case studied is referred to a real plant sited close to Bucharest, in Romania. The measurements were done using the device ION 7600 at the 110kV voltage level. At PCC the line-to-line voltage is 110 kV, the ac arc furnace has 60MVA (see Fig. 12).

The layout of flickermeter connection at the end-user monitored terminals is shown in Fig. 13.

The RMS variations of the voltages at PCC are shown in Fig. 14. As it can be seen, during the operation of the ac arc furnace, variations of the voltages at PCC occur.

Fig. 15 shows the real power adsorbed by the ac arc furnace for one week operating period. The initial and the final part of the waveform correspond to a not-operating period of the arc furnace, respectively Sunday. During its operation period, a variable real power necessary for the arc furnace functioning is adsorbed. The imaginary power variation is highlighted in Fig. 16.

In Fig. 17 the RMS variation of the short-term flicker severity P_{st} values on phase A, for the same one week monitoring period, is illustrated. The 95% value of P_{st} is 5.28, exceeding the 0.8 IEEE threshold established for the high voltage networks [7]. This value P_{st}=5.28 can be explained by the fact that the fault level at PCC is 1150MVA.

When voltage flicker sources with long and variable operating cycle (like the arc furnaces) are investigated, the evaluation of the long-term severity flicker is necessary. Fig. 18 illustrates the variation of the long-term flicker severity P_{lt} using the values P_{st} monitored on 120 minutes period, measured every consecutive 10 minutes. The 95% value of P_{lt} is 6.95, greater than the IEEE threshold of 0.6 settled for P_{lt}.

Figs. 19-20 show the phase *A* waveforms of the power quality indices THD_V and THD_I.

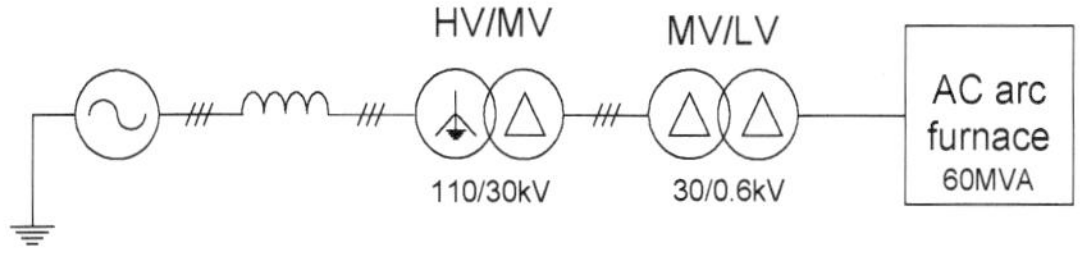

Fig. 12. Simplified diagram of the ac arc furnace power supply.

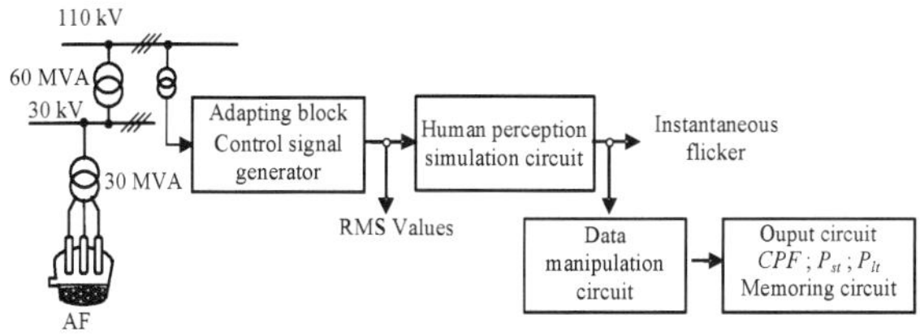

Fig. 13. Layout of flickermeter connection at end-user terminals

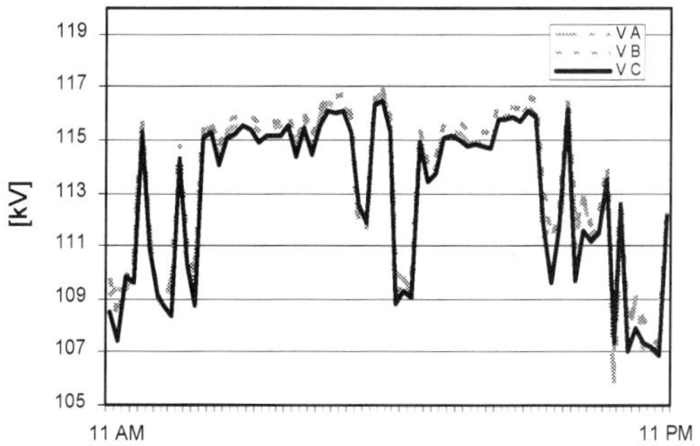

Fig. 14. RMS variation of the ac voltages at PCC (HV level).

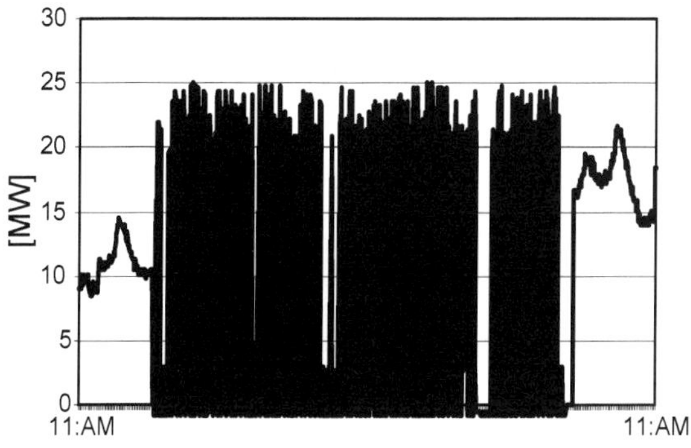

Fig. 15. Variation of the real power at PCC (HV level)

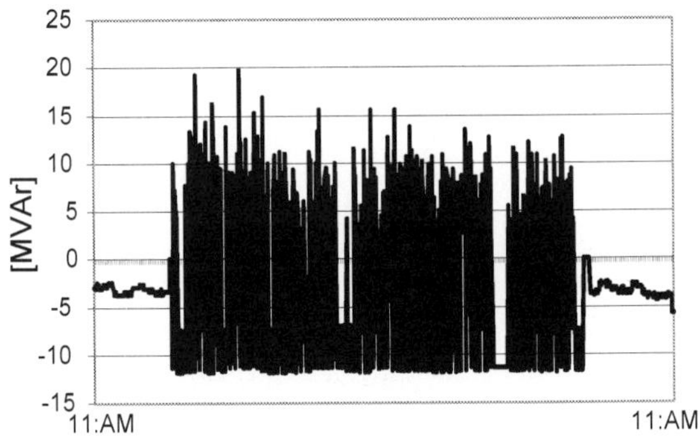

Fig. 16. Variation of the imaginary power at PCC (HV level)

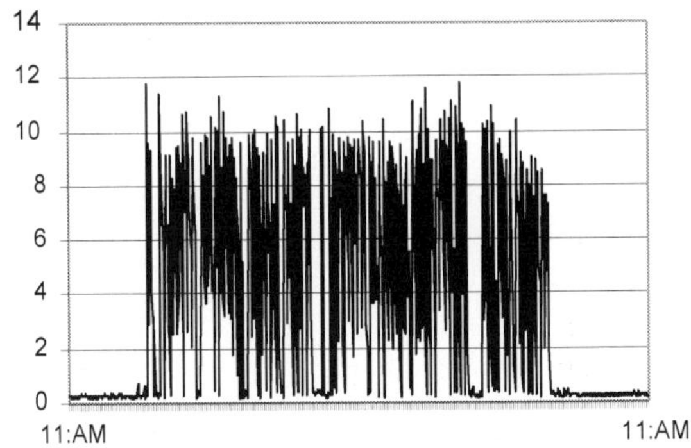

Fig. 17. Variation of P_{st} values on phase A.

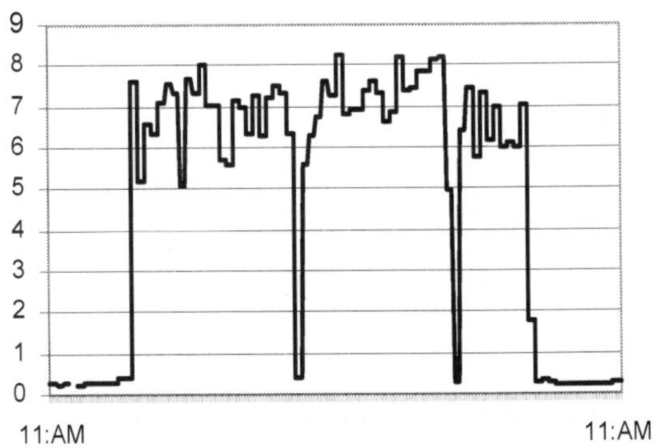

Fig. 18. Variation of P_{lt} values on phase A.

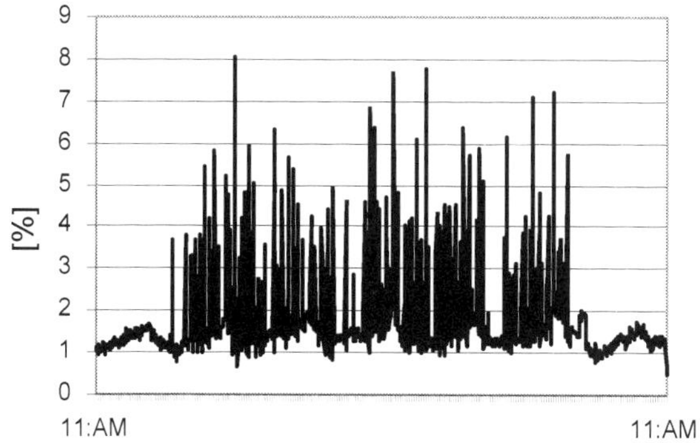

Fig. 19. Variation of *THDV* values on phase *A*

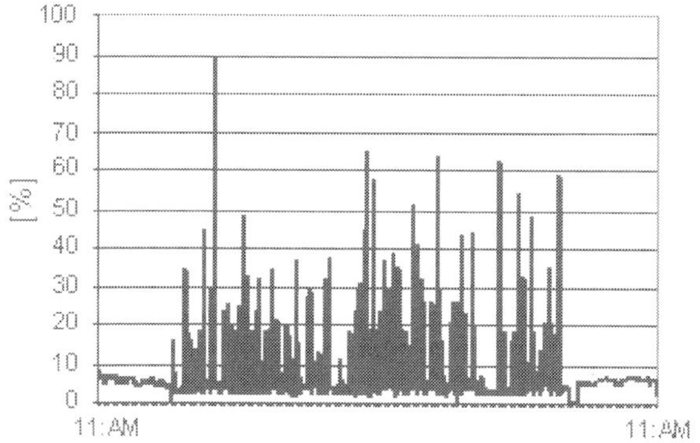

Fig. 20. Variation of *THDI* values on phase *A*.

Using the variation curve of the instantaneous flicker sensation the cumulative probability function (*CPF*) can be built [4]. The cumulative probability function at the 110kV buses for the ac arc furnace facility is illustrated in Fig. 21.

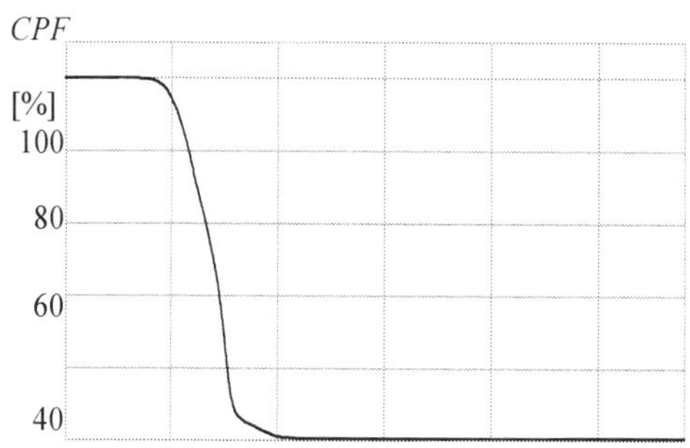

Fig. 21. Cumulative probability function CPF measured at the 110kV bus of the perturbing end-user

5. Conclusions

The high capacity ac arc furnaces connected to the HV level can determine power quality perturbations in the transmission networks.
To quantify the amplitude of such disturbance, reference is made to the instantaneous flicker sensation *S(t)*. The value of *S(t)* is measured by means of the so-called flickermeter. The setting-up of the flickermeter model is done in Simulink, the simulation tool, in the MATLAB environment.
For the real case studied investigated, the main power quality indices regarding harmonics and flicker levels are reported in the paper.
The aspects afore reported highlights the necessity to adopt efficient technical solutions for limiting the voltage variations until a threshold level. Thus, a precise knowledge of the perturbation characteristics and the choice of the best solution, from the technical and economical point of view, is necessary. For limiting the flicker level at PCC, the installation of a SVC will be implemented.

6. Biographies

Nicolae Golovanov received the Ph.D degree in Electrical Engineering from University Politehnica of Bucharest in 1975, and he is now Full Professor in the Department of Electrical Engineering, University Polytechnic of Bucharest. His areas of research include electrical end-use technologies and power quality. Dr. Golovanov is member of Energy Commission of The Romanian Academy, Chairman of Romanian Committee for Electrical End-use Technologies and member of Romanian Energy Policy Association.
George Cristian Lazaroiu received B. Sc and M. Sc. degrees from Department of Electrical Engineering, University Politehnica of Bucharest, in 2002 and respectively, 2003. He received the Ph.D. degree in Electrical Engineering, from Politecnico di Milano in 2006. His areas of research include distributed generation, power electronics and power quality.
Mariacristina Roscia, received M.S. degree (1999) and she obtained the degree PhD in Electrical Engineering at the University of Naples "Federico II". Her areas of research include Energy and Environment, Sustainable Development and renewable resources.
Dario Zaninelli, (SM'97) received the Ph.D degree in Electrical Engineering from Politecnico di Milano, in 1989, and he is now Full Professor in the Electrical Engineering Department of the Politecnico di Milano. His areas of research include power system harmonics and power system analysis. Dr. Zaninelli is a senior member of IEEE, a member of AEI and a member of the Italian National Research Council (C.N.R.) group of Electrical Power Systems.

7. References

[1] L. di Stasi, *Electric furnaces* (in Italian), Patron Editore, Bologna, Italy, 1976.
[2] E. Tironi, *Electrothermal Applications* (in Italian), CUSL, Milan, Italy, 2000.
[3] Electromagnetic compatibility (EMC), Part 4-30: Testing and measurement techniques - Power quality measurement methods, IEC 61000-4-30/2007
[4] *Flickermeter - Functional and design specifications*, IEC Standard 61000-4-15, Feb. 2003
[5] *Voltage characteristics of electricity supplied by public distribution systems*, SR EN 50160/2007
[6] Golovanov N., Postolache P. and Toader C., *Efficiency and quality of the electrical energy*, Bucharest, Ed. Agir, 2007.
[7] IEEE Recommended Practice for Measurement and Limits of Voltage Fluctuations and Associated Light Flicker on AC Power Systems, IEEE Standard 1453-2004, March 2005

5

Power Quality and Electrical Arc Furnaces

Horia Andrei[1], Costin Cepisca[2] and Sorin Grigorescu[2]
[1]*Valahia University of Targoviste*
[2]*Politehnica University of Bucharest*
Romania

1. Introduction

The chapter covers general issues related to power quality in Electric Arc Furnaces.
The use of electric arc furnaces (EAF) for steelmaking has grown dramatically in the last decade. Of the steel made today 36% is produced by the electric arc furnace route and this share will increase to 50 by 2030.
The electric arc furnaces are used for melting and refining metals, mainly iron in the steel production. AC and DC arc furnaces represent one of the most intensive disturbing loads in the sub-transmission or transmission electric power systems; they are characterized by rapid changes in absorbed powers that occur especially in the initial stage of melting, during which the critical condition of a broken arc may become a short circuit or an open circuit. In the particular case of the DC arc furnaces, the presence of the AC/DC static converters and the random motion of the electric arc, whose nonlinear and time-varying nature is well known, are responsible for dangerous perturbations such as waveform distortions and voltage fluctuations.
Nowadays, arc furnaces are designed for very large power input ratings and due to the nature of both, the electrical arc and the melt down process, these devices can cause large power quality problems on the electrical net, mainly harmonics, inter-harmonics, flicker and voltage imbalances.
The Voltage-Current characteristic of the arc is non-linear, what can cause harmonic currents. These currents, when circulating by the electric net can produce harmonic voltages, which can affect to other users.
In evaluation and limitation, there are some definitions and standards to quantify the disturbance levels, such as (***IEC, 1999), (***IEEE 1995), and (***IEEE, 1996). and. The total harmonic distortion (THD), short-term voltage flicker severity (Pst), and long-term voltage flicker severity (Plt) are used. However, sometimes it is desired to record voltage and current waveforms in the specified duration to track the disturbance levels.

2. Electrical arc furnaces

2.1 Construction and typical steelmaking cycle

An electric arc furnace (EAF) transfers electrical energy to thermal energy in the form of an electric arc to melt the raw materials held by the furnace. The arc is established between an electrode and the melting bath and is characterized by a low voltage and a high current. Arc

furnaces differ from induction furnaces in that the charge material is directly exposed to an electric arc, and the current in the furnace terminals passes through the charged material.
Sir Humphrey Davy conducted an experimental demonstration in 1810 and welding was investigated by Pepys in 1815. Pinchon attempted to create an electrothermic furnace in 1853 and, in 1878 - 79, William Siemens took out patents for an electric arc furnaces. The first electric arc furnaces were developed by Paul Héroult, with a commercial plant established in the United States in 1907. While EAFs were widely used in World War II for production of alloy steels, it was only later that electric steelmaking began to expand. Of the steel made today 36% is produced by the electric arc furnace route and this share will increase to 50 by 2030.
A schematic cross-section through an EAF is presented in figure 1: three electrodes (black), molten bath (red), tapping spout at left, refractory brick movable roof, brick shell, and a refractory-lined bowl-shaped hearth.

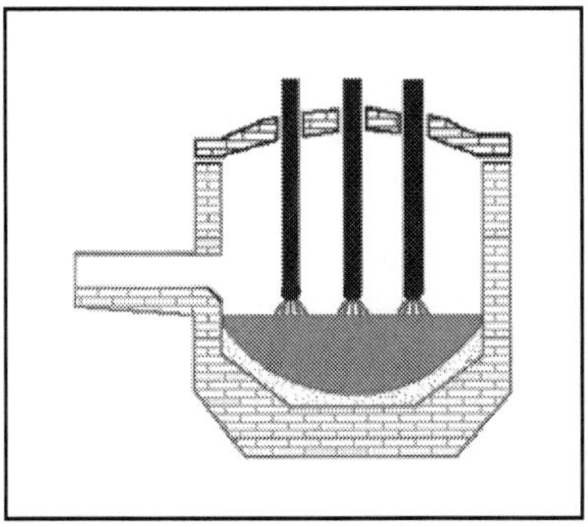

Fig. 1. Cross-section trough an EAF

The furnace is primarily split into three sections:

- the *shell*, which consists of the sidewalls and lower steel 'bowl';
- the *hearth*, which consists of the refractory that lines the lower bowl;
- the *roof*, which may be refractory-lined or water-cooled, and supports the refractory delta in its centre, through which one or more graphite electrodes enter.

Separate from the furnace structure is the electrode support and electrical system, and the tilting platform on which the furnace rests. Possible configurations: the electrode supports and the roof tilt with the furnace, or are fixed to the raised platform.
A typical alternating current furnace has three electrodes (Hernandez et al., 2007). The arc forms between the charged material and the electrode, the charge is heated both by current passing through the charge and by the radiant energy evolved by the arc. The electrodes are automatically raised and lowered by a positioning system and a regulating system maintains approximately constant current and power input during the melting of the charge, even though scrap may move under the electrodes as it melts. Since the electrodes move up and down automatically, heavy water-cooled cables connect the bus tubes/arms with the transformer located adjacent to the furnace.
The energy diagram shown in Figure 2 indicates that 70% of the total energy is electrical, the remainder being chemical energy arising from the oxidation elements such as carbon, iron, and silicon and the burning of natural gas with oxy-fuel burners. About 53 % of the total

energy leaves the furnace in the liquid steel, while the remainder is lost to slag, waste gas, or cooling.

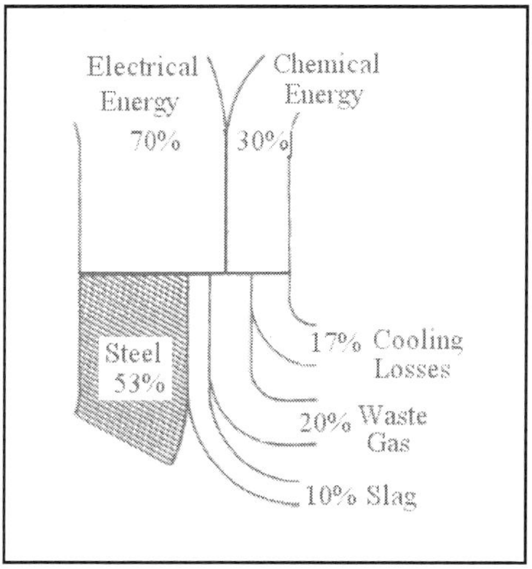

Fig. 2. Energy patterns in an EAF

A mid-sized modern steelmaking furnace would have a transformer rated about 60 MVA, with a secondary voltage between 400 and 900 volts and a secondary current in excess of 44,000 amperes. To produce a ton of steel in an EAF requires approximately 440 kWh per metric tone; the theoretical minimum amount of energy required to melt a tone of scrap steel is 300 kWh (melting point 1520°C).

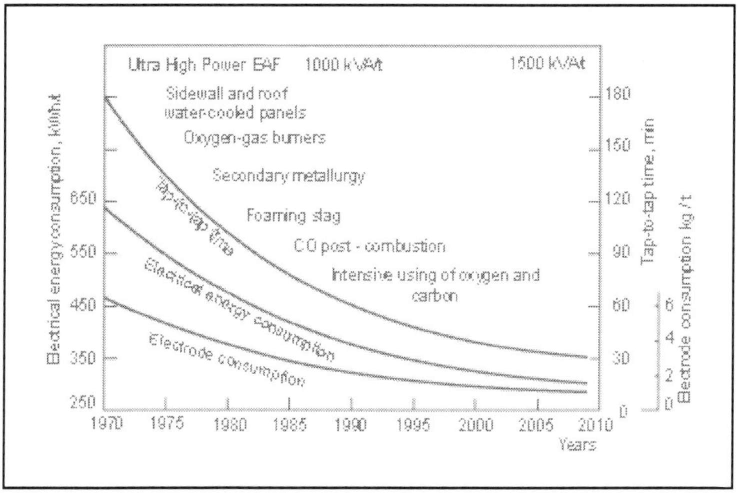

Fig. 3. Basic innovations and improvement in the 120-t EAF performances

Electric Arc Furnaces (EAF) are being greatly improved at a fast pace. Only 20–30 years ago today's EAF performance would be impossible to imagine (Hurst, 1994). Owing to the impressive number of innovations the tap-to-tap time has been shortened to 30–40 min. for the best 100–130 ton furnaces operating with scrap. Accordingly, their hourly and annual productivity increased. Electrical energy consumption got reduced approximately in half, from 580–650 to 320–350 kWh/ton. Electrical energy share in overall energy consumption per heat dropped to 50%. Electrode consumption was reduced 4–5 times - Figure 3.
Typical steelmaking cycles are:

- arc ignition period (start of power supply) - figure 4a
- boring period -figure 4b
- molten metal formation period - figure 4c
- main melting period - figure 4d
- meltdown period -figure 4e
- meltdown heating period - figure 4f

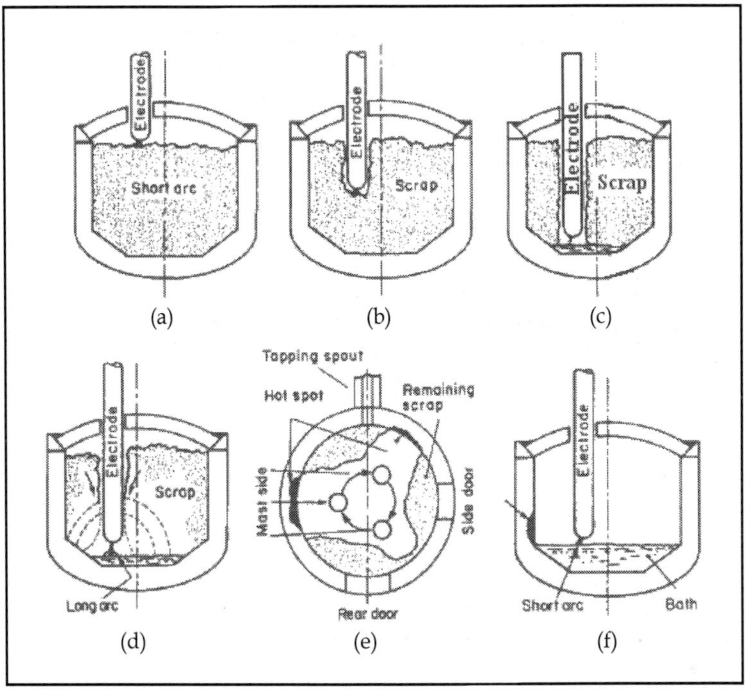

Fig. 4. Typical steelmaking cycle

Electrodes are initially lowered to a point above the material, the current is initiated, and the electrodes bore through the scrap to form a pool of liquid metal. The scrap itself protects the furnace lining from the high intensity arc. Subsequently, the arc is lengthened by increasing the voltage to maximum power. In the final stage, when there is a nearly complete metal pool, the arc is shortened to reduce radiation heat losses and to avoid refractory damage and hot spots.

After melt dawn, oxygen usually is injected to oxidize the carbon in the steel or the charged carbon. This process is an important source of energy; the carbon monoxide that evolves helps minimize the absorption of nitrogen and flushes hydrogen out of the metal. It also foams the slag, which helps minimize heat loss.
The random movement of the melting material has as consequence that no two cycles of the arc voltage and current waveforms are identical. The impact of these large, highly varying loads has a direct impact on the power quality of the interconnected power system.
The abrupt initiation and interruption of current flow provides a source of harmonic currents and causes considerable disturbance to high-impedance circuits. Voltage and current waves deviate considerably from symmetrical sinusoidal patterns. Disturbances are worst during early meltdown, and they occur at varying frequencies.
Generation of harmonics may result in further flicker problems, and equipment on the power system may also be damaged. If static capacitors are to be used to improve the power factor, an analysis to ensure that resonance does not exist at any of the harmonic frequencies should be made. Harmonics contribute to wave distortion and to the increase in effective inductive reactance. This increase is often in the 10 to 15% range and has been reported as high as 25%. Current into the furnace is therefore less than what would be expected from calculations based on sinusoidal wave shapes, and losses in frequency-sensitive equipment such as transformers are higher than the sinusoidal wave shape would produce.
Generally, the initial period of melting causes the most electrical disturbances. As the scrap temperature begins to rise, a liquid pool forms, and disturbances begin to diminish. This is generally about 10 minutes or so after power-on and can vary depending on power levels and practices.
After about 20 minutes, most electric furnaces will have begun converting scrap to liquid metal. Hence, wide swings in disturbances will diminish considerably. When sufficient molten metal exists the arc is shortened by an adjustment to the electrode regulators. The current will rise since overall resistance is reduced, and the power factor and arc power will decline.

2.2 Perturbations

The majority of electric and electronic circuits (arc welders and furnaces, variable speed controllers, PC's, medical equipment, etc) use switch mode techniques which act as a non-linear load or disturbance generator which degrades the quality of the electricity supply.
In these electro energetic steady state circuits, the importance of the inconvenience caused by the non sinusoidal system of running is directly correlated to the amplitude of the harmonics. Also, it is of utmost importance to determine the variation of the apparent power at non defined node, in accordance with the presence of the current and voltage harmonics. Understanding the current harmonics and voltage harmonics is of utmost scientific importance both to the beneficiaries, who thus can prevent the undesirable effects of non sinusoidal steady state in a given network, and to the possible consumers as for as the corresponding measurement and pricing are concerned. Hence the elaboration of certain rules and prescription as regards the influence of the harmonics upon the fundamental component (first harmonic).
Such combinations of traditional and non-traditional loads, coupled with fluctuating loads, causes problems often classified as "random" or "sporadic" (problems with sensitive

devices), annoying (light flickering) or as "strange" or "without apparent reason" (problems with cabling, capacitor banks, tripping, signaling etc.).

The electric arc furnace produces strong disturbing effects featured by non-symmetries of currents and voltages, harmonics, flickers, voltage drops and over-voltages, characteristic parameters of power quality.

Many ways exist to reduce the effects of the arc disturbances. These are determined by the utility system to which the furnace or furnaces are to be connected, and they are influenced mainly by the size and stability of the power grid. Some sizable shops require no particular flicker control equipment. It is quite possible that, if a furnace shop is fed from a 220 kV or higher system with a short-circuit capacity of 6500 MVA or more, the utility will experience very little load disturbance, and the steelmaker can have considerable flexibility in configuring his internal plant power system.

Most utilities require power factor correction. Shops with large electric furnaces would more than likely use static capacitors; synchronous condensers of sufficient capacity would be prohibitively expensive for a multi-furnace shop. Before such systems are installed, transient analysis is required to determine:

- Capacitor bank configuration
- Need for harmonic tuning of sections
- Switching procedure

If additional regulation is needed, VAR control equipment would probably be required. However, if plans have already been made for power factor capacitors, including tuning reactors, then the thyristors and main reactor are the only further additions required.

The perturbations caused by electric arc furnaces are of random nature and encompass a frequency range from DC to a few hundreds of Hz. Depending on whether AC of DC is used to supply the electric arc furnace there are unbalances, harmonics, inter-harmonics or voltage flicker.

2.3 Arc furnace models

For the design of EAF is necessary to utilize a suitable model. In this regard, numerous models have been presented to describe the electric arc (Lazaroiu & Zaninelli, 2010); (Math et al., 2006); (Hooshmand & Esfahani, 2009); (Sankaran, 2008).

In general the models can be classified into:

a. Time domain analysis methods:

- *Nonlinear Resistance Model*: The approximation on the V-I characteristic of the arc, performed by piecewise linearization, neglect of the voltage rising time or nonlinear approximation. This method uses the numerical analysis method to solve the differential equation which is used to describe the furnace system with the assumed V-I characteristic.

  However it is a primitive model and does not consider the time-varying characteristic of arc furnaces;
- *Current source models*: An EAF is typically modelled as a current source represented by the Fourier series where the coefficients may change randomly during every period. This model is perfectly suited to size filter components and to evaluate voltage distortions resulting from the harmonic current injected into the system.
- *Voltage Source Models*: The voltage source model for an EAF is a Thévenin equivalent circuit where equivalent impedance of the furnace load impedance including the

electrodes. The voltage source can be modelled in different ways. One possibility is to form it by major harmonic components that are known empirically. This method loses the stochastic characteristics of arc furnaces like the nonlinear resistance model does.

- *Nonlinear Time Varying Voltage Source Model*: The arc voltage is defined as a nonlinear function of the arc length. The time variation of the arc length is modeled with deterministic or stochastic laws.
- *Nonlinear Time Varying Resistance Models*: Arc furnace operation can be described by three basic states: open circuit, short circuit and normal operation. During normal operation the arc resistance can be modelled following an approximate Gaussian distribution. The random fluctuation in arc resistance accounts for the short-term perceptibility flicker index P_{st}.

b. Frequency domain analysis methods represent the arc voltage and current by their harmonic components (Key & Lai, 1997). The Harmonic Voltage Source Model first applies the Fourier transform to the arc voltage to obtain its harmonic components. Then the current harmonic components are calculated through the arc voltage harmonic components. Calculations provide an equivalent circuit for the fundamental frequency component consisting of an equivalent arc resistance and a reactance. The equivalent circuit for the calculation of the different order harmonics consists of a harmonic voltage source and the system impedance for that harmonic frequency. The model is simple, but suitable for steady-state iterative harmonic analysis.

c. Power balance method.
This model provides a harmonic domain solution method of nonlinear differential equation. The arc furnace load model is developed from the energy balance equation, which is actually a nonlinear differential equation of arc radius and arc current. This model uses some experimental parameters to reflect the arc furnace operation, but it neglects the influence of its supply system.

3. Basic principles for the power quality analysis

3.1 Power quality and harmonic distortion

One of the most important problems in nowadays consumers power supply is to ensure the power quality. Together with the power suppliers, the consumers are interested to use, to produce and to transport the electrical power as clean as possible. Any perturbation produced in the power system by any of its elements (components) may seriously affect the power quality consumed by the other elements especially those closely situated to the perturbing component (Filipski, et al., 1994).

The Power Quality has concerned the experts from power engineering area as far back as first years of using the energy, in a large amount of applications, the alternating current; during the last decade, we can observe several ascertainments to the involvement for this domain, owing to development based on power electronics.

Institute of Electrical and Electronic Engineers (IEEE) Standard IEEE 1100 define power quality as "a concept of powering and grounding sensitive electronic equipment in a manner suitable for the equipment". But this is not the only interpretation. Another simple and more concise definition might state: "Power quality is a set of electrical boundaries that allows equipment to function in its intended manner without significant loss of performance or life expectancy", definition that embraces two things that we demand from electrical equipment: performance and life expectancy. Another definition of power quality, based on the principle of EMC, is as follows: power quality refers to a wide variety of electromagnetic phenomena that characterize voltage and current at a given time and at a given location on the power system. IEC 61000-4-30 defines power quality as "the characteristics of the electricity at a given point on an electrical system, evaluated against a set of reference technical parameters" (Toulouevski & Zinurov, 2010); (***IEEE, 1995).

Power quality can be interpreted by the existence of two components:

- Voltage quality. It expresses the voltage deviation from the ideal one and can be interpreted as the product quality delivered by the utilities.
- Current quality. It expresses the current deviation from the ideal one and can be interpreted as the product quality received by the customers.

The main Power quality disturbances are:

- harmonics;
- under-voltages or over-voltages;
- flicker;
- transients;
- transients and voltage sags;
- voltage sags;
- interruptions.

Among the greatest electrical perturbations in a power system is the electrical arc furnace. Its perturbations are visible upon the reactive power flow, the load unbalance and the harmonics injected in the supply network. Also the random variation of the EAF electrical load, leads to the "flicker" phenomena characterized by variation in the field of 0.3-0.5% of the rated voltage and frequencies variations of 6 up to 10 Hz. Physically, the flicker phenomena is visible for the electrical bulbs that are rapidly changing the light intensity. Also, the side effects of the flicker are visible for the modern computation technique that could be damaged by the voltage variations.

At this moment we cannot talk about a united standardization of electrical energy quality on an international level and sometimes on national one. Currently, several engineering organizations and standard bearers in several parts of the world (IEEE, IEC, ANSI,...) are spending a large amount of resources to generate power quality standards. Some of them classify the events as steady-state and non-steady-state phenomena, in some regulations the most important factor is the duration of the event, other guidelines use the wave shape (duration and magnitude) of each event to classify problems and other standards (e.g., IEC) use the frequency range of the event for the classification. These documents come in three levels of applicability and validity: guidelines, recommendations and standards. In almost all the countries, the directives system of electrical energy quality is composed by several quantitative characteristics of slow or rapid variations of effective voltage value, the shape or symmetry as well as characteristics of slow or rapid frequency variations (***IEEE-WG, 1996); (***PE, 2004) (***SREN, 1998); (***CMP, 1987).

As it can be seen in Figure 5 there are presented the main causes of an improper electrical energy quality.

For the measurements of disturbances, IEC 61000-4-7 describes testing and measurement techniques for harmonics and inter-harmonics measurements and instrumentation, for power supply systems and equipment connected thereto.

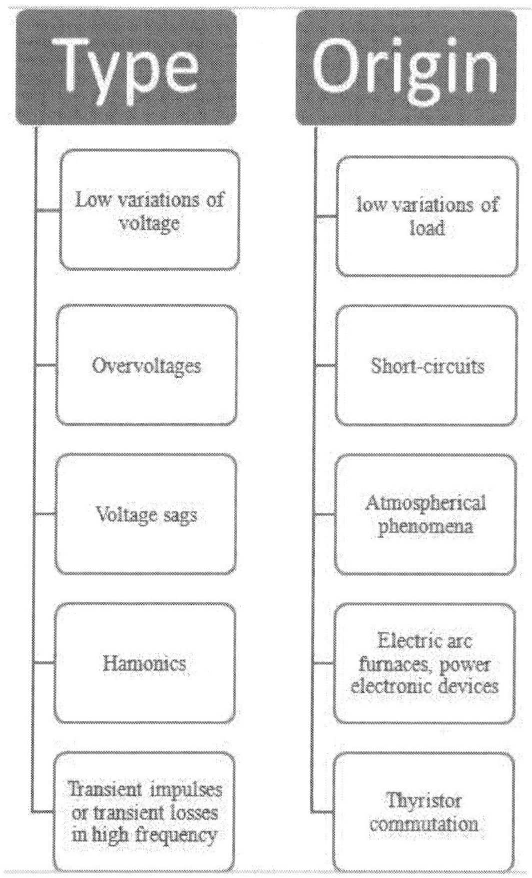

Fig. 5. Causes of an improper electrical energy quality

3.2 The prominent power quality aspects

The prominent power quality aspects considered are the following:

a. Voltages and currents are non sinusoidal quantities, and can be expressed by relations:

$$u(t) = \sum_{k=1}^{N} U_k \sqrt{2} \sin(k\omega t + \gamma_k)$$

$$i(t) = \sum_{k=1}^{N} I_k \sqrt{2} \sin(k\omega t + \gamma_k - \varphi_k)$$

where U_k, I_k are the RMS of each k-harmonic of voltage, respectively current, ω is the angular frequency, γ_k is the phase angle or each k-harmonic of voltage, k-harmonic of voltage, φ_k is difference of each phase angle of k-harmonic of voltage and current, t is the time.

- the active power: $P = \sum_{k=1}^{N} U_k I_k \cos\varphi_k$

- the reactive power: $Q = \sum_{k=1}^{N} U_k I_k \sin\varphi_k$

- the apparent power: $S = \sqrt{\sum_{k=1}^{N} U_k^2}\sqrt{\sum_{k=1}^{N} I_k^2}$

- the power factor: $K_P = \dfrac{P}{S} = \dfrac{P}{\sqrt{P^2 + Q^2 + D^2}}$

- the reactive factor: $\rho = \dfrac{Q}{P}$

- the deforming factor: $\sigma = \dfrac{D}{\sqrt{P^2 + Q^2}}$

where $D = \sqrt{S^2 - P^2 - Q^2}$ is the Budeanu distortion (deforming) power.

b. The presence of voltage and current harmonics is evaluated through a relative quantity, the total harmonic distortion (THD). Voltage harmonics are asserted with THD_U, the ratio of the RMS value of the harmonic voltage to the RMS value of the fundamental, calculated by relation:

$$THD_U = \sqrt{\sum_{n=2}^{N}\left(\frac{U_n}{U_{(1)}}\right)^2}$$

Everything presented for voltage harmonics is also valid for current harmonics and THD_I, the ratio of the RMS value of the harmonic current to the RMS value of the fundamental, calculated by relation:

$$THD_I = \sqrt{\sum_{n=2}^{N}\left(\frac{I_n}{I_{(1)}}\right)^2}$$

Total harmonic distortion is the ratio between deforming residue and effective value of fundamental waveform:

$$\delta_u = \frac{U_d}{U_1}100(\%) = \frac{\sqrt{\sum_{h=2}^{40} U_h^2}}{U_1}100(\%)$$

Harmonic level is the ratio between effective value of the considered harmonic and the effective value of the fundamental:

$$\gamma_u = \frac{U_n}{U_1} 100(\%)$$

c. Voltage imbalance. Applying the theory of symmetrical components, an unbalanced three-phase sinusoidal voltage system [Va, Vb, Vc] can be decomposed into a positive-sequence three-phase balanced system V_+, a negative-sequence system V_-, and a zero sequence system V_0

d. Disturbance transiting among voltage levels: Rapid voltage changes, Transient over-voltages and voltage fluctuation and flicker.

3.3 Power quality measurements

A simple way for a technician to determine power quality in their system without sophisticated equipment is to compare voltage readings between two accurate voltmeters measuring the same system voltage: one meter being an "averaging" type of unit (such as an electromechanical movement meter) and the other being a "true-RMS (rms)" type of unit (such as a high-quality digital meter). Remember that "averaging" type meters are calibrated so that their scales indicate volts RMS, based on the assumption that the AC voltage being measured is sinusoidal. If the voltage is anything but sine wave-shaped, the averaging meter will not register the proper value, whereas the true-RMS meter always will, regardless of wave-shape.

The rule of thumb here is this: the greater the disparity between the two meters, the worse the power quality is, and the greater its harmonic content. A power system with good quality power should generate equal voltage readings between the two meters, to within the rated error tolerance of the two instruments.

Measurement and testing of supply voltage quality, according to EN 50160, requires specialized apparatus and measuring methods.

This arrangement enables continuous monitoring, short time and long time, over 7 days, of the following parameters:

- voltages and currents in three phases;
- frequency;
- total harmonic distortion factor THD_U and THD_I;
- voltage unbalance factor, which is a multiple of positive and negative sequence voltage components;
- fast and slow voltage variations, which are defined as short term (P_{st}) and long term (P_{lt}) flicker;
- severity factors.

This type of equipment, named digital power analyzer also enables measurement of voltage dips and outages, its frequency and duration.

The RMS values of voltages and currents can be determined correctly by digital methods in any harmonic content of waveforms. Also, with the results of RMS voltage and current can calculate the apparent power. The active power may be calculated and accurately measured in any circumstances of harmonic pollution. Unfortunately this is not the case for reactive power. For reactive power can be used different definitions and methods (Arrillage et. al., 2001); (Czarnecki, 1987); (Emmanuel, 1995); (Emmanuel, 1999); (Katic, 1994):

- reactive power measurement (Budeanu definition);
- Hilbert transform method;
- power triangle method;
- quarter period time delay method;
- low-pass filter method.

Table 1 presents the test conditions, voltage and current, used to test the measurement performances of the reactive power measurement solutions. Table 2 presents the errors obtained for different tests using notations: H- for Hilbert transform, LPF- for low pass filter, PT-power triangle, CTD- compensated time delay.

The traditional measurement methods, like Power triangle and the Time delay, comply with international standards but show limitations in the presence of harmonics or line frequency variation.

One can observe that Hilbert method give the best results, followed by the low pass filter method and then power triangle method. So, different analyzers implemented with different formulas can give discrepancies measuring the same loads.

Type of Test	Current and Voltage (Normalized to rms values) i(t) u(t)
IEC1268 -Reference Voltage and current input: sin wave, f_{line}; $PF = 0$	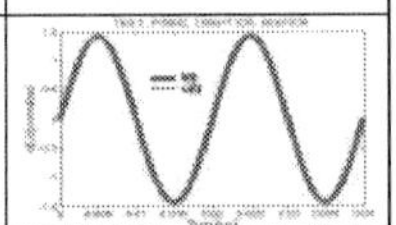
IEC1268 - Frequency variation (±2%) sin wave; $PF = 0.87$	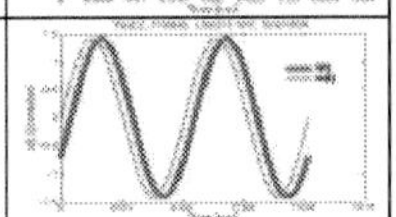 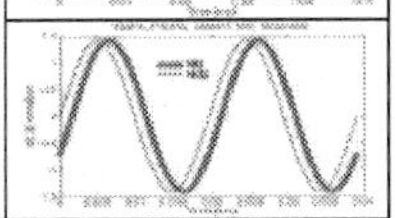
IEC1268 -Harmonic Reference test + 10% of the third harmonics on the current signal	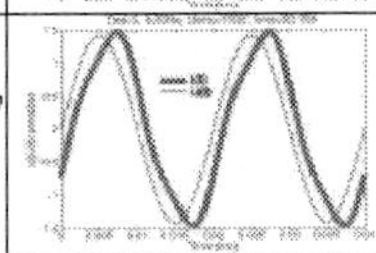
Reference test + 10% of the third harmonics on voltage input and 20% of the third harmonics on current input ($\varphi_1 = \varphi_3 = 30^0$)	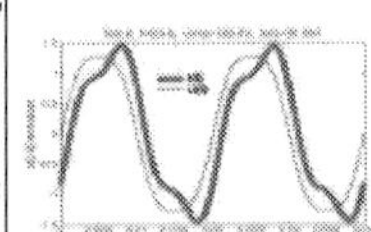
Additional Test	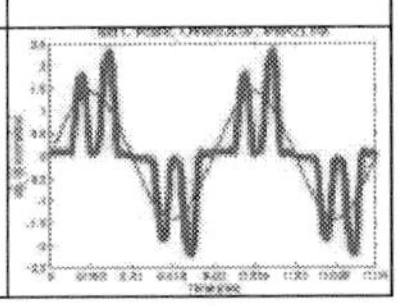

Table 1.

Type of test	H	LFF	FT	CTD
IEC 1268 - Reference Voltage and current input sin wave, f_{line}; $PF = 0$	~0	~0	~0	~0
IEC 1268 - Frequency variation (±2%) sin wave; $PF = 0.87$	~0	~0	~0	~0
	~0	~0	~0	~0
IEC 1268 - Harmonic Reference test + 10% of the third harmonics on the current signal	~0	~0	2	~0
Reference test + 10% of the third harmonics on voltage input and 20% of the third harmonics on current input ($\varphi_1 = \varphi_3 = 30^0$)	~0	-1.3	1.9	-3.9
Additional Test	~0	2.83	3.90	3

Table 2.

4. Numerical simulations for energy calculation in power measurements

The model presented in (Vervenne et. al., 2007) is based on exponential-hyperbolic form which causes many problems in the power system quality. Also the model can describe different operations of the EAF and it does not need specific initial conditions.

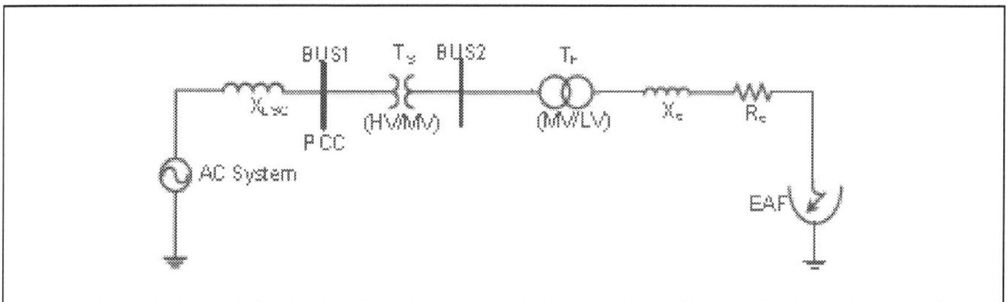

Fig. 6. EAF connected to supply system

The electric diagram of a electrical circuit supplying an EAF is illustrated in Figure 6. In this figure, bus 1 is the point of common coupling (PCC) which is the supplying bus of the EAF

transformer. The arc furnace is also connected to the PCC through the transformer TS, (HV/MV). In this figure, X_C and R_C are the reactance and resistance of the connecting cable line to the furnace electrodes, respectively. Also, X_{Lsc} is the short circuit reactance at bus PCC.
The electric arc is modeled by the following equations:

$$V(i)=\begin{cases} V_{at}+\dfrac{C}{D+i}, & \dfrac{di}{dt}\geq 0, \quad i>0 \\ V_{at}\left(1-e^{-i/I_0}\right), & \dfrac{di}{dt}<0, \quad i>0 \end{cases}$$

where V and i are arc voltage and current of the EAF, respectively. Also V_{at} is the voltage threshold magnitude to which voltage approaches as current increases. Furthermore, I_0 is the current time constant in kA. It should be noted that the voltage V_{at} depends on the arc length.
The constants C and D are corresponding to the arc power and arc current, respectively. These constants can take different values which depend on the sign of the derivative of the arc current.
As it can be seen in electric arc modeled equation, for the positive current and regarding the hysterias property of the arc, there are two cases. In the increasing current case, the hyperbolic equation and in the decreasing current case exponential equation is used. Hence, this model is called exponential-hyperbolic model. The proposed method has the capability of describing the EAF behavior in time domain using differential equation. In addition, it is able to analyze the behaviors in the frequency domain without solving the sophisticated differential equations.
Moreover, the proposed model can describe different operating conditions of the EAF such as initial melting (scrap stage), mild melting (platting stage) and refinement of the EAF.
With the parameters of the system:

$$X_{Lsc} = 9.4245\Omega, \; Xc = 2.356 \; m\Omega, \; Rc = 0.4 \; m\Omega, \; fsys = 50 \; Hz$$

and:

$$Vat = 200 \; V, \; Ca = 190 \; kW, \; Cb = 39 \; kW, \; Da = Db = 5 \; kA, \; Io = 10 \; kA$$

the voltage-current characteristic of the arc is obtained and shown in Figure 7. The voltage and the current of the arc are illustrated in Figure 8.
The characterization of flicker produced by an arc furnace is an extremely difficult operation (Alonso & Donsion, 2004); (Beites et. al., 2001); (Webster, 2004). The flicker is variable from one cycle to another and during melting stage very high peaks are produced. It depends on following parameters: quality and quantity of used scrap, reference operating points, quantity of injected oxygen, unpredictable consequences due to crumbling of the scrap during melting.
Consequently it is recommended to evaluate the level of flicker produced during at least one week of operation, representing several tens of operation cycles. LabVIEW and MATLAB software are used for simulation on EAF (Andrei et. al., 2006); (Andrei et. al., 2006); (Andrei et. al., 2006); (Beites et. al., 2001); (Bracale et. al., 2005); (Buzac & Cepisca, 2008).

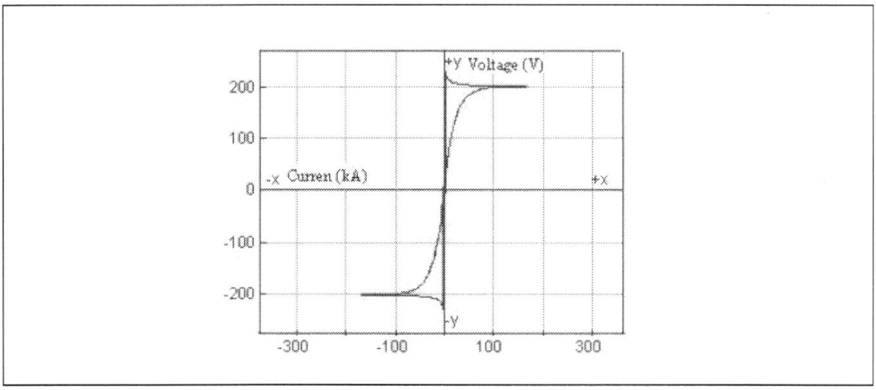

Fig. 7. Voltage-current characteristic for the exponential-hyperbolic model

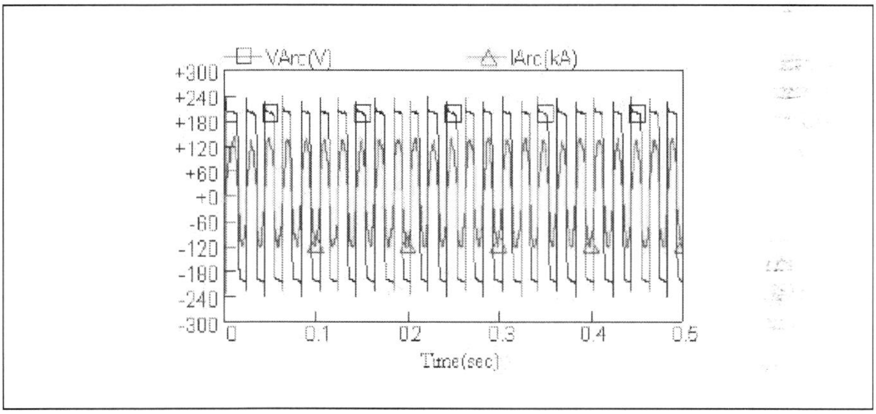

Fig. 8. Waveforms in the exponential-hyperbolic mode

5. Results of measurements in a real electric installation of arc furnace

5.1 Measurement method and equipment

The three-phase power analyzer is used for the analysis of power quality with compatible software analysis. The following quantities are necessary to be measured: voltage, current, flicker (IEC 68, IEC 61000-4-15-P_{ST} and P_{LT}), THD, waveform snapshots and harmonics up to the minimum order of 64, frequency, transient events (Chi-Jui Wu & Tsu-Hsun Fu, 2003); (Pretorius et al., 1998).

The strategy of measurements was to carry out recordings on EAF with all electrical quantities: RMS voltage, RMS current, flicker, frequency, THD voltage, THD current, current and voltage waveforms, powers kW, kVAR, kVA, power factor, voltage and current vectors for the short and long time (Cepisca et al., 2004); (Cepisca et al., 2006).

One example of measurement equipment is a multifunctional Power Quality Analyzer METREL, shown in Figure 9, one advanced instrument for measuring quality of electrical power in compliance with the EN60150. It incorporates a number of different measurement instruments for calculating various electrical parameters which is based on current and voltage measurements.

Fig. 9. Measurement equipment METREL

5.2 Results of the measurements in a real electric installation of EAF

The electrical power networks of arc furnaces are presented in Figure 10 (Cepisca et al., 2008).

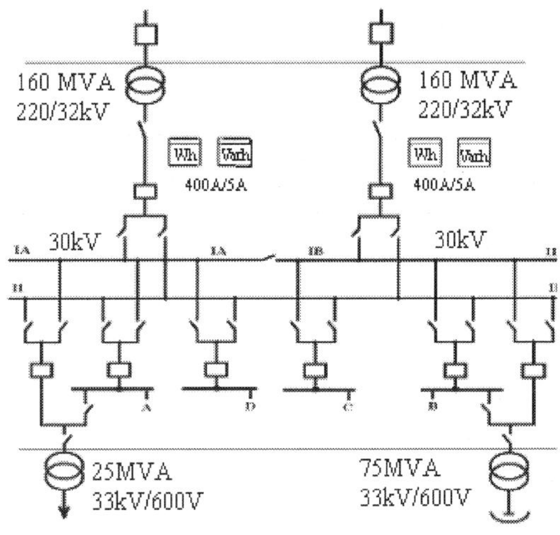

Fig. 10. Electrical power supply networks for arc furnaces

5.2.1 The real measurements of voltage and current harmonics, and of the powers

Figure 11 presents the current (a), the voltage (b) and Figure 12 presents the powers for a technological cycle of arc furnace. This cycle presents two phases: melting phase (6-8 minutes) and phase of stable arc burning (12-15 minutes). The electrical quantities are strong variation in the melting phase, with an important voltage fall. In the phase of stable arc burning the variation of electrical quantities are more reduced (Cepisca et. al., (2007); (Grigorescu et al., 2006); (Grigorescu et al., 2009); (***PE, 2004).

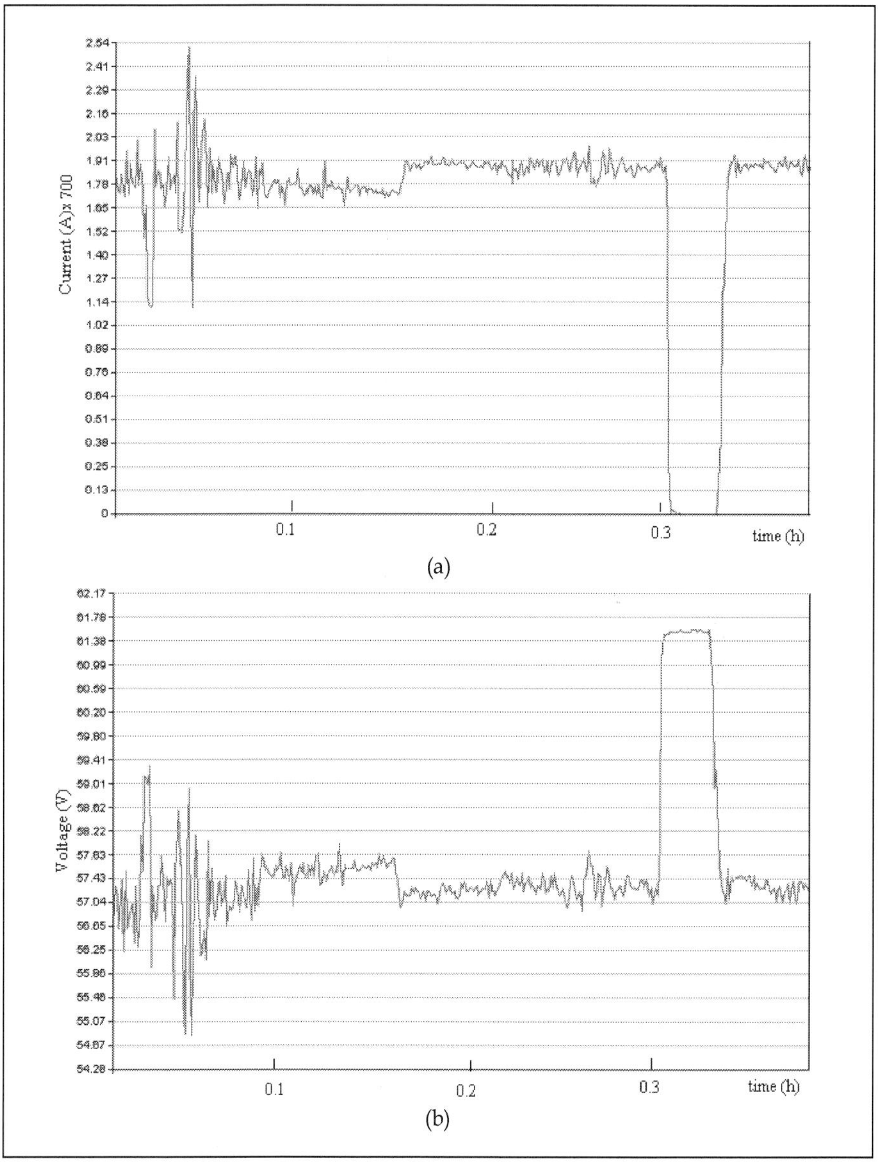

Fig. 11. The real measurements for a technological cycle of EAF: a) current b) voltage

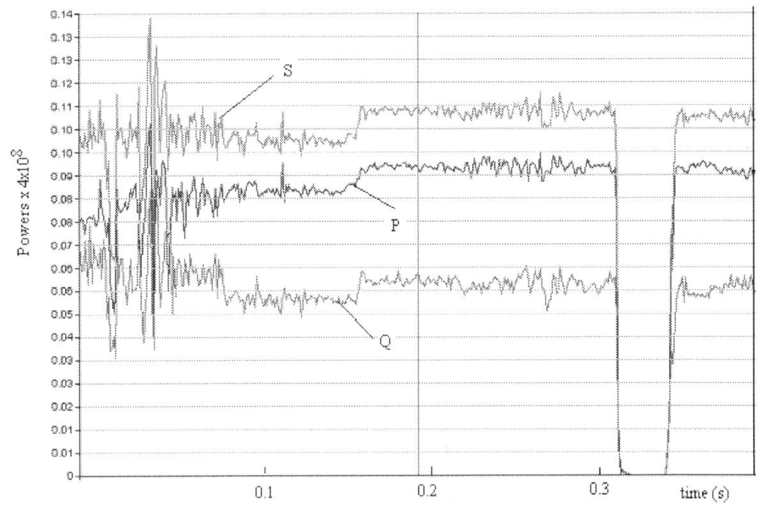

Fig. 12. The real measurements of powers (P, Q, S) for a technological cycle of arc furnace

5.2.2 The real measurements of wave forms of voltage and current, and of the THD_U and THD_I for melting phase of the technological cycle of arc furnace

As regard to the wave forms of the voltages, shown in Figure 13, a, and, respectively the wave forms of the currents shown in Figure 13, b, on the 30 kV voltage supply line in the melting phase is found a strong distortion of currents.

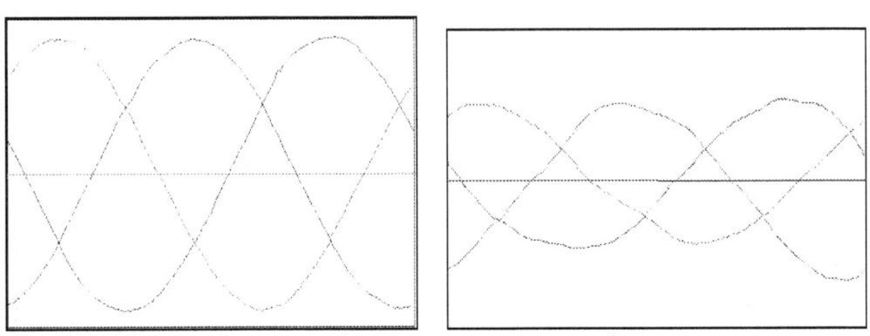

a. The wave forms of voltages b. The wave forms of currents

Fig. 13. The wave forms of voltages and currents in the melting phase

The Figure 14 presents: (a) the total harmonic distortion calculated for voltages (THD_U, 2,8…3%), and (b) the total harmonic distortion calculated for the currents (THD_I, 10.…11%).

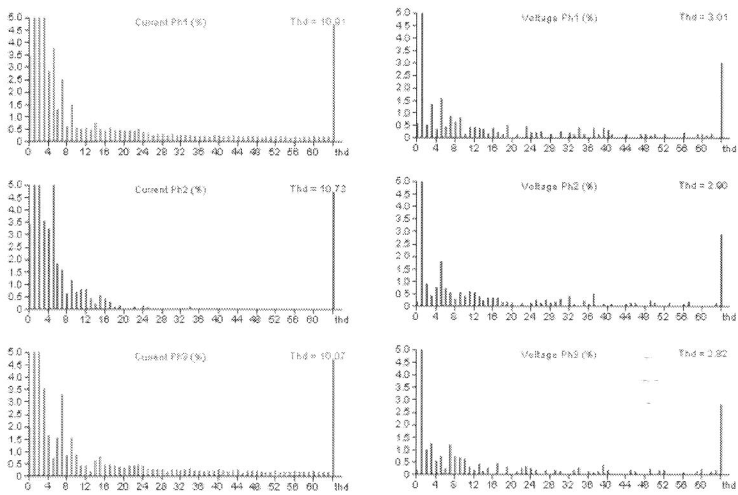

a. THD_I b. THD_U

Fig. 14. The total harmonic distortion calculated for voltages (THD_U) and currents (THD_I) in the melting phase

5.2.3 The real measurements of wave forms of voltage and current, and of the THD_U and THD_I in the phase of arc burning of the technological cycle of arc furnace

In the phase of the electric arc stable burning (Figure 15, a, and b), that appears towards the final of the heat's making, is found that the distortion that appear in the currents and voltages wave forms are more reduced. In this phase, the amplitude of the three phase currents and voltages are closer as value, fact which shows that the load impedance is more balanced.

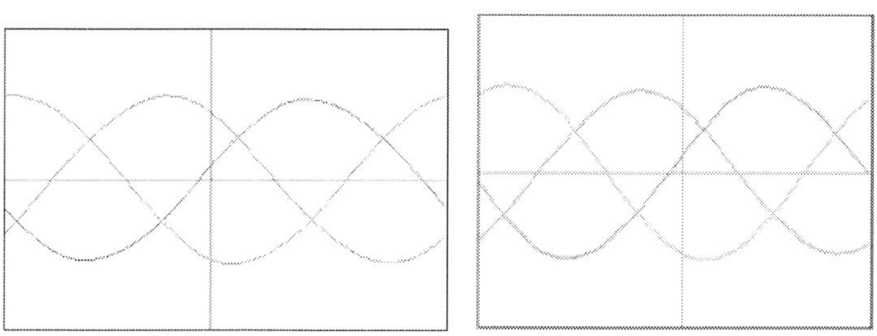

(a) (b)

Fig. 15. The wave forms in the arc stable phase: a) voltages ; b) currents .

The TDH for voltages and for currents in the arc stable phase are presented in Figure 16, a, and b. We observe that in the arc stable phase the THD_U is reduced (1...2%) and THD_U are an acceptable value (4...5%). One can reach to the conclusion that the deformation of the current and voltage waves is smaller in the stable burning phase also by the fact that the distorting power is smaller in this phase, in conditions where the apparent, active and reactive power is higher.

As regard the voltage on the 30 kV line, in the melting phase one can observe the presence of the important harmonics while in the oxidation phase is found practically only the presence of the fundamental. In the current's case, the important values of harmonics demonstrate that in this phase the current is strongly deformed.

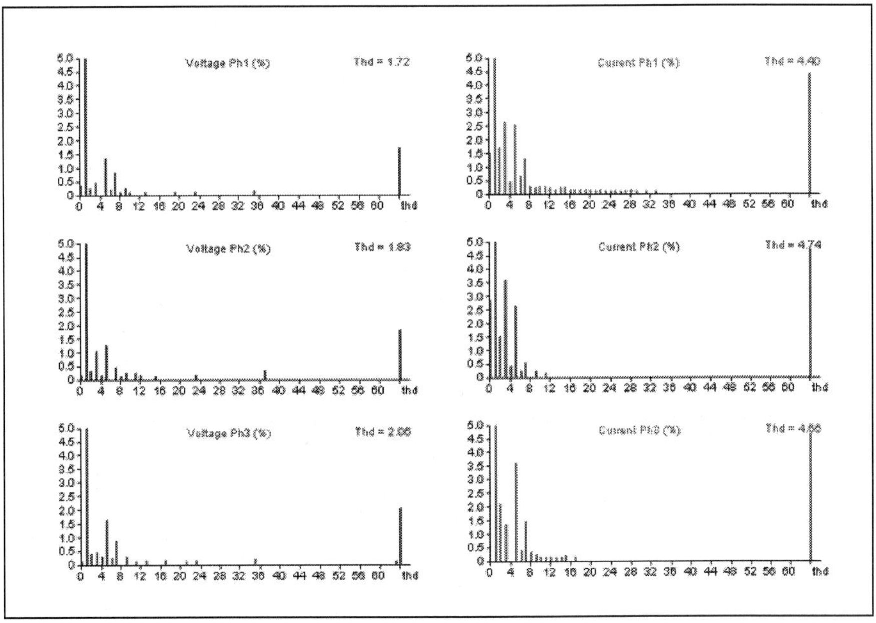

Fig. 16. The total harmonic distortion calculated for voltages (THD_U) and currents (THD_I) in the arc stable phase

The variation form of powers measured values presented on the heat time presents in the first period, corresponding to the melting phase, a smaller apparent power. The electrodes are more lifted-up, in order to ensure protection against breaking and this determining a smaller value current. In the stable phase the apparent power is approximately constant and higher than in the melting phase. The variation of the voltage, as well as of the arc current, is reflected partially in the variation of active and reactive powers during the heat.

5.2.4 The variation of the THD_U and THD_I, and the variation of the power factor

The THD_U and THD_I (Figure 17) are higher in the melting phase than in the stable burning phase, bat the reactive power is higher in the stable phase than in the melting phase.

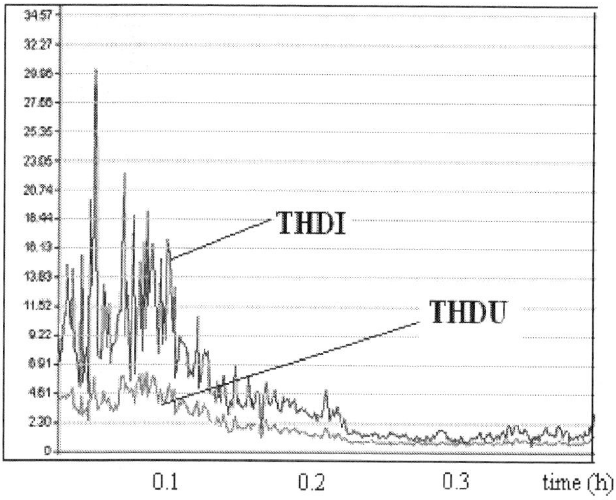

Fig. 17. The variation of THD_I and THD_U

The power factor value (Figure 18) is higher in the stable arc phase and lower during the melting phase. For this reason results that on the 30 kV line the currents wave is more distorted than the voltages wave.

In different moments of technological process, following the measurements, were obtained values for THD_I within 1-21% for current and 1-6% for voltage. Comparing these values with the standard results that the furnace is not matched in the national and international standards.

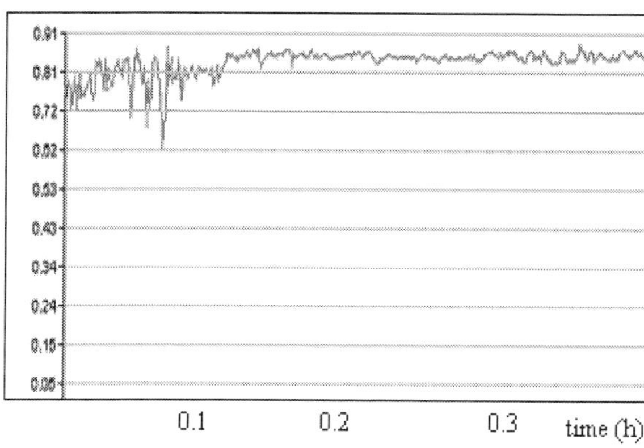

Fig. 18. The variation of power factor

6. References

Alonso, M. & Donsion, M. (2004). An Improved Time Domain Arc Furnace Model for Harmonic Analysis, *IEEE Transaction on Power Delivery*, 19(1), 2004, 367-373

Arrillage, J.; Watson, N. & Chen, S. (2001). *Power System Quality Assessment*, John Wiley & Sons, New York

Andrei, H.; Spinei, F.; Cepisca, C. & Caciula, I. (2006). 3-D mathematical model of the power factor in electro-energetical systems, *Proceedings of VI World Energy System Conference*, pp. 257-261, Torino, Italy, July 10-12, 2006

Andrei, H.; Cepisca, C.; Chicco, G.; Dascalescu, L.; Dogaru, V. & Spinei, F. (2006). LabVIEW measurements in steady state nonsinusoidal regime, *WSEAS Transactions on Circuits and Systems*, 11 (5), 2006, 1682-1687

Andrei, H.; Cepisca, C. & Spinei, F. (2006). The modelling of the power factor in steady state non sinusoidal regime with Mathcad techniques, *Proceedings of IEEE-TTTC International Conference on Automation, Quality and Testing, Robotics AQTR THETA 15- Tome I*, pp. 58-62, Cluj Napoca, Romania, May 25-28, 2006

Beites, L. F.; Mayordomo, J. G.; Hernandes, A. & Asensi, R. (2001). Harmonics, Inter harmonic, unbalances of arc furnaces: a new frequency domain approach, *IEEE Transactions on Power Delivery*, 16(4), 2001, 661-668

Bracale, A.; Carpinelli, G.; Leonowicz, Z.; Lobos, T. & Rezmer, J. (2005). Waveform distortions due to AC/DC converters feeding, *Proceedings of International Conference Electrical Power Quality and Utilization*, pp. 130-136, Cracow, Poland, June, 1-2, 2005

Buzac, E. & Cepisca, C. (2008). The importance of accurate measurement of electrical energy and the performance of modern electricity meters, *OIML Bulletin*, vol. XLIX, no.1, 2008, Paris, 5

Cepisca, C; Andrei, H.; Ganatsios, S. & Grigorescu, S. (2008). Power quality and experimental determinations of electrical arc furnaces, *Proceedings the 14th IEEE Mediterranean Electrotechnical Conference - MELECON*, vol. 1 and 2, pp. 546-551, Ajaccio France, May 5-7, 2008

Cepisca, C.; Ganatsios, S.; Andrei, H.; Cepisca, C. I.; Dogaru, V. & Lefter, E. (2004). The measurements of electrical nonsinusoidal signals, *The Scientific Bulletin of University of Pitesti, Romania, Metrology, Measurements system and quality*, 1, 2004, 22 -26

Cepisca, C.; Grigorescu, S. D.; Seritan, G.; Banica, C. & Argatu, F. (2006). Experimental results of harmonic pollution in the electric networks by the electric arc furnaces, *Proceedings of the 5th International Symposium Advanced Topics in Electrical Engineering-ATEE*, pp. 34-40, Bucharest, Romania, November 14-16, 2006

Cepisca, C.; Covrig, M.; Grigorescu, S. D.; Predescu, C.; Banica, C. & Argatu, F. (2007). Harmonic pollution in the electric networks by the electric arc furnaces. Experimental results, *Proceedings of 7th WSEAS/IASME International Conference on Electric, Power Systems, High Voltage, Electric Machine - POWER'07*, pp.233-238, Venice, Italy, April 20-22, 2007

Chi-Jui Wu & Tsu-Hsun Fu (2003). Data compression applied to electric power quality tracking of arc furnace load, *Journal of Marine Science and Technology*, vol.11, no.1, 2003, 39-47.

Czarnecki L. S. (1987). What is wrong with Budeanu's concept of reactive and distortion power and why it should be abandoned. *IEEE Transaction on Instrumentation and Measurement*, IM-36 (3), 1987, 345-352

Emmanuel, A.E. (1995). On the assessment of harmonic pollution, *IEEE Transaction on Power Delivery*, vol. 10 (3), 1995, 1693-1698

Emmanuel, A. E. (1999). Apparent power definition for three-phase systems. *IEEE Transactions on Power Delivery*, 14 (3), 1999, 762-772

Filipski, P.S.; Baghzouz, Y. & Cox, M.D. (1994). Discussion of power definitions contained in the IEEE dictionary, *IEEE Transaction on Power Delivery*, vol. 9 (3), 1994, 1237-1244

Fuchs, E. F. & Masoum, M.A.S. (2008). Power Quality in Power Systems and Electrical Machines, *Elsevier Academic Press*, Amsterdam

Grigorescu, S. D.; Cepisca, C.; Potirniche, I.; Ghita, O. & Covrig, M. (2009). Numerical simulations for energy calculation in power measurements, *Proceedings of the European Computing Conference (ECC09) and Proceedings of the 3rd International Conference on Computational Intelligence (CI09)*, pp.152-158, Tbilisi, Georgia, June, 20-22, 2009

Grigorescu, S.D.; Cepisca, C. & Ghita O. (2009). Strategies for energy quality monitoring in decision-making networks nodes, *Proceedings of 5th International Conference Metrology&Measurement Systems – METSIM*, pp. 174-178, Bucharest, Romania, 14-15 November, 2009

Hernandez, A.; Mayordomo, J.G.; Asensi, R. & Beites, L.F. (2005). *A Method Based on Interharmonics for Flicker Propagation Applied to Arc Furnaces, IEEE Transactions on Power Delivery*, 20(3), 2005, 2334-2342

Hurst, R. (1994). Power Quality and Grounding Handbook, *The Electricity Forum*, Toronto, Canada

Katic, V. (1994). Network harmonic pollution - A review and discussion of international and national standards and recommendations, *Proceedings of Power Electronic Congress-CIEP*, pp. 145-151, Paris, France, October 24-26, 1994

Key, T.S. & Lai, J.S. (1997). IEEE and international harmonic standard impact on power electronic equipment design, *Proceedings of International Conference Industrial Electronics, Control and Instrumentation -IECON*, pp. 430-436, London, England, May, 25-27, 1997

Lazaroiu, C. & Zaninelli, D. (2010). DC arc furnace modeling for power quality analysis, *Scientific Bulletin of Politehnica University of Bucharest*, Serie C, vol.72, Issue 1, 2010, 56-62

Math, H. J.; Bollen, Irene. & Yu-Hua Gu. (2006). Signal Processing of Power Quality Disturbances, *Wiley-Interscience*, New York

Pretorius, J.H.C.; Van Wyk, J. D. & Swart, P.H. (1998). An evaluation of some Alternative Methods of Power Resolution in a Large Industrial Plant, *Proceedings of the Eights International Conference on Harmonics and Quality of Power-ICHQP-VIII*, pp. 331-336, Athens, Greece, vol. I, October, 7-8, 1998

Hooshmand, R.A. & Esfahani, M.T. (2009). Optimal Design of TCR/FC in Electric Arc Furnaces for Power Quality Improvement in Power Systems, *Leonardo Electronic Journal of Practices and Technologies*, Issue 15, 2009, 31-50

Sankaran, C. (2008). Power Quality, *CRC Press*, London

Toulouevski, Y.N. & Zinurov, I.Y. (2010). Innovation in Electric Arc Furnaces, *Springer-Verlag*, Berlin, Heidelberg

Vervenne I.; Van Reuse K. & Belmans R. (2007). Electric Arc Furnace Modeling from a Power Quality Point of View, pp. 1-6, *Proceedings of IEEE Conference on Electrical Power Quality and Utilisation*, Lisbon, Portugal, September, 21-23, 2007

Webster, J. G. (2004). Electrical measurement, Signal processing and Displays, *CRC Press*, New York

***IEC. (1999). IEC 61000-4-30, Testing and measurement techniques-power quality measurement method

***IEEE. (1995). IEEE 1159:1995, IEEE recommended practice for monitoring electric power quality

***IEEE-WG. (1996). IEEE Working Group on Nonsinusoidal Situations. (1996). Practical Definitions for Powers in Systems with Nonsinusoidal Waveforms and Unbalanced Loads, *IEEE Transactions On Power Delivery*, vol. II, no. 1, 1996, 79-101

***IEEE. (1996). IEEE 100-1996 The IEEE Standard Dictionary of Electrical and Electronics Terms, Sixth Edition

***PE. (2004). PE 143/2004, Romanian norm for limitation of harmonic pollution and unbalance in electrical networks

***SREN. (1998). SREN 50160, Characteristics of supplied voltage in public distribution networks, October, 1998

***EN. (2004). Standard EN50160-Power Quality Application Guide, Voltage Disturbance, July 2004.

***CMP. (1987). Understanding electric arc furnace operations for steel production, Center for Metals Production-CMP, vol.3, no.2, 1987

Part 3

Power Quality and Monitoring

6

Power Quality Monitoring and Classification

Sjef Cobben
University of Technology, Eindhoven
Netherlands

1. Introduction

Experts generally agree on the need for standardized power quality indices allowing to monitor and to report power quality on a common basis. Values for the quality indices represent a few numbers that are the result of characterizing, reducing or extracting from a large volume of power quality measurement data. The systems for power quality data acquisition, storing and reporting are mostly separated stand alone applications, which have its drawbacks. However, distribution automation reveals new possibilities for continuous power quality monitoring. Advanced computer systems with open architecture make it possible to integrate power quality data with the normal network operation data.
In this chapter an overview of power quality monitoring is given. The power quality indices will be discussed and the compatibility levels as stated in the European standards and as example the National Dutch grid code will be described. Then possible methods for normalizing indices for power quality phenomena are introduced. An example of how power quality phenomena can be classified and translated into easy to understand and to analyze data is described. The end of this chapter gives details of some applications.

2. Power Quality indexes, levels and limits

For several reasons, there is a need for common power quality indices. With indices it is possible to report quality in a consistent and harmonised manner, either to customers, regulators or within the grid operator. The values for these indices have to be compared to the limits described by the national regulator. Because there is still a lot of discussion about the acceptable levels and the corresponding time limits (95% or 99% or 100%), converting power quality data into indices has to be done in a flexible way, so changing the limits must be possible without a lot of consequences. A number of international standard documents define the limits and the measurement process, including EN 50160 (Cenelec 2009) and IEC 61000-4-30 (IEC 2009). The last one is a standard that explains exactly how power quality instruments should work.

2.1 Power Quality indexes

In general the responsibility for Power Quality is a shared responsibility between grid operator, manufacturer and customers, as shown in Fig. 1.
The grid operator is responsible for the quality of the supply voltage. The manufacturers are responsible for the immunity and emission of the equipment and devices. The customer is responsible for the installation connected to the installation. The current at the point of

connection (POC) shall influence the quality of the supply voltage. That means that there should be good limits and indexes, not only for the supply voltage but also for the current at the POC. In table #.1 some of the existing power quality indices are given. Also is indicated where indices are still not available or which indices are subject of discussion in international committees. The table has indices for the quality of voltage (grid operator's responsibility) and quality of current which is (mostly) the responsibility of the customer.

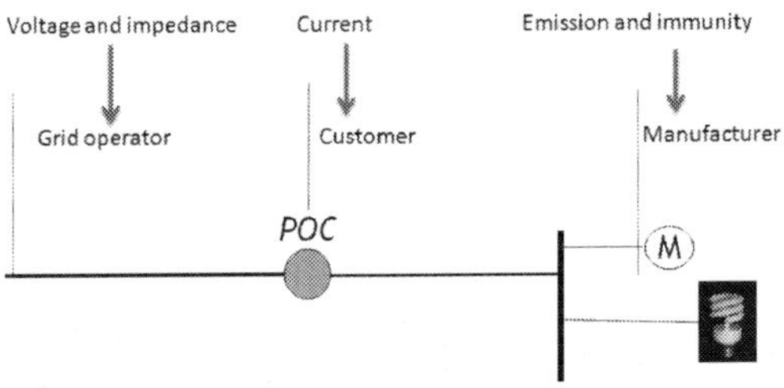

Fig. 1. Shared "Power Quality" responsibility along several parties

The current capacity of the point of connection is limited by the protection device at the POC. Certainly this is more an economic limit, because the grid operator includes a certain diversity factor for the grid design. There is only a weak relation with the voltage indices. Harmonic current limits are defined for devices, but at the point where there is a contract between customer and grid operator no harmonic current limits exist in European standards or grid codes. For the grid operator, limits at the point of connection are needed. The IEEE 519 (IEEE, 1992) aims to state harmonics for a whole installation. For flicker there is an additional requirement for flicker severity variation implemented in the Dutch National grid code. In most European countries this is not the case.

Power Quality phenomena	Indices grid operator	Indices customer
Magnitude of voltage	U_{nom}, U_c	current capacity at the point of connection 2)
Voltage level	$\Delta U_{nom}, \Delta U_c$ 1)	-
Harmonic voltage	THD_U, U_h	THD_I, I_h 3)
Flicker severity	$P_{lt}, \Delta U$	$\Delta P_{st}, \Delta P_{lt,}, \Delta U$ 4)
Voltage dips	- 5)	-

1)Discussion about the acceptable levels
2) Only as indication, no time restrictions
3) Harmonic currents are only regulated for devices and not yet at the POC
4) Regulated in the Dutch grid code, in most countries no indices at the POC
5) No regulation yet, discussion about acceptable number of dips is ongoing. Some restrictions in National grid codes
General remark: All power quality indices are defined for a certain percentile within a measuring period of one week. Measured are the 10-minutes average values. There is discussion in international committees concerning power quality about this measuring period of 10 minutes.

Table 1. Indices for power quality phenomena

Phenomenon	Criteria
Frequency	50 Hz +/-1% during 99.5% of each year 50 Hz +4%/-6% during 100% of the time
Magnitude of voltage (Voltage level)	Low voltage : U_{nom}=230V U_{nom} +/-10% for 99% of all 10 min. average measured values during 1 week U_{nom} +15/-15% for all 10 min. average values Medium voltage Un<35 kV U_c=nominal voltage as stated in contract with client U_c +/-10% for 99% of all 10 min. average measured values during 1 week U_c +15/-15% for all 10 min. average values *+15% is still under consideration on time of printing this book
Flicker	Low voltage and medium voltage Un< 35kV $P_{lt}\leq 1$ during 95% of the time In the Dutch grid code for example is stated: $P_{lt}\leq 1$ during 99.5% of the time $P_{lt}\leq 5$ during 100% of the time
Unbalance	Low voltage and medium voltage U_c< 35kV The negative sequence voltage is smaller then 2% of the positive sequence voltage during 99.5% of the time. In the Dutch grid code is stated: The negative sequence voltage is smaller then 2% of the positive sequence voltage during 99.5% of the time. The negative sequence voltage is smaller then 3% of the positive sequence voltage during 100% of the time.
Harmonic distortion	Low voltage and medium voltage U_{nom}<35 kV Under normal operating conditions, during each period of one week, 95 % of the 10 min mean r.m.s. values of each individual harmonic voltage shall be less than or equal to the values given in Table 3. $THD\leq 8\%$ for all harmonics up to the 40th during 95% of the time In the Dutch grid code is additional stated: For harmonics that are not mentioned the smallest value referred to in the standard counts. $THD\leq 12\%$ for all harmonics up to the 40th during 99.9% of the time The relative voltage per harmonic is smaller than the percentage stated in EN-50160 multiplied by 1.5 for 99.9% of the ten-minute values.

Table 2. Requirements for supply voltage, EN50160

2.2 Voltage quality levels

The limits set by the national Regulator are used as the starting point for defining the quality of the supply voltage. As a minimum requirement, these limits have to be fulfilled at each POC (the point of connection of a customer). The European Standard EN50160 is always used in European countries as a basis for the quality of the supply voltage, so as basis for the different national grid codes. The quality of the voltage is characterised by the following aspects:

- Magnitude of voltage
- Voltage changes (Dips and flicker)
- Harmonic distortion
- Unbalance
- Frequency

Table 2. gives an overview of the power quality levels as stated in the EN50160 on June 2009.

Odd harmonics				Even harmonics	
Not multiples of 3		Multiples of 3			
Order h	Relative voltage U_h	Order h	Relative voltage U_h	Order h	Relative voltage U_h
5	6,0 %	3	5,0 %	2	2,0 %
7	5,0 %	9	1,5 %	4	1,0 %
11	3,5 %	15	0,5 %	6 ... 24	0,5 %
13	3,0 %	21	0,5 %		
17	2,0 %				
19	1,5 %				
23	1,5 %				
25	1,5 %				
NOTE No values are given for harmonics of order higher than 25, as they are usually small, but largely unpredictable due to resonance effects.					

Table 3. Limits for harmonic voltages

At the time of printing there were no requirements in the EN50160 for the amount, depth and duration of voltage dips. Because a dip is an event and not a continuous appearing phenomenon a different classification method is necessary. In section 5 this classification method for dips is described

3. Power Quality monitoring

The need for monitoring the Power Quality (voltage and current) is growing. Industrial processes in installations are more and more depending on a sufficient quality of the voltage. The (over)loading of equipment as transformers, motors, conductors is depending on the quality of the current. Due to the shared responsibility of power quality it is important to monitor the voltage and currents at the POC and in the low-, medium and high voltage grid.

3.1 Monitoring in installations

There can be several reasons to monitor the power quality in installations. Measuring at the point of connection can be useful to check the quality of the supply voltage. Power quality

monitoring can also be used for analyzing a problem. The time scale of the measurement is totally different in both cases. The quality of the supply voltage has to be measured during at least one week. Analysing a problem for example with inrush currents and voltage deviations could be done during the moments of switching on and off the equipment involved. Also the data needed for doing these analyses is different. In the first case we need average values; in the second case even waveforms could be needed.
Fig.3 shows an example of voltages measured at a low voltage POC. Measured are according the EN 50160, 10 minutes averages during a week. The voltages are within the given limits, so no problem with voltage level is measured.

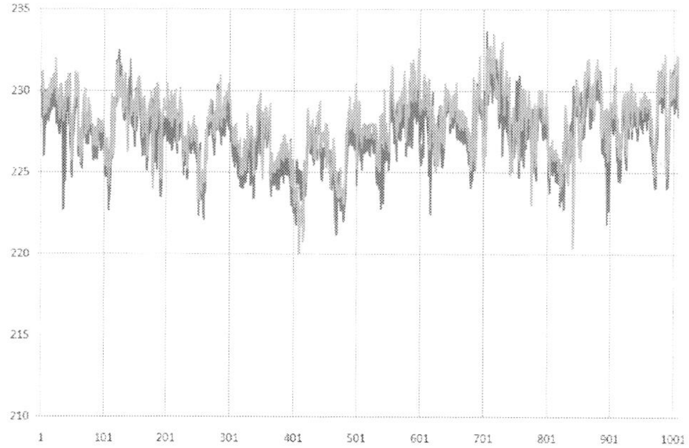

Fig. 2. Voltages measured on LV-POC (1008 measured 10-minutes averages)

Another example of a measured power quality phenomena is the flicker index Plt as shown in fig.3. The 95% limit of this index should be below 1. A serious problem is measured on this POC.

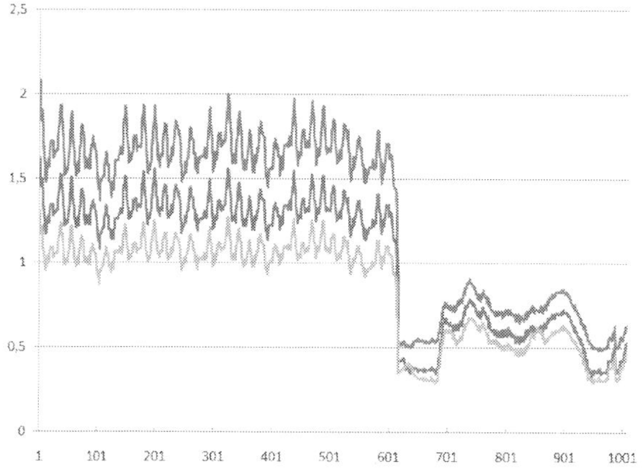

Fig. 3. Flicker index Plt measured at LV-POC

Most PQ-measurement devices will give a quick overview of several power quality phenomena to check of all phenomena are within the given limits. In fig.4 is a picture given of a possible overview screen.

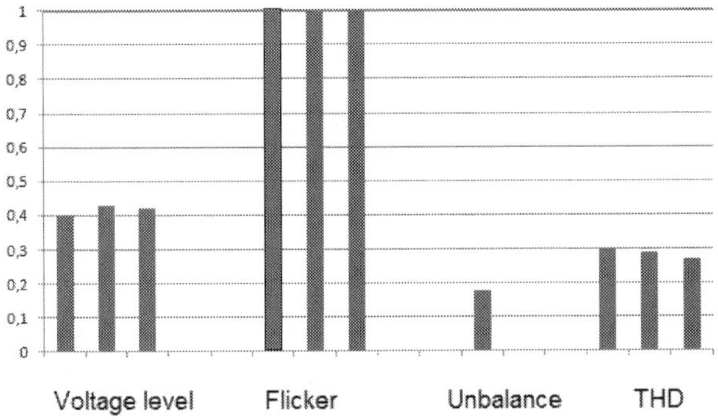

Fig. 4. Overview screen of PQ-measurement device

For a single installation it could be important to measure all power quality phenomena in voltage and current for the following reasons:

- Checking the quality of the supply voltage.
- Analyzing the harmonics in the current (overloading most of the components).
- Analysing the power flow (contract with the grid operator, overloading transformer, cables, switchgear, ...).
- Controlling trends in the power quality phenomena.
- Analyzing problems within the installation.

For most of the measurements a fixed installed measurement device is needed. For analyzing problems a portable device could be used. Measuring with portable devices could give a need for additional safety measures (see fig. 5)

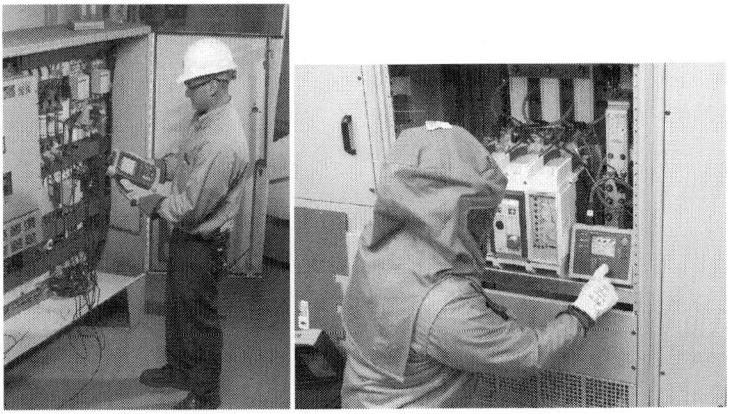

Fig. 5. Measuring in an installation requires several safety measures (Source: Fluke).

3.2 Monitoring in grids

Monitoring in grids could be done for the same reasons as for monitoring in installations. Nevertheless, there are more reasons to measure in (more) points in the grid.

First of all is it important for analyzing the quality of the grid (supply voltage) and the operation of the grid. Also, more and more information about the quality of supply voltage is required, for individual customers but also for the National regulator.

The interest of the regulator is not only on the quality of one POC but the general level of voltage quality. This could be given by measuring the quality at a random chosen number of POC's and report about it. In Fig. 6 an example of the result of measuring the THD-level in the low voltage grid in the Netherlands is given.

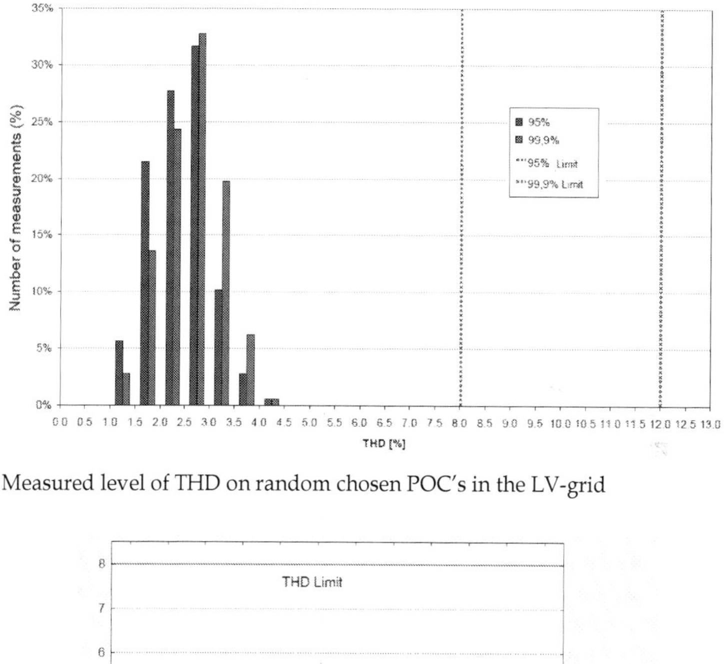

Fig. 6. Measured level of THD on random chosen POC's in the LV-grid

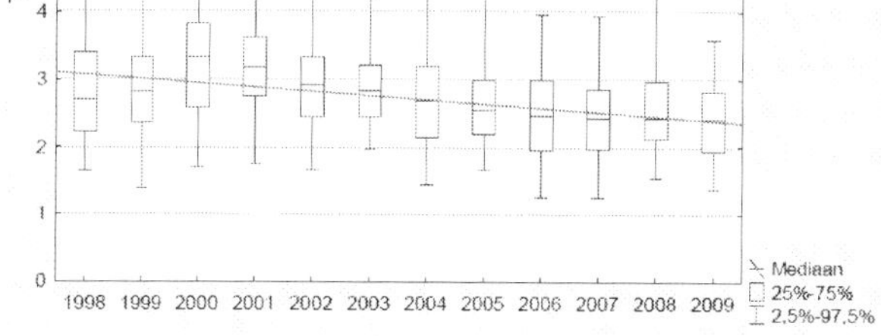

Fig. 7. Trend over the years 1998-2009, LV-grid in the Netherlands

When this is done for several years the trend over the measured years can be given. In Fig. 7 this is done for the measurements over the years 1998-2009. In spite of the connection of more non linear loads we can conclude that the level of the THD in the low voltage grid is decreasing.
In the future there will be more PQ-measurements in the grids, installed as PQ-devices in the substations (see Fig. 8) or as sensors measuring voltage and current which will also be used for analyzing the PQ-level.

Fig. 8. PQ measurement devices installed in a substation (Source: Metrum)

The advantage of having more information is more control and knowledge of the grid. The measured data can be combined with other data for fault localisation, protection, control of loads or dispersed generation and lot of other applications. The disadvantage is a lot of data what has to be analysed. So, automatic processing of data and application of data is needed. Also, for the grid operator, the customer and the regulator an easy to understand classification is needed. In the next chapter a possible method is described.

4. Normalizing and classifying PQ-data

Making a classification we have to realise that most people that will use this classification are not familiar in detail with power quality and all the different limits of each phenomenon. This classification can be used for several purposes as:

- Customer's information, important for the image of the grid operator but also for industrial customers.
- To report about the quality of the voltage to the management of the grid operator, the grid owner or the regulator
- To use the data in network operation and planning

Monitoring the quality of the voltage will lead to an enormous amount of data which has to be reduced to a format that can be analysed quickly and easily. An important element in the communication with the customer is that the results should be easy to interpret. The first step in reaching a suitable classification is to normalise all power quality aspects (Meyer, J., Schegner, 2005). The second step is to make the appropriate classification (Cobben, J.F.G. 2007). Choices made here by realizing a classification are proven to be practical but can be discussed. Other choices can be made depending on purpose, planning levels, etc.

Nevertheless, the handling of data into such visual classification is a promising development.

4.1 Normalising

For each continuous phenomenon we can calculate the normalised power quality level using the formula:

$$r_{(v,q,p)} = 1 - \frac{m_{(v,q,p)}}{l_{(q)}} \tag{1}$$

$r_{(v,q,p)}$ = the normalised power quality aspect q, on location v, for phase p
$m_{(v,q,p)}$ = the actual level of phenomenon q, on location v, for phase p
$l_{(q)}$ = the compatibility level of phenomenon q

When there is no disturbance the normalised value will be 1 (m=0). If the disturbance level is equal to the accepted compatibility level, the normalised value will be zero. If the disturbance level exceeds the specified limit, the performance index r becomes negative.
The plane from "no disturbance" to a level of "twice the acceptable disturbance level" is divided into six areas ranging from very high quality (A) to extremely poor quality (F), as shown in Fig. 9. Although the number of areas is disputable, there are good arguments for choosing six. These arguments are:

- For each voltage disturbing phenomenon a compatibility level and a planning level can be recognised. So three areas could be defined by these levels (the levels of 0.33 and 0.66 could be fitted to these levels).
- More areas will define more levels of quality than necessary but also will lead to more switching results week by week to a different quality level.
- By analysing complaints it is of interest to have some differentiation in not acceptable quality, which makes it easier to make trends visible. But in the "not acceptable quality" part it is not helpful to have more areas than three for the same reason as mentioned before.

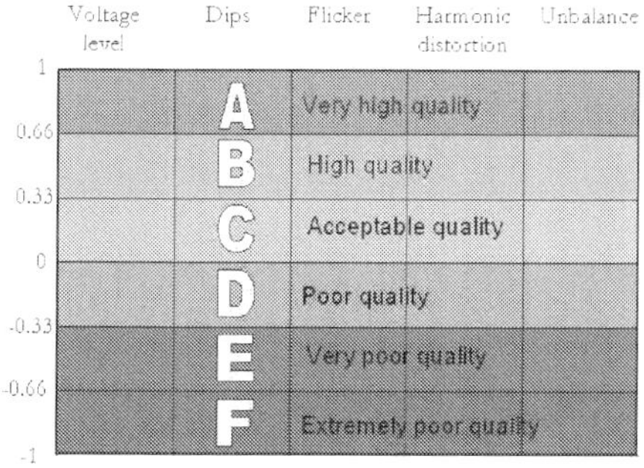

Fig. 9. Classification of power quality phenomena

The classification from very high quality to extremely poor quality is made on basis of a technical judgement of the voltage. Of course, for some customers poor quality can be acceptable quality too, certainly when they can pay less for their energy. These economic aspects are not introduced into this classification.

4.2 The percentile method

The most accurate method for classifying a certain power quality aspect is the so-called "Percentile Method". For each continuous existing characteristic the EN 50160 uses a certain percentile and level as a limit for the measured average values. For example, the flicker level (P_{lt}) has to be below 1 for 95% of all measured average ten-minute values.

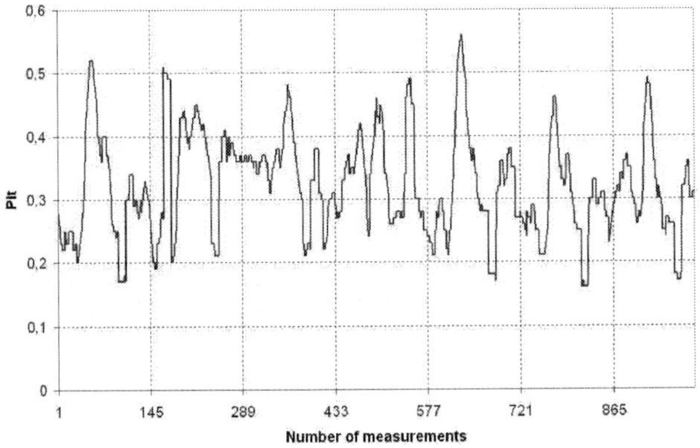

Fig. 10. Flicker level at a random chosen POC

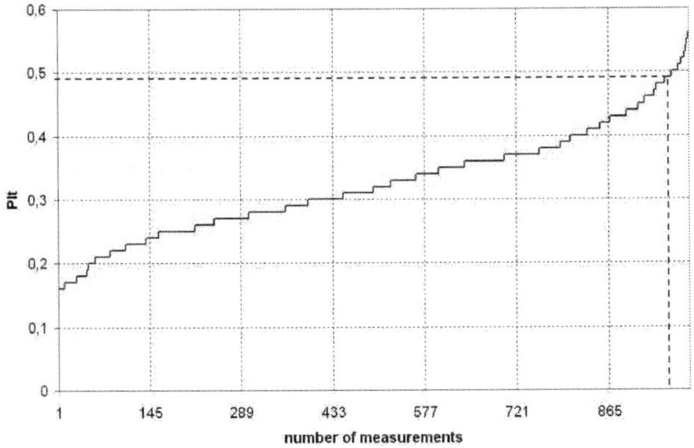

Fig. 11. Sorted data flicker level

The measured flicker level for a point of connection of a customer (POC) chosen at random is shown in Fig. 10. By sorting the data the 95% percentile value can be established, which in the given example is 0.48 (see Fig. 11). Classification of the POC with respect to flicker can be done by normalising the flicker level, the result being:

$$r_{(v,q,p)} = 1 - \frac{0.48}{1} = 0.52$$

This corresponds to a high quality classification (B) for this POC with respect to flicker. The disadvantage of this method is that the end result does not give a lot of information about the actual flicker level. Fig. 12 shows two extreme distributions with the same 95% percentile result, nevertheless the distribution in a) presents a better performance with respect to flicker than situation b).

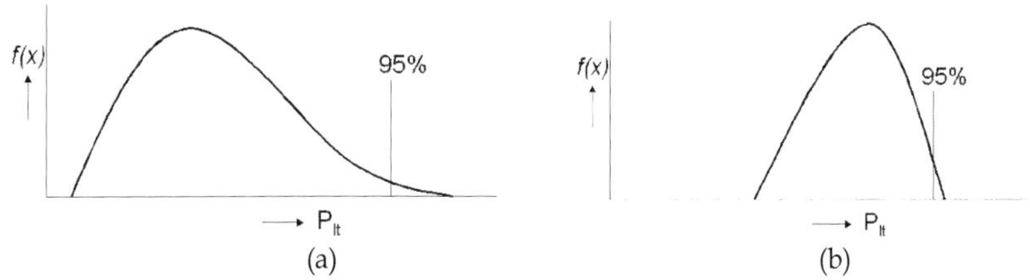

Fig. 12. Two different distributions of flicker levels

Grid operators also need good general information about the power quality level, so the 95% percentile value alone does not satisfy their needs. However, for checking the quality in relation to the standard (usually set by the regulator), this method is the most accurate.

4.3 The STAV method

Classification according the "STAV (standard deviation, average value) Method" explores more information about the distribution of the aspect involved. Take for example the flicker level measured over the course of a week, as shown in Fig. 10. Calculating the average value and the standard deviation of this distribution results in:

$$P_{lt,av} = \frac{\sum_{i=1}^{n} P_{lt,i}}{n} = 0,323 \quad \text{and} \quad \sigma = \sqrt{\frac{\sum_{i=1}^{n} \left(P_{lt,i} - P_{lt,av}\right)^2}{n-1}} = 0,082$$

Fig. 13. Shows how the data, as shown in Fig. 10, is distributed. Fig. 13 also includes the best fitted normal distribution.

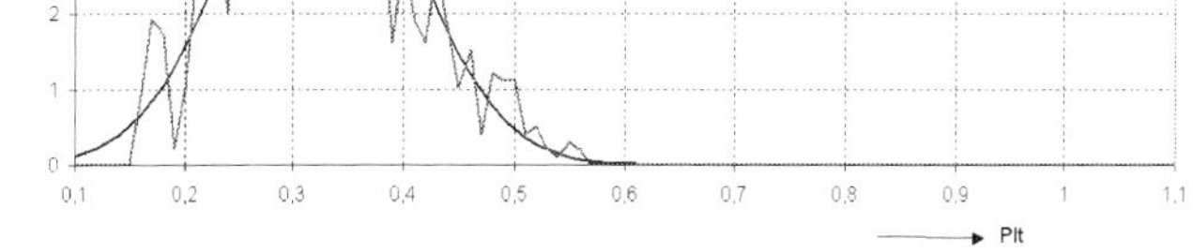

Fig. 13. Distribution of flicker level and best fitted normal distribution

Assuming that the flicker level is normally distributed, a classification method based on the same basic principles as used for the percentile method can be made. The most important requirements for the flicker level according to the national grid code are:

- 95% of all ten-minute average values of P_{lt} within a period of 1 week have to be ≤1.
- All ten-minute average values of P_{lt} within a period of 1 week have to be ≤5.

The relation between the standard deviation and the average value of all ten-minute values can be calculated using the following formula (#.2).

$$P_r\{X \le x\} = P\left\{Y \le \frac{x - P_{lt,av}}{\sigma}\right\} = P\left\{Y \le \frac{1 - P_{lt,av}}{\sigma}\right\} = 0.95 \quad (2)$$

P_r = chance of occurrence
X = normal distribution variable
x = normal distribution value
Y = standard normal distribution variable
$P_{lt,av}$ = average value of P_{lt} within given distribution
σ = standard deviation of P_{lt} within given distribution

This is the case where $(1-P_{lt,av})/\sigma=1.65$, based on the table for a standard normal distribution. According to formula (2.1) and the borders 0, 0.33, 0.66, 1, 1.33, 1.66 and 2 as explained before, six planes can be made with the classification A through F, as shown in Fig. 14. Advantages of the STAV method are the quick overview of the quality of the considered aspect, a simplified result that is easy to communicate and less influence of some of the extreme measuring points on the weekly data. A disadvantage is that it does not answer the question of whether the measured value was within the limits required by the regulator in a certain week.

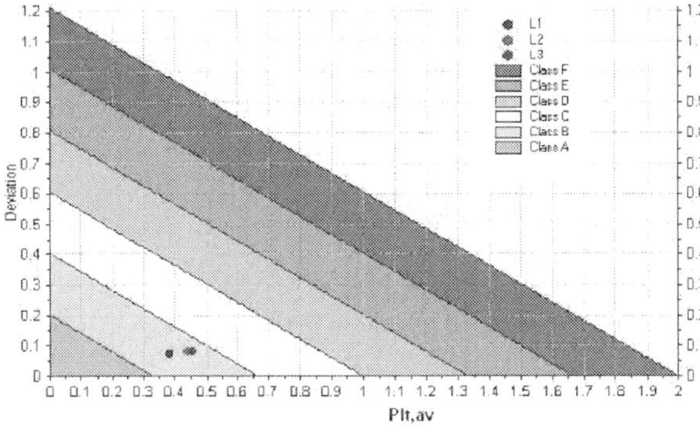

Fig. 14. Classification according STAV method (flicker)

All other continuous power quality phenomena can be classified in the same way as for the flicker phenomenon.
A particular point in relation to voltage level is that there are two limits. Take as an example the voltage measured during a week, as shown in Fig. 15.
Calculating the average value and the standard deviation of this distribution results in:

$$U_{av} = \frac{\sum_{i=1}^{n} v_i}{n} = 225.3 \text{ V} \text{ and } \sigma = \sqrt{\frac{\sum_{i=1}^{n}(v_i - U_{av})^2}{n-1}} = 2.43 \text{ V}$$

Assuming that the voltage is normally distributed, a classification method based on the same basic principles as before can be used. Fig. 16 shows how the data, as shown in Fig. 15, is distributed. Fig. 15 also includes the best fitted normal distribution. The normally distributed fitting curve in Fig. 15 shows that the measured voltage is indeed close to a normal distribution. The limits given by the national Dutch regulator (as still valued during the printing of this book) are taken to create the border between classification C and D. The requirements for the low voltage level are:

- The nominal voltage is 230 V.
- 95% of all measured average ten-minute values of the voltage within a period of 1 week have to be within ±10% of the nominal voltage.
- All average ten-minute values of the voltage within a period of 1 week have to be between -15% and +10% of the nominal voltage.

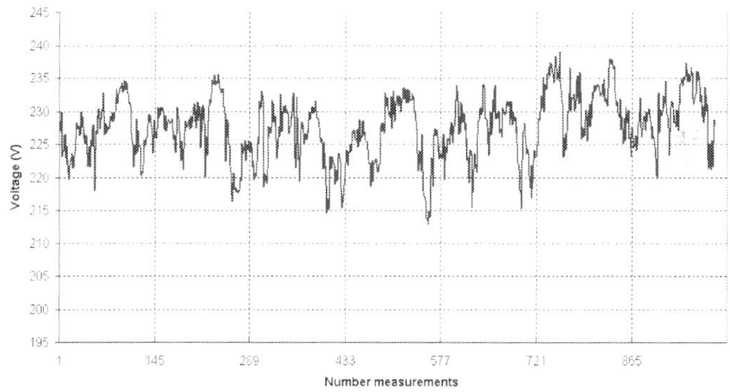

Fig. 15. Voltage measured at random POC

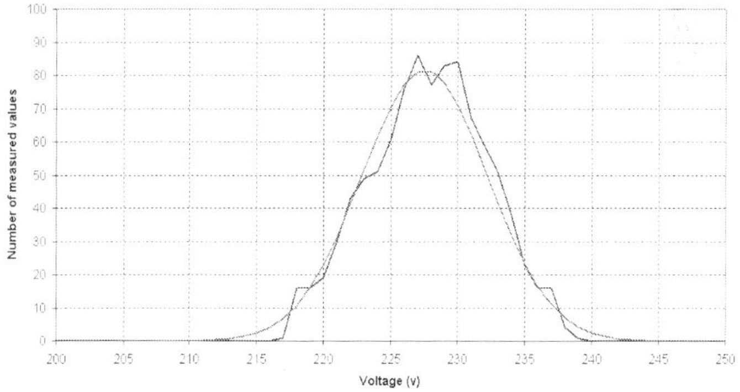

Fig. 15. Comparing measured distribution with normal distribution

For the voltage level the two theoretical extreme distributions, which just fulfil the requirements of the national Dutch grid code, are shown in fig.16. This figure plots the ±10% limits of the voltage. The upper limit may not be exceeded. Here we suppose that 99.9% is acceptable. The limit on the left (lowest voltage) may be exceeded for 5% of all measured values.

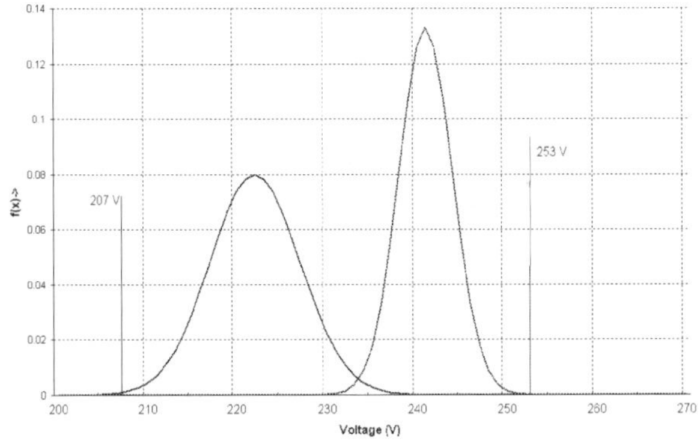

Fig. 16. Extreme distributions of low voltage level

For the upper limit the relation between the deviation and the average of all measured values is represented by:

$$P\{X \le x\} = P\left\{Y \le \frac{x - U_{av}}{\sigma}\right\} = P\left\{Y \le \frac{253 - U_{av}}{\sigma}\right\} = 99,9\%^{1} \tag{3}$$

Looking at the statistical table of the normal distribution (see appendix A), this is the case where:

$$\frac{253 - U_{av}}{\sigma} < 3.1 \,.$$

For the lower limit the following formula has to be used:

$$P\{X \le x\} = P\left\{Y \le \frac{x - U}{\sigma}\right\} = P\left\{Y \le \frac{207 - U_{m}}{\sigma}\right\} = 5\% \tag{4}$$

This is the case where $(207\text{-}U_{av})/\sigma$ = -1.65. The relation between the deviation and the average voltage for both extreme distributions are shown in Fig. 17.
The lines drawn in fig.17 are the borders of the normal quality C.
Again formula (#.3) and the borders from 0 to 2 as described before are used for normalising (in this case) the voltage level. The following borders can be defined for classification A, B, D, E and F.

$$\text{Border A / B} \;: r = \frac{1}{3} = 1 - \frac{\left|\dfrac{253 - U_{A/B}}{230}\right| \bullet 100}{10} \rightarrow U_{A/B} = 253 - \frac{2}{3} \bullet \frac{230}{10} = 237.66\,V$$

$$\text{Border B / C} \;: r = \frac{2}{3} = 1 - \frac{\left|\dfrac{253 - U_{B/C}}{230}\right| \bullet 100}{10} \rightarrow U_{B/C} = 253 - \frac{1}{3} \bullet \frac{230}{10} = 245.33\,V$$

---

[1] For practical use, the 99.9% value is used instead of the 100% value

A normalised border on the minimum voltage level side is calculated as follows:

$$\text{Border B / C} : r = \frac{2}{3} = 1 - \frac{\left|\frac{U_{B/C} - 207}{230}\right| \bullet 100}{10} \rightarrow U_{B/C} = 207 + \frac{1}{3} \bullet \frac{230}{10} = 214.66\ V$$

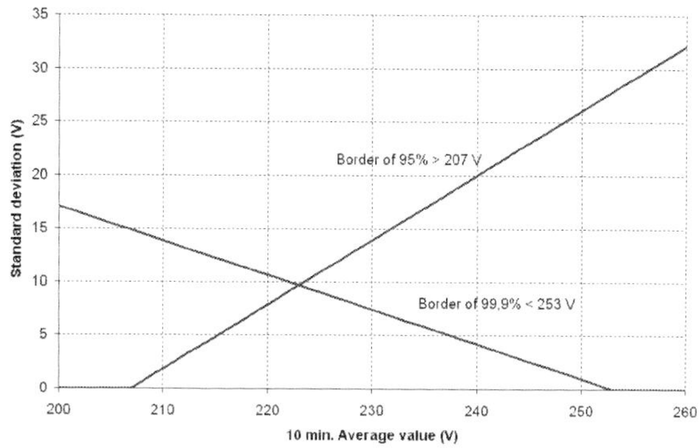

Fig. 17. Border of normal classification, low voltage level

Similarly, all other borders can be calculated, resulting in the classification areas as shown in Fig. 18.

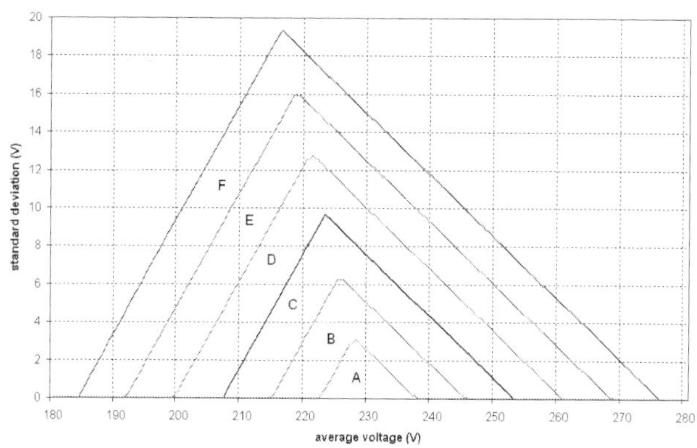

Fig. 18. Classification low voltage level

Implementation of a power quality monitoring system using this classification can help a grid operator to have a general overview of the quality of the grid without an additional need for analysing a lot of data.

5. Classification of dips

There are several causes for dips. The most important are faults in the MV or HV grid. Three-phase faults will have the greatest impact on the installation or processes of customers if they are sensitive to voltage dips. Mitigation of dips will be very difficult or expensive if a lot of active and/or reactive energy is required to keep the voltage on an acceptable level. If the remaining voltage falls below 50% of the nominal voltage, a critical border is reached.

The reasons for classifying dips are:

- To make it easy to inform customers about the expected dip profile, without details which are irrelevant or even unjustified.
- To make it easy to inform the national regulator about the quality of the grid, in relation to this power quality phenomenon.
- To make it easy to gather information about dips in the grid and to draw some conclusions about the quality of the grid, taking into consideration trends over a period of several years.

Converting the dip table into a system indices or the presented "ABCDEF" classification makes it very suitable as a management tool to control the overall quality of the grid and to communicate about this power quality phenomenon in the same way as for other phenomena.

It is also possible to make the general classification more tailor-made for a specific customer with a sensitive process or installation. A good insight is then needed into the cost arising from a dip and the possible cost of mitigating a dip.

By developing a classification for dips there are some general applications where this classification can be used:

- Giving information about the background (sources) of the dips.
- Giving some guidance about the way to solve possible dip problems.
- Giving information about the quality of the grid.
- Showing responsibilities (manufacturer, grid operator, customer).

A distinction can be made between the following sources of the dip problem:

- Faults in the HV grid.
- Faults in the MV grid (primary section).
- Faults in the MV grid (secondary section, own feeder).
- Low voltage problems.

Faults in the HV grid are mostly of short duration, so the duration of the dip is rather short too. Faults in the MV grid are disconnected in times varying from 0.3 seconds to a few seconds. The depth of the dip is mostly not very big, but sometimes the depth can be large if it is a fault in the same feeder after a secondary protection device.

Low voltage problems leading temporarily to a voltage below 90% of the nominal voltage can last a longer time but these dips are never deep. In fact it should not be counted as a real dip.

Another way of looking to the problem is through the "how to solve the problem" debate. Most severe dip problems are the three-phase dips, with a remaining voltage less then 50% (Didden M., 2003). Customers can protect their installation against these kinds of dips, only with high costs. This can be a reason to divide the table into dips with remaining voltages below and above 50% of the nominal voltage. Seen from the manufacturer's point of view the ITIC curve is an important curve. All dips that occur with a remaining voltage above this curve should not give any problem in the installation. Furthermore, we can conclude that dips with a remaining voltage below 70%, in combination with duration of longer than five seconds, seldom occur.

All these considerations make it advisable to divide the table in six parts, as shown in Table 4. The green part lies above the ITIC curve and is the responsibility of the manufacturer. All devices have to work properly within this area. Furthermore the following parts can be recognised:

- S1: Short duration 1, mostly having its origin in the HV grid, but with a remaining voltage above 50% of the nominal voltage. Easy to solve in customer's own installation.
- S2: Short duration 2, mostly having its origin in the HV grid, but with a remaining voltage below 50% of the nominal voltage. Difficult to solve in customer's own installation.
- M1: Short duration 1, mostly having its origin in the MV grid, but with a remaining voltage above 50% of the nominal voltage. Easy to solve in customer's own installation.
- M2: Short duration 2, mostly having its origin in the MV grid without coils or with secondary protection, but with a remaining voltage below 50% of the nominal voltage. Difficult to solve in customer's own installation.
- L1: Long duration 1, occurs due to low voltage problems, mostly with high remaining voltage which makes it easy for the customer to solve in customer's own installation.
- L2: Long duration 2, will lead to severe problems but will in practice only occur in very extreme situations.

ΔU t_d	0.01-0.02	0.02-0.1	0.1-0.5	0.5-1	1-2	2-5	5-10	10-20	20-60s
80-90						ITIC-curve			
70-80									
60-70									L1
50-60		S1			M1				
40-50									
30-40									
20-30		S2			M2				L2
10-20									
1-10									

Table 4. Classification of dips

This is a similar classification of dips as is described in the South African standard NRS 048 [Nrs01]. The areas chosen in this standard are given in Table 5.

Retained voltage U	Duration t		
	20=t<150 (msec)	150=t<600 (msec)	0,6 = t < 3 (s)
90 > U = 85	Y	Y	Y
85 > U = 80	Y	Y	Z_1
80 > U = 70	Y	S	Z_1
70 > U = 60	X_1	S	Z_2
60 > U = 40	X_2	S	Z_2
40 > U = 0	T	T	Z_2

Table 5. Dip-classification according South African standard

Although it is also a dip depth and time presentation, it is a little bit more complex and not completely according to the EN50160 dip definition. Furthermore it defines some small dips areas, what increases the problem of unstable values. Also the important border of 50% is not used.

The way to develop an "ABCDE" classification in the same sense as done in this research for other Power Quality phenomena is shown in Fig. 19.

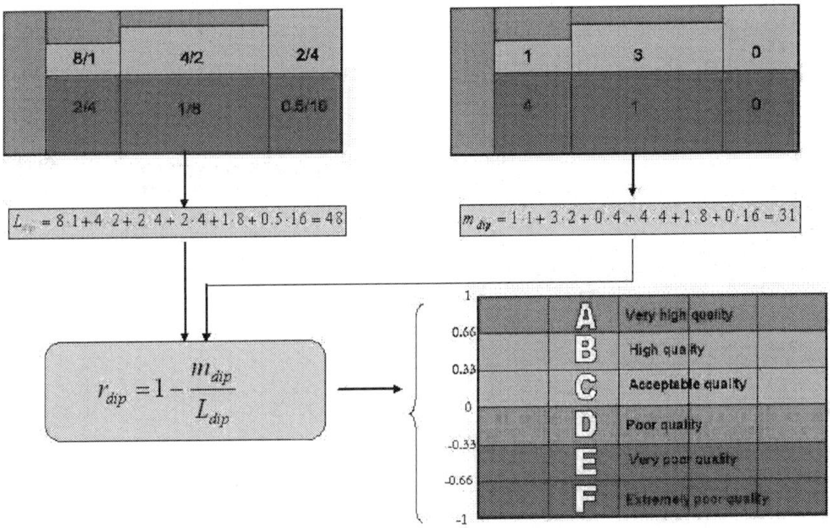

Fig. 19. Converting measured dips to "ABCDE" classification

First we have to define L_{dip}, being the limit for dips, as for example a P_{lt} value of 1 was the limit for the flicker phenomena. In each defined dip zone two parameters has to be given. These parameters are:

- The amount of dips, which normally should not be exceeded. This value could be determined by taken the 95% value of measured dips in a substation. In this example it is taken as 8 for the dip zone in the upper left corner of the table.
- The weighting factor. A certain weighting factor must be given to each defined dip zone. A deep dip with long duration will have a much higher impact on the installation than a shallow dip with short duration. These weighting factors should have a relation with the costs of mitigation of the corresponding dip. Research as described in (Wang, J., Chen, 2007) or (Driesen, J., Belmans, R., 2006) could be helpful to determine these weighting factors. In fig.19 this relation is not yet made, but the values of 1 (zone S1), 2 (zone M1) 4 (zone L1 and S2), 8 (zone M2) and 16 (zone L2) are given as example.

The maximum limit value for dips L_{dip} can be calculated using the given limits for the number of dips and the weighting of each dip zone (see fig.19). When on a POC or a substation the amount of dips are measured as is displayed in the right table, the dip value m_{dip} can be calculated. In this case it is 31, resulting in classification B, high quality but near to acceptable quality. Due to the different distribution of dips each year, the measured value

to be used should be an average value over a period of five years. A gliding window of five years could be used, giving a more stable and reliable classification of the grid.
More research has to be made after the optimal level for voltage dips objectives. The total costs for network and customer has to be considered, for determine the weighting factors. Measurements have to be performed and published to estimate the existing performance level and to determine the maximum amount of dips for each zone. This can be done for several type of grids (cables, lines, urban grids, rural grids). These measurements also see the "hidden dips" which are not included in the calculating or prediction process. With "hidden dips" is mend the dips due to faults in the grid, lasting a very short time and do not lead to interruptions

6. Application of Power Quality data

Power Quality monitoring can be defined as the process of gathering data about voltages and currents including time, transporting that data and converting it into useful information. The ideal power quality monitoring system should have the following characteristics:

- It gathers all of the data that is required.
- It moves the data to a certain location where it can be saved and processed.
- It converts the data into information that can be used to take action or to inform people involved.
- It combines the power quality data with other sources of data.

Power quality monitoring as stand-alone function on certain points in the network has its drawbacks because:

- It will not give a total overview of the voltage quality in the grid.
- An additional communication system is needed.

Monitoring these devices is difficult to integrate in normal operation. Therefore, power quality monitoring will in future be integrated in substation automation systems. Two examples of integrated systems which are under development within Liander (Dutch grid operator) are described in the following sections.

6.1 PQ-monitoring in HV/MV substations

Fig. 20 shows the integration of a power quality monitoring system in a substation, as already is implemented in several substations end 2006, beginning 2007 (Riet van, M. J., Baldinger, F. L. 2005).
The several components in the systems are:

- VIM (Voltage Interface Module); measuring the voltage.
- CIM (Current Interface Module); measuring the current.
- BIM (Breaker Interface Module); for status information about the breaker or giving commands to the breaker.

In every incoming and outgoing feeder the currents are measured with a high sampling frequency to gather information about the current waveform. This information can be used for several purposes such as:

- Power quality issues (for example harmonic distortion).
- Short circuit currents for calculating the location of the fault in combination with measured voltage.

- Remaining capacity of the feeder.

All the measured data can be stored on a local computer. Transmitting all data to a central operating room of the network operator is not efficient. Only in faulted situations it will be needed to get on line detailed information. For power quality phenomena, the data can be processed locally to determine power quality indices. These indices can be further analysed centrally. When needed, for instance in case of complaints, detailed information can be sorted out and examined centrally. The system is able to measure harmonics up to the 50th and all other Power Quality aspects stated in EN 50160, such as voltage dips, sags, swells, flicker, etc.

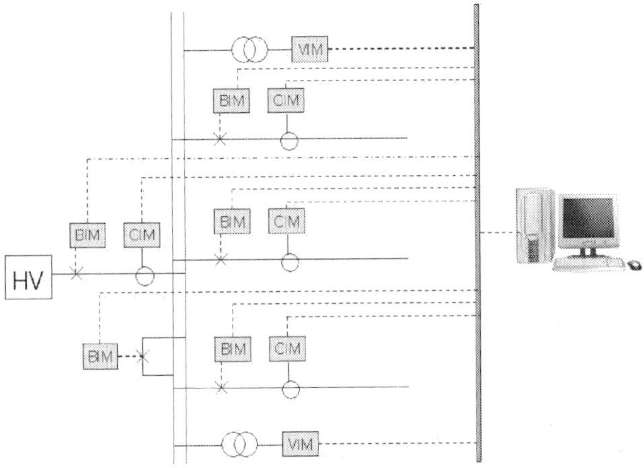

Fig. 20. Distribution automation system, implemented in a HV/MV substation

Gathering a lot of power quality data with this system is possible but analyzing all the data will take an enormous amount of time. For this reason the percentile or STAV method is used.

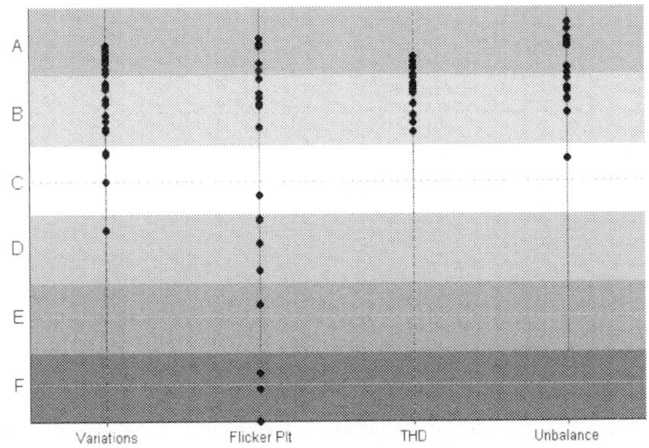

Fig. 21. Classification of supply voltage in substations (example, not based on real data)

First of all there is the advantage of sending only three values of a weekly measured power quality phenomenon: the average, the standard deviation and the 95% percentile value, whatever is important for the regulator. Using this method makes it possible to analyse the data quickly. All weekly data from the substations and transforming substations measured is plotted on a screen, as shown in Fig. 21. This gives an overview of the quality of the supply voltage for each measured busbar in the different substations and for the different power quality phenomena.

Studying Fig. 21., we can conclude that there is one location with a voltage level problem and several locations with a flicker problem. To get more information about the grid situation a more detailed screen relating to voltage level can be selected. Fig. 22 shows this screen, made using the STAV method. There is indeed one bullet across the border of plane C, placed in plane D. This means that the limit of the grid code (border between plane C and D) is exceeded.

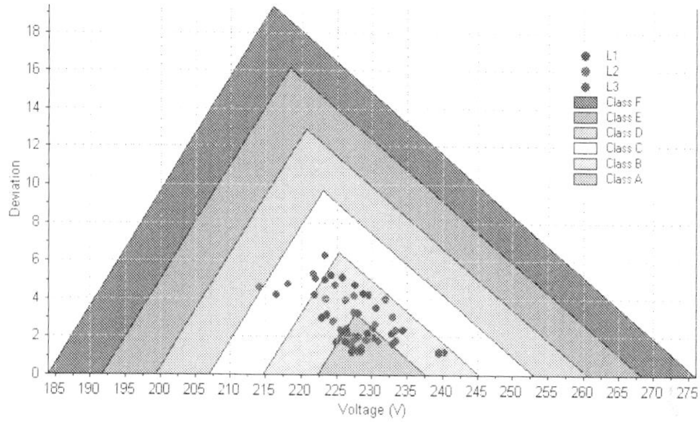

Fig. 22. Measured voltage in transforming substations (LV side), using STAV method

The flicker problem can be analysed using the screen shown in Fig. 23.

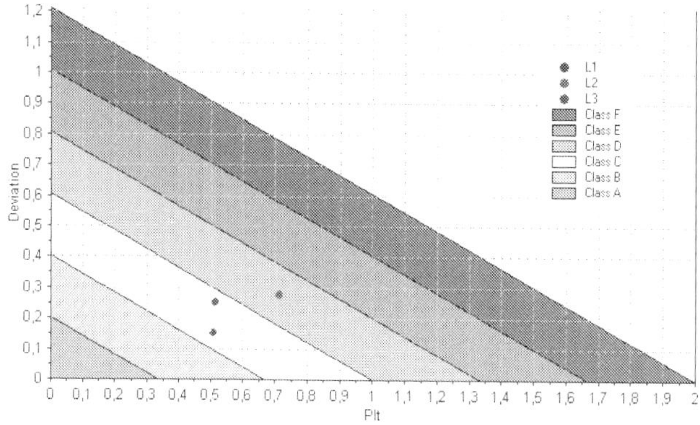

Fig. 23. Flicker data for one selected point in the grid (one phase is exceeding the limit)

Only when more detailed information is needed can all the data of this measurement be downloaded and a graph of the measured values of P_{lt} made according to fig.10. In most cases this will not be necessary because the supply voltage will generally be well within the limits. Presenting the data in the way described above makes it possible to implement data of several weeks, months or even years in the same plot. Trends in the different power quality phenomena can, with all selected data in one graph very easily be analysed.

6.2 PQ monitoring in MV/LV substation

A second example of integrated distribution automation is the MV/LV transforming substation as shown in fig.24. With increasing DG there is a need to have more control on the low voltage network. Therefore more information about the voltage and the currents in the low voltage network must be gathered. This automation will be developed and realised in the coming years.
In the coming 5 to 10 years at all Dutch consumers premises an energy meter with additional functionality will be installed. The energy meter can be remotely managed and read. This could be based on power line communication using the existing low voltage electricity network.

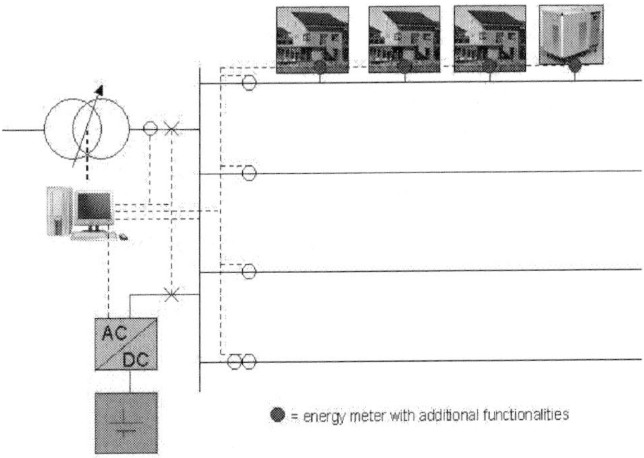

Fig. 24. Development of advanced MS/LS substation with additional functionalities

In the MV/LV transforming substation possibilities for energy storage could be present. This approach enables the grid operator to control the low voltage level, to improve power quality (harmonics flicker) and to optimise the loading of the network. Again, concentrating on power quality phenomena, the amount of data available will be enormous, so it has to be converted locally and in normal situations only indices has to be communicated to a central location.

7. Conclusions

For several reasons there is a need for common power quality indices. With indices it is possible to report quality in a consistent and harmonised manner, either for customers, regulators or for use within the grid company. However, due to the growing importance of power quality, monitoring the present power quality is needed and more monitoring programs are done in the last years and will be done in the coming years. A new development is integrating power quality measurements into substation automation. However, this will lead to enormous amounts of data, even using indices. Therefore an additional classification will be helpful.

In this chapter a promising and already implemented classification is explained. The first step in reaching this classification is to normalise all power quality aspects. The second step is to make the appropriate classification. The most accurate method for classifying a certain power quality aspect is the so-called "Percentile Method". This method can be used for controlling the power quality on a single connection point of a customer. The disadvantage of this method is that the end-result does not give a lot of information about the actual power quality level. To overcome this disadvantage the "STAV Method" is introduced. Advantages of the "STAV Method" are the quick overview of the considered quality aspect, a simplified result that is easy to communicate and with less influence of some of the extreme measuring points on the weekly data. A disadvantage is that it does not answer the question of whether the measured voltage was within the limits required by the regulator in a certain week. Nevertheless, for the grid operator it is a very useful tool for controlling the performance of the grid and to inform customers and regulators about the general voltage quality in the network.

Measuring events as voltage dips and transient are even more complex. Nevertheless for each power quality phenomena an easy to use classification can be made. The data available and the need for detailed information is depending on the application. For analyzing a problem detailed information as waveforms could be needed. For monitoring the quality of the grid more average and general information can be used.

In the coming years, the need of monitoring the quality of voltage and current will rise and depending on the application data handling will become more important.

8. References

EN 50160: Voltage characteristics of electricity supplied by public distribution systems, (2009): CENELEC, Brussels, Belgium.

IEC 61000-4-30: Testing and measurement technigues - Power quality measurements methods. (2003). : IEC, Geneva, Switzerland.

IEEE 519-1992: Recommended Practices and requirements for harmonic control in Electrical Power Systems. (1992). :IEEE

Meyer, J., Schegner, P., & Winkler, G. (2005, June). Efficient method for power quality surveying in distribution networks. CIRED 2005, Turin, Italy.

Cobben J.F.G., Dissertation Eindhoven University of Technology, Power Quality, Implications at the Point of Connection, June 2007.

Didden, M., Techno-Economic Analysis of Methods to reduce Damage due to Voltage Dips (Doctoral dissertation, KU Leuven, 2003).

Wang, J., Chen, S., & Lie, T. T., A systematic approach for evaluating economic impact of voltage dips. Electric Power Systems Research 2007.

Driesen, J., & Belmans, R. Distributed generation in Future Grids. Retrieved August , 2006, from http://www.leonardo-energy.org

Riet van, M. J., Baldinger, F. L., Buijtenen, W. M., Erp van, F. T., Volberda, F., & Provoost, F. (2005). Alternative approach for a total integrated secondary installation in MV substations covering all possible and required functions. CIRED 2005, Turin, Italy.

7

Management, Control and Automation of Power Quality Improvement

Heidarali Shayanfar[1] and Ahad Mokhtarpour[2]

[1]*Center of Excellence for Power System Automation and Operation, Electrical Engineering Department, Iran University of Science and Technology, Narmak, Tehran*

[2]*Department of Electrical Engineering, Islamic Azad University, Science and Research Branch, Tehran, Iran*

1. Introduction

Electrical energy is a type of existing energy in the nature. Because of its capability of changing to different kinds of energy and possibility of transmission; electrical energy has become one of the major energy sources. Electrical energy from generation to consumption passes from three stages, generation, transmission and distribution. Distribution systems composed of a large scale area of power system with many nodes. Distribution networks transmit electrical energy from high voltage power towards low voltage customer's devices. Electrical companies have the duty of preparing good quality electrical energy in the consumer side. For properly work of a device it is essential to pay attention to the quality of the energy. In bad power quality condition, electric devices may fail completely. Thus, electrical companies should control and limit power quality problems. Also, consumers should not use devices that generate power quality problems. If not, electrical companies should find them. Fig. 1 shows a simple distribution system (Mulhausen et al., 2010). If a fault is occurred in jth node of network short circuit current will effect in ith node voltage such as Eq. (1).

$$\Delta v_i = -Z_{ij} I_f \tag{1}$$

where, I_f is the fault current. Voltage problems are generally voltage sag and swell, unbalance, zero sequence and harmonics. Current problems are generally zero sequence, unbalance, reactive power and harmonics. For the improvement of voltage problems, series active filters such as Distributed Voltage Regulator (DVR) that are from the family of Custom Power (CP) can be used in the distribution systems and for the improvement of current problems, parallel active filters such as Distribution STATic COMpensator (DSTATCOM) can be used. Nowadays, for the instantaneous improvement of both voltage and current problems, efficient combination of the above mentioned CP devices are being used (Mokhtarpour et al., 2009). This device is called Unified Power Quality Conditioner (UPQC). The scope of this chapter is the investigation and analysis of the control approaches for power quality improvement, power quality management, automation of distribution systems and analytical review of the power quality control papers. A typical distribution system power quality control will be attached as the case study.

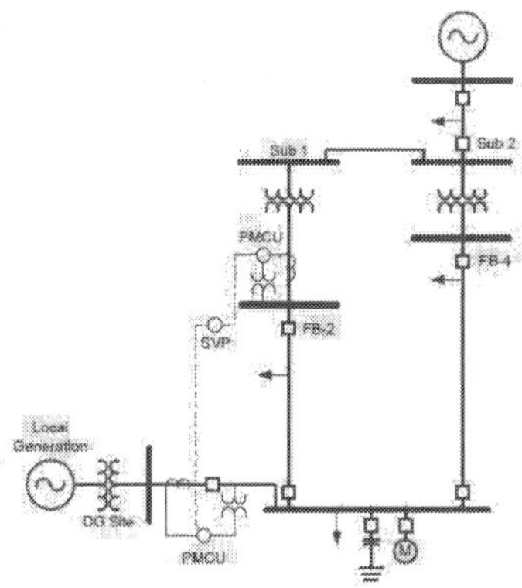

Fig. 1. A simple distribution system

2. Power Quality

How can we determine quality of electrical energy and what is a good power quality? Suppose that:

$$v = V_m \cos(\omega t) \tag{2}$$

$$i = I_m \cos(\omega t - \varphi) \tag{3}$$

Then, the instantaneous power can be fine as:

$$p = vi = V_m \cos(\omega t) \times I_m \cos(wt - \varphi) = \frac{V_m I_m}{2}\cos(\varphi)(1+\cos(2\omega t)) + \frac{V_m I_m}{2}\sin(\varphi)\sin(2\omega t) \tag{4}$$

Eq. (4) shows that the electrical power is related to voltage, current and frequency. It is obvious from Eq. (4) that the instantaneous power is composed of two part, active and reactive power (Grainger & Stevenson, 1994). Active power has a dc and an ac part. Based on Eq. (5) it is obvious that the instantaneous active power oscillates around a dc amount by frequency of 2ω and can be found as:

$$\text{Active power } = \frac{V_m I_m}{2}\cos(\varphi)(1+\cos(2\omega t)) \tag{5}$$

Also, reactive instantaneous power can be written as:

$$\text{Reactive power } = \frac{V_m I_m}{2}\sin(\varphi)\sin(2\omega t) \tag{6}$$

Fig. 2 shows these powers variation versus time. Based on Eq. (6), the second part of the instantaneous power which is named reactive power has zero dc amount and oscillates with frequency of 2ω. Obviously, the phase difference between ac amount of active and reactive powers is $\pi/2$. This means that the reactive power has zero amounts when active power has maximum amount. Based on Eq. (7) the average of the total power that has been named active power is constant and can be found as:

$$P=\frac{1}{2\pi}\int_{0}^{2\pi}p(\omega t)d(\omega t)=\frac{V_m I_m}{2}\cos\varphi=V_e I_e\cos\varphi \tag{7}$$

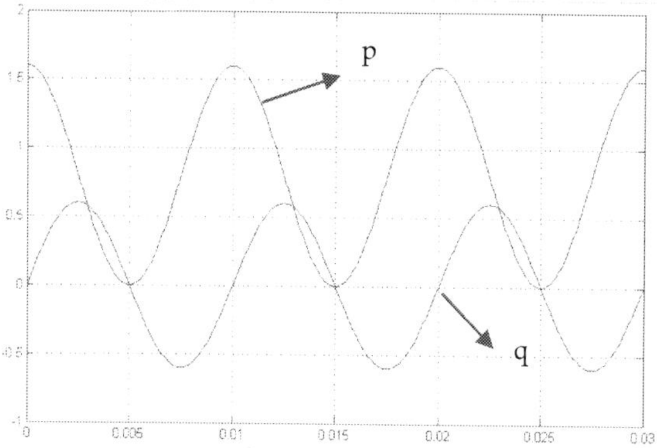

Fig. 2. Typical active and reactive powers

Obviously, for good quality of energy, voltage, current and frequency of distribution system should be proper. Thus, the investigation of energy quality can be divided into three parts, voltage quality, current quality and frequency. When the voltage quality is good? In other words, when the voltage is proper for a good performance of the device?

The magnitude of voltage is a quality problem in the first part. In good condition, the voltage of distribution system is changed by a sinusoidal function with the frequency of power system. The effective amount of voltage should be constant in all times and conditions. But, nonlinear loads and short circuit faults can produce voltage harmonics, unbalance voltages and negative and zero sequences of the voltage. This is the power quality problem which is produced by the utility side.

In the second case, the quality of the current is considerable. But, it is obvious that the load current is not controllable. In other words the characteristics of the load determine load current waveform. If there is a linear load, then the relationship between its voltage and current is linear. Thus, if voltage is a fixed magnitude sinusoidal function, then the current will be a sinusoidal function, with constant magnitude. This problem has been shown in Fig. 3. But, if characteristic of the load not be a linear function, then a sinusoidal voltage results in a nonsinusoidal current. This problem has been shown in Fig. 4. Based on the ohms law, $v=Ri$ and a nonsinusoidal current will cause nonsinusoidal voltage drop and finally nonsinusoidal voltage in all nodes. Therefore, the current quality can change the voltage quality. The load current is not controllable, but source current can be controlled that will be explained in the parallel active filter section.

In the third case, the quality of frequency is considered. The frequency of voltage and current in all conditions should be constant. Major cause of frequency change is linearity or nonlinearity of the load. If we calculate the Fourier transform of a nonlinear load current it can be fund that there are frequencies f, 2f, 3f and …in its spectrum. A nonlinear load current spectrum has been shown typically in Fig. 5. In other words the nonlinear loads can produce other frequency components in the current and voltage. Also, voltage frequency change can be produced by electrical source, because of non proper design of generator.

Power quality problems that are related to the voltage are voltage unbalance and negative sequences and zero sequences of the voltage. Power quality problem that are related to the current are harmonics and reactive power.

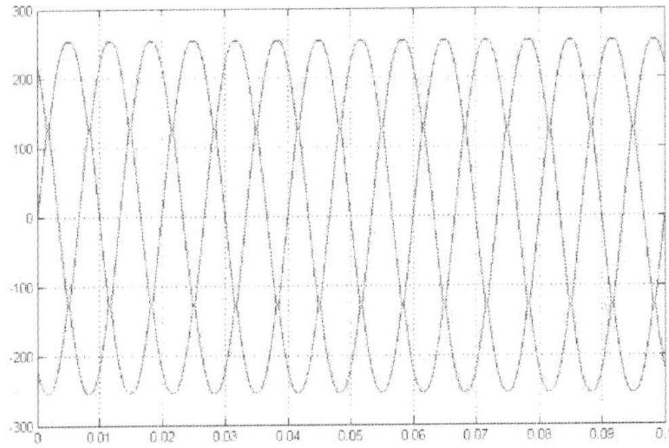

Fig. 3.(a) Voltage waveform;

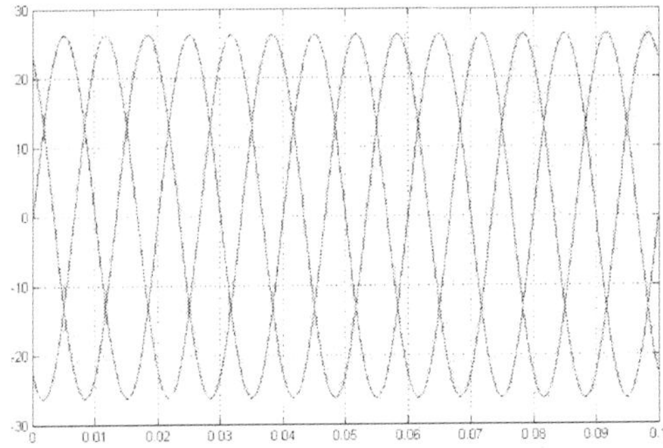

Fig. 3.(b) Linear load current waveform

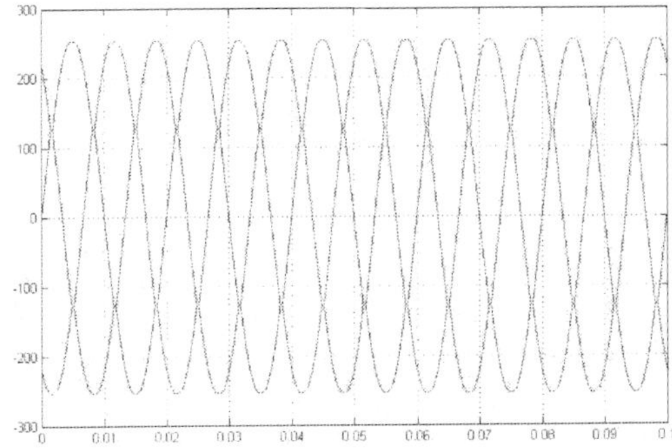

Fig. 4.(a) Voltage waveform;

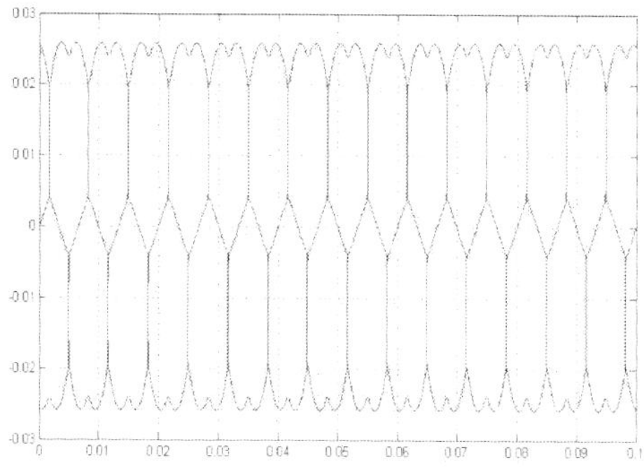

Fig. 4.(b) Nonlinear load current waveform

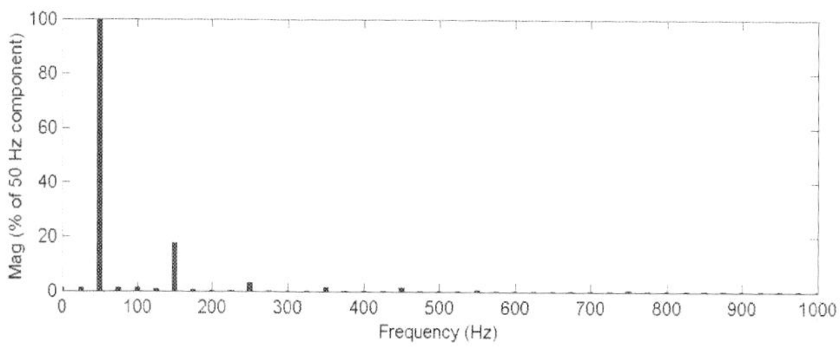

Fig. 5. Spectrum of a nonlinear load current

3. Power Quality control

Now, it is important to investigate how can we control these energy quality problems and improve them. Different approaches have been presented for the improvement of energy quality. Passive filters can compensate voltage and current harmonics. Passive filter has composed of an inductor in series or parallel by a capacitor that in the resonance frequency of $1/\sqrt{lc}$ acts as a short circuit. Thus, relative harmonic can be neglected. In the design of passive filters their resonance frequency are generally equal to third or fifth harmonic; because these harmonics are very important in the distribution systems. But, passive filters are designed for the predetermined conditions. Therefore, in the case of changing the operating point for a dynamic power system, passive filter can not compensate the quality of the energy. Also, passive filters can compensate only one or two harmonic of the voltage and current. Thus, they are not good devices for the improvement of the power quality. So, for the control of all power quality problems, it is better to use active filter. These filters can be divided into two parts of parallel and series active filters. Based on the mentioned problems, power quality problems can be related either to the load side or to the source side. Nonlinear load current should not appear in the source side current. This problem can be solved by the use of a parallel active filter. A parallel active filter has been composed of a current inverter fed by a DC link that is connected to the grid via a parallel transformer (Shayanfar et al., 2005a). Fig. 6 shows the structure of a parallel active filter. Parallel activefilters are used for compensating negative sequence, zero sequence and harmonics of the

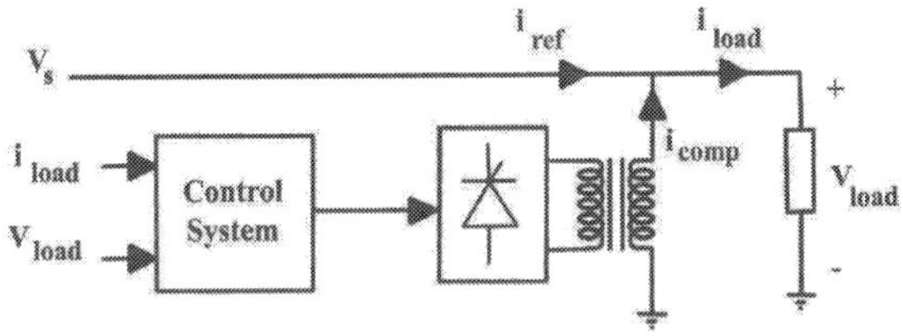

Fig. 6. Structure of a parallel active filter

load current. Major principle of its control is the measuring of load voltage and currents and then computing the reference and compensators current. The scope of using the parallel active filter is the injection of unwanted components of the load current via the inverter and transformer to the load instead of using main source. Reference current is a sinusoidal current that is extracted by the control circuit of the parallel active filter. Compensator current is the component of load current that has composed of negative sequence, zero sequence and harmonics of it. It is known that the power of electrical system has two part of active and reactive. Also, the average of electrical power is equal to the active power. This problem is explained in Eq. (7). From Eq. (6) reactive power will oscillate by frequency of 2ω. This means that the amount of it in a time of period is positive and in another time of period is negative. So, the direction of the reactive power in a time is from source to the load and in other time is from load to source. It says that the reactive power have the average of zero and only limits the capacity of the distribution line. It is known that the inductor current has 90 degree lag to the voltage. Eqs. (8) and (9) show the active and reactive powers of an inductor, as:

$$\text{Active power} = \frac{V_m I_m}{2}\cos(90)(1+\cos(2\omega t)) = 0 \tag{8}$$

$$\text{Reactive power} = \frac{V_m I_m}{2}\sin(90)\sin(2\omega t) = V_e I_e \sin(2\omega t) \tag{9}$$

But, capacitor current has 90 degree lead to the voltage. Eqs. (10) and (11) show the active and reactive powers of a capacitor, as:

$$\text{Active power} = \frac{V_m I_m}{2}\cos(-90)(1+\cos(2\omega t)) = 0 \tag{10}$$

$$\text{Reactive power} = \frac{V_m I_m}{2}\sin(-90)\sin(2\omega t) = -V_e I_e \sin(2\omega t) \tag{11}$$

From Eqs. (9) and (11) it seams that the signs of reactive power of inductor and capacitor are opposite. In other word the sum of reactive powers in inductor and a capacitor is zero in all times and can be said that the capacitors reactive power cancels that of an inductor. Load reactive power can be compensated by parallel active filter. In other words the reactive current is a component of load current that has 90 degree phase difference with the voltage and it can be concluded that the unwanted component of the load current. Thus, if the reactive current be included in the compensation current of parallel active filter, then the reactive power of load will be compensated. In other words, compensating current is the unwanted component of the load current. After calculation of compensating current, it should be fed to the network in parallel with load. This problem has been shown in Fig. 6. For injection of the compensation current by using of an inverter, there are some approaches for the generation of the fire pulses of the inverter and the best of them is PWM approach. For minimum error between calculated compensating current signal that has calculated from parallel control system and the injected current that is generated from an inverter, generally a PI controller is being uesd.

A typical two arm and 6 pulse thyristor based inverter has been shown in the Fig. 7. Power quality problems can be related to the source side. These are voltage problems such as

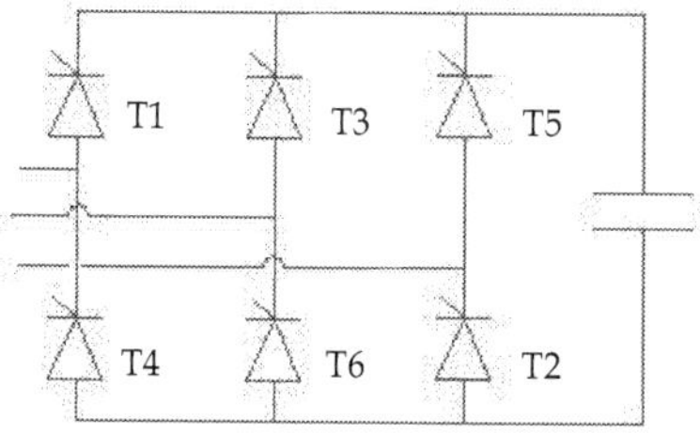

Fig. 7. Two arm and 6 pulse thyristor based inverter

unbalance, negative sequence, zero sequence and harmonics of the voltage (Shayanfar et al., 2006a). The cause of the voltage problems can be non symmetric short circuit faults, non symmetric distribution of loads between three phases and nonlinear load. Passive filters could improve a little voltage harmonics only in one point of the work and they can not improve all voltage problems completely. To improve this problem, series active filters are being used instead of passive filters. These filters have composed of a voltage inverter that is fed from a DC link (Mokhtarpour et al., 2009). This inverter is connected to the grid via a series transformer. Fig. 8 shows structure of a series active filter.

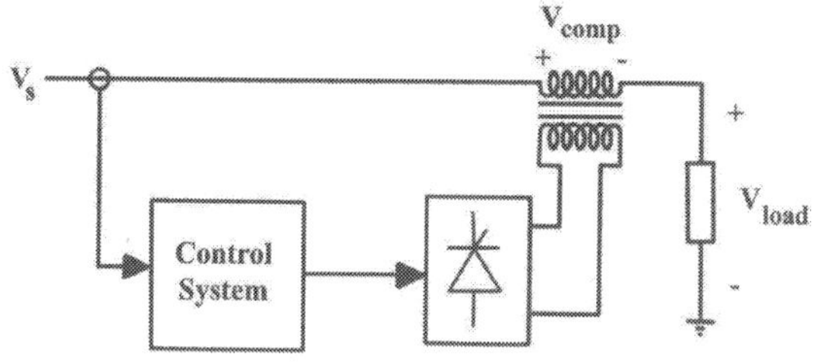

Fig. 8. Structure of a series active filter

Series active filter extracts the reference voltage from the source voltage. Reference voltage is a sinusoidal voltage. Compensator voltage is the difference between reference and source voltages. Compensator voltage is injected to the grid via an inverter and series transformer. Based on the mentioned problems it is obvious that the parallel active filters are being used for the current quality problems improvement such as negative sequence, zero sequence and harmonics and reactive power and series active filters can be used to improve the voltage quality problems such as unbalance, negative sequence, zero sequence and harmonics.

A typical harmonized, compensator and reference signals have been shown in the Figs. 9, 10 and 11. Obviously compensator signal include unwanted components of harmonized signal.

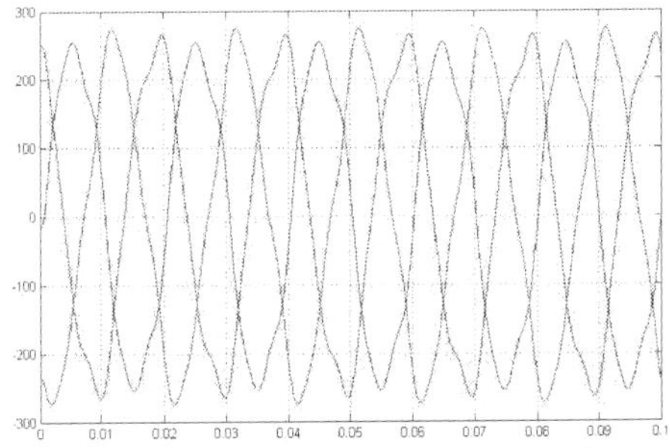

Fig. 9. A typical harmonized signal

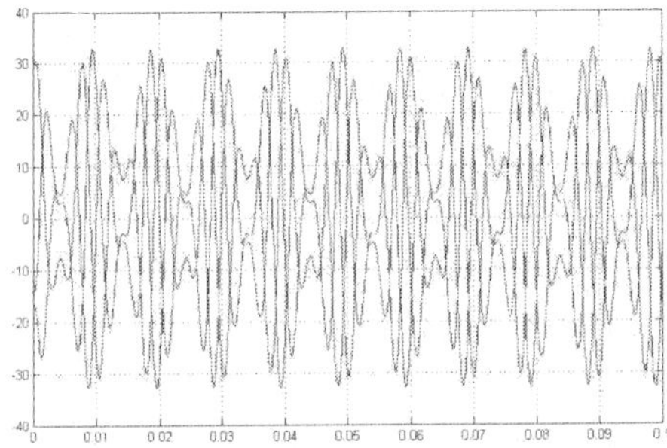

Fig. 10. Compensator signal

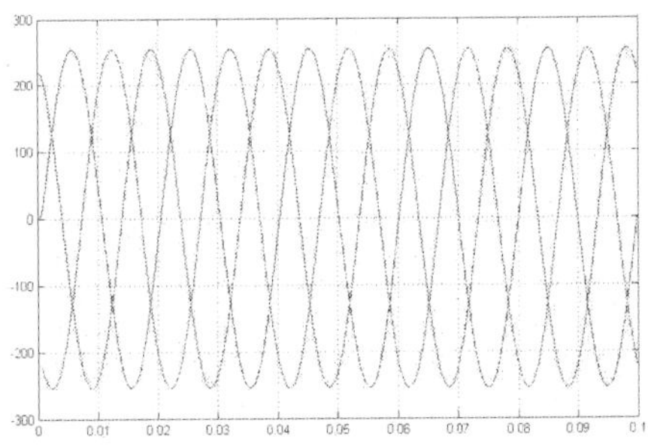

Fig. 11. Reference signal

4. Parallel active filter control

Parallel active filter compensates and improves current quality. As mentioned in the text, the current problems are related to the load and consumer side. They are including negative sequence, zero sequence, reactive power and harmonics. Design of the control circuit of the parallel active filter is based on the extraction of a sinusoidal current which has the same phase with compensated load voltage as the reference current and determination of the compensation current. The compensation current will be injected to the grid via an inverter and a parallel transformer. Generally, generation of the inverter pulses are based on the PWM approach. First, the load side voltage and current are measured, then the control circuit extracts a reference current from the measured voltage and current (Shayanfar et al., 2006b). The difference between different approaches is in the parallel active filter control circuits which is the way of reference current extraction. In this section a very simple control approach for learning the major principle is explained and in the case study of the chapter a full control approach of the parallel active filter will be described. Usually load current is non sinusoidal. After measurement of the voltages and currents and use of a low pass filter a reference current will be extracted. Compensator current (I^{*}_{comp}) is determined from the difference between load side current and the reference current. Block diagram of a parallel active filter has been shown in Fig. 12.

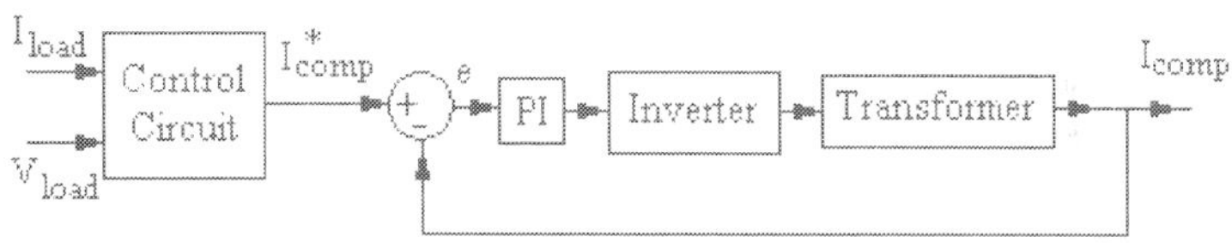

Fig. 12. Block diagram of a parallel active filter

5. Series active filter control

Series active filter compensates and improves voltage quality. As mentioned above the voltage problems are related to the source side and include unbalance, negative sequence, zero sequence and harmonics. The design of the control circuit of series active filter is based on the determination of a sinusoidal voltage as the reference voltage and the difference between that and the source side voltage as the compensation voltage (Shayanfar et al., 2006a, 2006c). Compensation voltage will be injected to the grid via an inverter and a series transformer.

Usually, generation of the inverter pulses are based on the PWM approach. First the source side voltage is measured, then the control circuit extracts a reference voltage from the measured voltage. Block diagram of a series active filter has been shown in Fig. 13. The difference between different approaches are in the series active filter control circuits which is a function of the reference voltage extraction. In this section a very simple control approach for learning the major principle is explained and in the case study of the chapter a full control approach of the series active filter will be described. Usually, source voltages are non sinusoidal. Using a low pass filter can extract a sinusoidal voltage as the reference voltage. Compensator voltage (V^{*}_{comp}) is determined from the difference between source side voltage and the reference voltage.

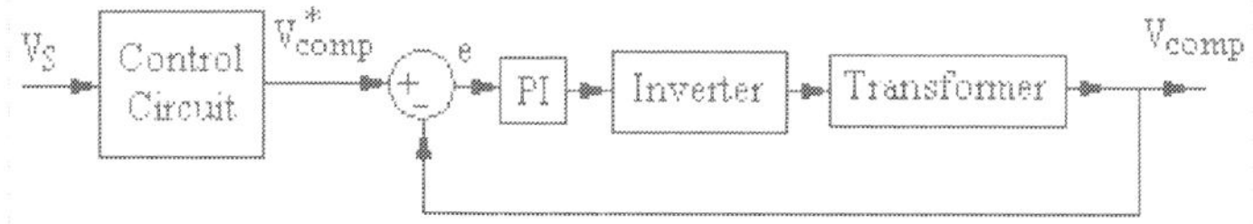

Fig. 13. Block diagram of a series active filter

6. Unified Power Quality Conditioner

Proper composition of parallel active filters as Distribution STATic COMpensator (DSTATCOM) and series active filters as Distributed Voltage Regulator (DVR) can compensate current problems such as reactive power, negative sequence, zero sequence and harmonics as well as voltage problems, such as unbalance, negative sequence, zero sequence and harmonics. This composition is named Unified Power Quality Compensator or UPQC. UPQC has composed of two inverters that are connected back to back. One of them is connected to the grid via a parallel transformer and can compensate the current problems. Another one is connected to the grid via a series transformer and can compensate the voltage problems. These two inverters are controlled for the compensation of all power quality problems instantaneously. Fig. 14 shows the schematic of a UPQC.

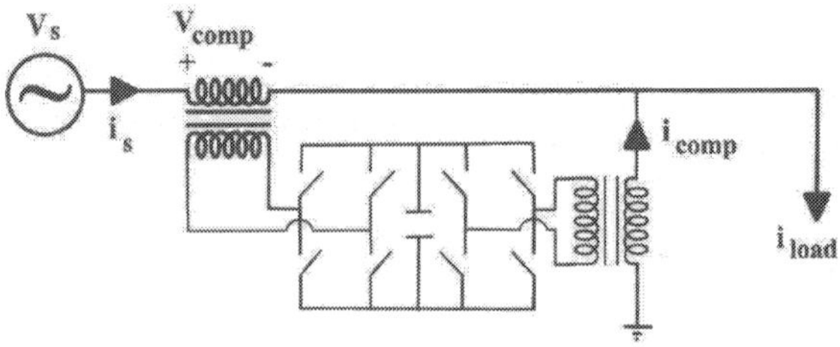

Fig. 14. Schematic of a UPQC

For control of this device there are different approaches in which their principles of them are the same. Parallel active filter control circuit extracts a sinusoidal current as the reference current (Kazemi et al., 2006). The difference between reference current and load current is the compensator current and will be injected via an inverter and a parallel transformer to the grid for the improvement of the power quality problems that are related to the load side. Series active filter control circuit extracts a sinusoidal voltage as the reference voltage. The difference between reference voltage and source voltage is the compensator voltage and is injected via an inverter and a series transformer to the grid for the improvement of the power quality problems that are related to the source side.
A case study for the power quality compensation by the UPQC and their mathematical functions will be described in the next section.

7. Automation and control of distribution systems

In the power system application, there are two layers; measurement and control. In Supervisory Control And Data Acquisition systems (SCADA) application, before the control action, measurement of essential parameters has been done in the measurement centers by relative devices. Then, these signals are sent to the control centers via communication channels. In the control centers proper process will be done and finally controller signals are generated. In automation and control of the distribution systems the need to measure parameters, communicate and control them exists. This process can be done in the SCADA systems, but different from each other. It will be excellent in the automation of power system to do the measurement and control of signal together. Integrated Object Network (ION) devices can be used in automation of distribution systems and do measurement and control actions in one the units (Shayanfar et al., 2005b). Fig. 15 shows a typical ION device.

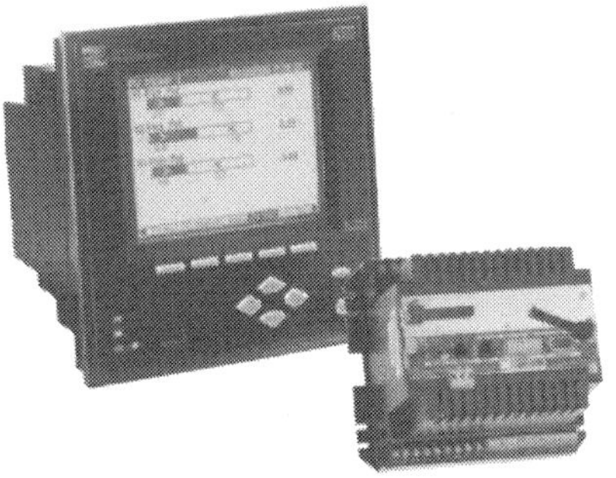

Fig. 15. Measurement and control device (ION) (Shayanfar et al., 2005b)

IONs as an intelligent device with the capability of measuring and controlling the required parameters of compensation can improve power quality disturbances such as negative sequences of voltage and reactive powers. Intelligent devices of ION analyze the retrieved data from PTs and CTs and then determine compensating currents intelligently regarding the programmed control algorithm (Shayanfar et al., 2006a, 2006b). The process of measurement and control system is divided to the following sections.

7.1 Data acquisition

First stage is voltage and current measurements. The measurement transformers are being used to produce the applicable signals to the ION in high voltage and current cases. If we use the devices in low voltage and current systems it is not required to use PTs and CTs.

7.2 Data analysis

The acquired data will be analyzed and all of the required parameters for the power quality compensation will be calculated. If we use and determine the proper process function in these devices, reference voltages and currents for the control of UPQC can be determined. These signals will be transacted via the communication ports; such as: RS232, RS485 or modem for triggering blocks and inverters to the other ION's and computers. Using ION has the following benefits:

1. Intelligent operation and control.
2. The ability of easy communication.
3. Possibility of connection to several IONs.
4. Increasing of speed and resolution.
5. Capability of new control function determination.

Obtaining the mentioned advantages in the distribution systems, automation and control of the UPQC can be done via an ION device. The major principle of the distribution systems power quality management is data acquisition and analysis. This can be done by using power quality meters and controller such as ION. Fig. 16 shows a typical small intelligent distribution control system which uses meters data for the automation. Data are gathered from essential points of a distribution system. These data can be analyzed there or can be sent to the major control unit for other application. Distribution systems have many of nodes that are connected together. Based on the visibility and controllability of the system can be said that meters should be settle where and which of the parameters can be controlled in a typical network.

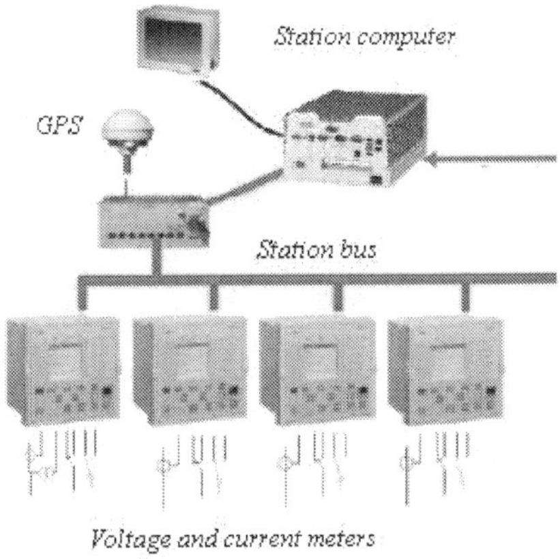

Fig. 16. A typical intelligent distribution control system (Valtari et al., 2009)

8. Case study

For display of the UPQC performance, different control strategies for the unified power quality conditioner are investigated. In the first case, the control strategy of Parallel Active Filter (PAF) and Series Active Filter (SAF) are based on Fourier transform theory. In the second case, the control strategy of PAF is based on the PQ theory and control strategy of SAF is based on the Fourier theory. In the third case, the control strategy of PAF is based on the Fourier transform theory and control strategy of SAF is based on the positive sequence detection (Shayanfar et al., 2007).

8.1 First control strategy

8.1.1 SAF control

First measure the load voltages and then, calculate their 1st order harmonics in magnitude and phase. This operation is based on the Fourier transform theory which can give us 1st order harmonics magnitude and phase. In Eq. (12), details and the theory have been illustrated. For detection of n'th order harmonic magnitude and phase, number of n should be replaced in Eq. (12). T is the major period of signal that is 0.02 s in this simulation. For detection of 1st order harmonics magnitude and phase, n=1 is replaced in the Eq. (12).

$$
\begin{aligned}
A(n) &= \frac{2}{T}\int_0^T f(t)\cos(\frac{2n\pi}{T}t)dt \\
B(n) &= \frac{2}{T}\int_0^T f(t)\sin(\frac{2n\pi}{T}t)dt \\
Amplitude(n) &= A(n)^2 + B(n)^2 \\
Phase(n) &= Arc\tan(\frac{A(n)}{B(n)})
\end{aligned}
\qquad (12)
$$

The control strategy of SAF is based on considering the Fourier transform of utility voltages. Block diagram of parallel active filter control circuit has been shown in Fig. 17.

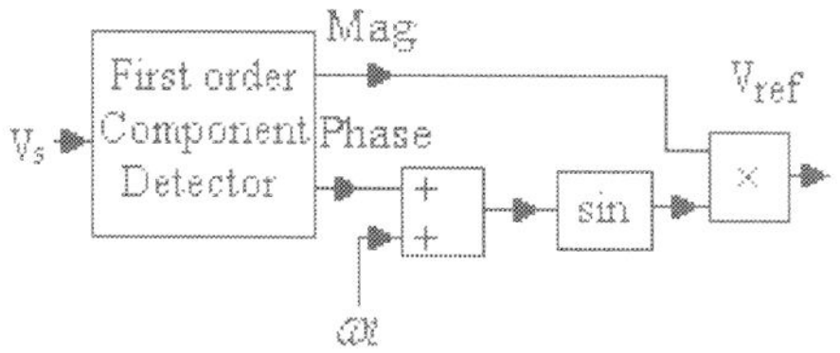

Fig.17. The control circuit of SAF

8.1.2 PAF control

Block diagram of parallel active filter control circuit is similar to Fig. 17. But tangent component of load current to the load voltage will be determined as reference current. Eq. (13) shows parallel active filter control method (Shayanfar et al., 2005a, 2006a).

$$\varphi = \theta_v - \theta_i \qquad I_{ref} = I_1 \times \cos\varphi \tag{13}$$

Where, θ_v and θ_i are 1st order harmonic of voltage and 1st order harmonic of current phases, respectively. I_1 is the magnitude of 1st order harmonic of current.

8.2 Second control strategy

8.2.1 SAF control

Series active filter control in this case is like first case because strategy of SAF control is based on the Fourier theory.

8.2.2 PAF control

For control of parallel active filter, one phase PQ theory is being used. We can show that active and reactive powers can be written as Eqs. (14) and (15):

$$p(t) = v_{la+}(t)i_a(t) + v_{lb+}(t)i_b(t) + v_{lc+}(t)i_c(t) \tag{14}$$

$$q(t) = v'_{la+}(t)i_a(t) + v'_{lb+}(t)i_b(t) + v'_{lc+}(t)i_c(t) \tag{15}$$

Where, $v'_{la+}(t)$ has 90 degree phase shift to $v_{la+}(t)$. Writing in matrix form of the above Eqs. can be done as:

$$\begin{bmatrix} p(t) \\ q(t) \end{bmatrix} = \begin{bmatrix} v_{la+}(t) - v_{lc+}(t) & v_{lb+}(t) - v_{lc+}(t) \\ v'_{la+}(t) - v'_{lc+}(t) & v'_{lb+}(t) - v'_{lc+}(t) \end{bmatrix} \begin{bmatrix} i_a(t) \\ i_b(t) \end{bmatrix} \tag{16}$$

$$\begin{bmatrix} i_a(t) \\ i_b(t) \end{bmatrix} = \frac{1}{\Delta} \begin{bmatrix} v'_{la+}(t) - v'_{lc+}(t) & -(v_{lb+}(t) - v_{lc+}(t)) \\ -(v'_{la+}(t) - v'_{lc+}(t)) & v_{lb+}(t) - v_{lc+}(t) \end{bmatrix} \begin{bmatrix} p_s(t) \\ q_s(t) \end{bmatrix} \tag{17}$$

It should be indicated that $q_s(t)$, compensated reactive power is zero. Then, Eq. (18) can be found as:

$$\begin{bmatrix} i_{refa}(t) \\ i_{refb}(t) \end{bmatrix} = \frac{1}{\Delta} \begin{bmatrix} v'_{la+}(t) - v'_{lc+}(t) & -(v_{lb+}(t) - v_{lc+}(t)) \\ -(v'_{la+}(t) - v'_{lc+}(t)) & v_{lb+}(t) - v_{lc+}(t) \end{bmatrix} \begin{bmatrix} p_s(t) \\ 0 \end{bmatrix} \tag{18}$$

8.3 Third control strategy

8.3.1 SAF control

The control strategy of SAF is based on considering the instantaneous positive sequence of utility voltages as the reference load side voltages. In this way, it is enough to subtract $V_{la+}(t)$, $V_{lb+}(t)$ and $V_{lc+}(t)$ from V_a, V_b and V_c, respectively, for computing the reference compensating voltages of SAF as follows:

$$\begin{bmatrix} V_{compa}(t) \\ V_{compb}(t) \\ V_{compc}(t) \end{bmatrix} = \begin{bmatrix} V_a(t) \\ V_b(t) \\ V_c(t) \end{bmatrix} - \begin{bmatrix} V_{la+}(t) \\ V_{lb+}(t) \\ V_{lc+}(t) \end{bmatrix} \tag{19}$$

8.3.2 PAF control

Block diagram of parallel active filter control circuit has been shown in Fig.18.

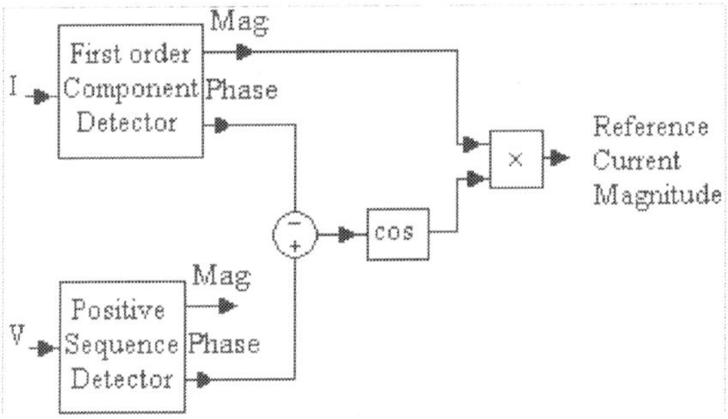

Fig. 18. The control circuit of PAF

9. Simulation results

Operation of three control circuits have been simulated using the power circuit of Fig. 19. This power system consists of a three phase 20 kV (RMS, L-L), 50 Hz utility, a three phase balanced R-L load, a nonlinear three phase load connected to the circuit in 0.05 s (t_{on} = 0.05 s) and one phase load connected to the circuit in 0.1s (t_{on} = 0.1 s) . For investigation of the proposed control strategy, operation under dynamic condition, magnitude of the first phase of the voltage is reduced to the 0.75 pu in 0.15 s. A number of selected simulation results for three control strategies are shown differently.

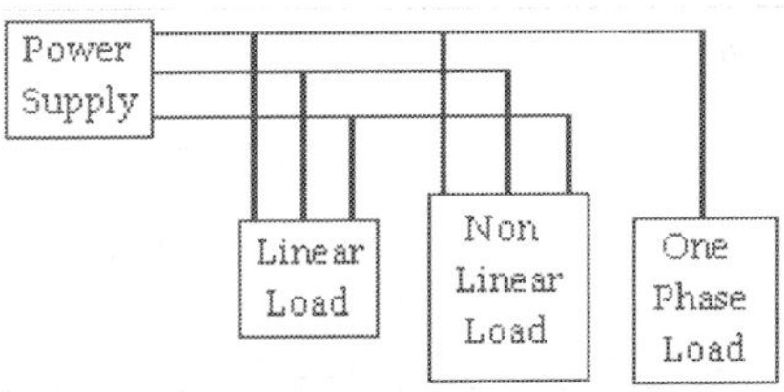

Fig. 19. Test circuit

Table 1 shows test circuit parameters. Fig. 20 shows the load voltages before compensation which are harmonized and unbalanced. Fig. 21 shows compensator voltages and Fig. 22 shows load side voltages in cases (1) and (2). Fig. 23 shows compensator voltages and Fig. 24 shows load side voltages in case (3). It is shown that the proposed control strategies can solve unbalance and reduction of voltage magnitude in the load side and balance them. Fig. 25 shows load side current before compensation. Figs. 26, 27 and 28 show compensator currents in cases (1), (2) and (3), respectively. Figs. 29, 30 and 31 show source side current in cases (1), (2) and (3), respectively. Fig. 32 shows load side current spectrum. Fig. 33 shows source side current spectrum in cases (1) and (3). Fig. 34 shows source side current spectrum in cases (2), respectively. It is seen from the results that control strategy in case (2) can not solve current harmonic problems as well as cases (1) and (3). Fig. 35 shows uncompensated power factor. Figs. 36, 37 and 38 show compensated power factor in case (1), (2) and (3), respectively. Finally Fig. 39 shows source side active and reactive powers after the compensation. Table 2 shows relative speed of three control strategies.

	Load 1	Load 2	Load 3	System Impedance
R	5 Ω	15 Ω if $\|V\| \le 0.5$ pu 2 Ω if $\|V\| > 0.5$ pu	5 Ω	0.0002 Ω
X	8 mH	10 mH if $\|V\| \le 0.5$ pu 5 mH if $\|V\| > 0.5$ pu	10 mH	4.5 mH
t_{on}	0 sec	0.05 sec	0.1 sec	-

Table 1. Test circuit parameters

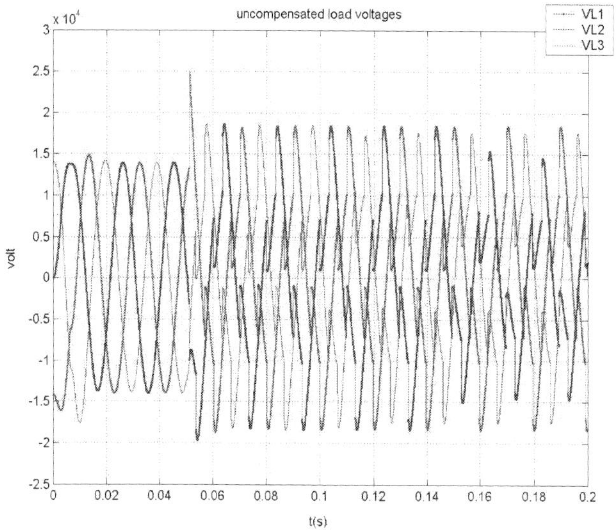

Fig. 20. Harmonized and unbalanced load side voltages (magnitude of phase 1 voltage is reduced to 0.75 pu in t=0.15 s)

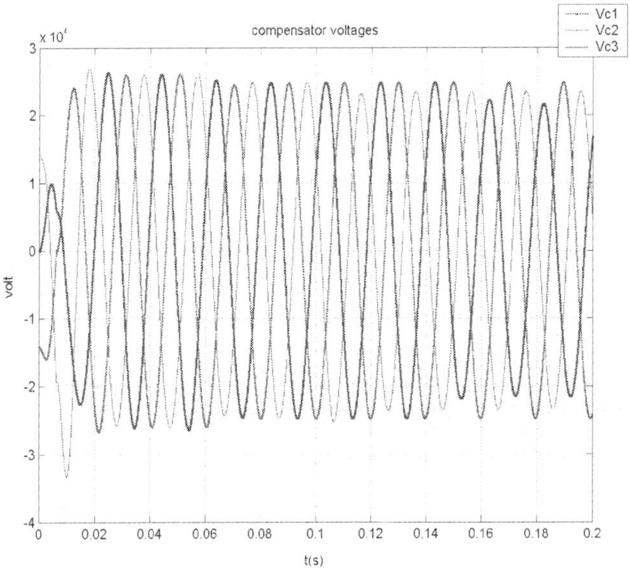

Fig. 21. Compensator voltages (these voltages have been injected to the power system to solve source harmonic and unbalance voltages) in case (1) and (2)

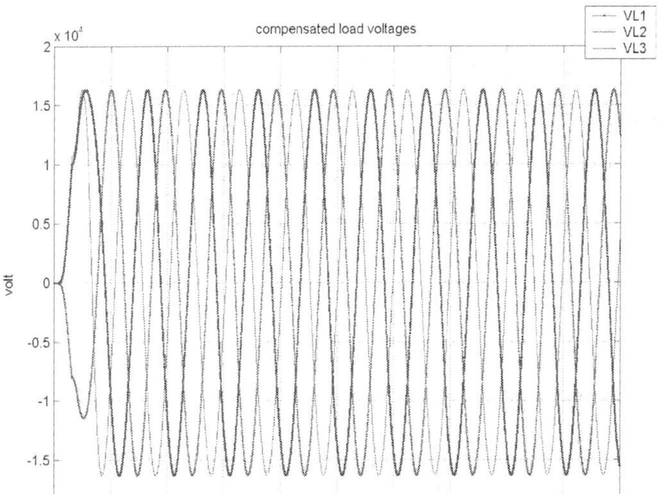

Fig. 22. Load side voltages (after compensation, load side voltages are sinusoidal and balance) in case (1) and (2)

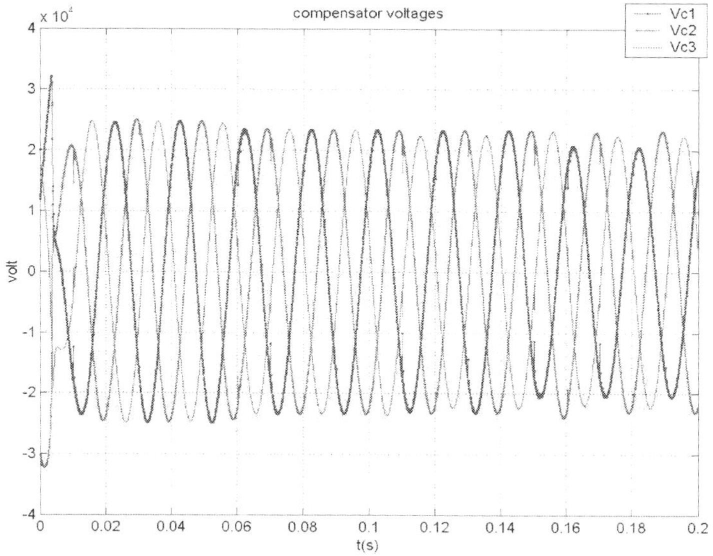

Fig. 23. Compensator voltages (these voltages have been injected to the power system to solve source harmonic and unbalance voltages) in case (3)

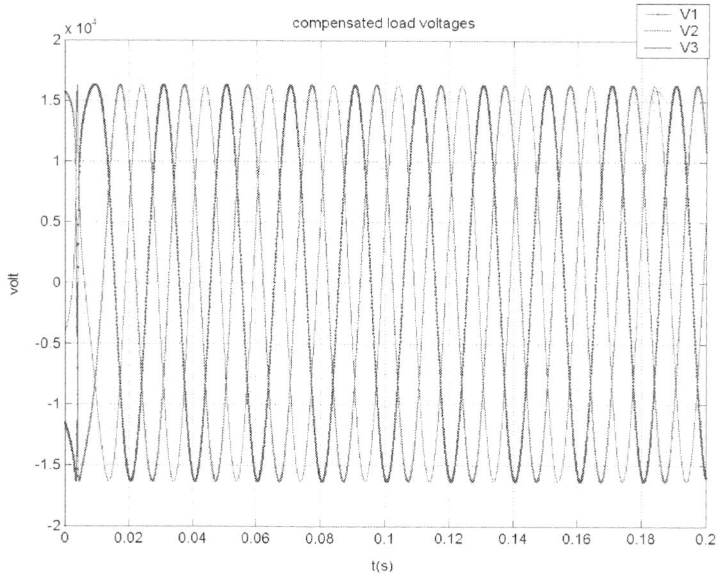

Fig. 24. Load side voltages (after compensation, load side voltages are sinusoidal and balance) in case (3)

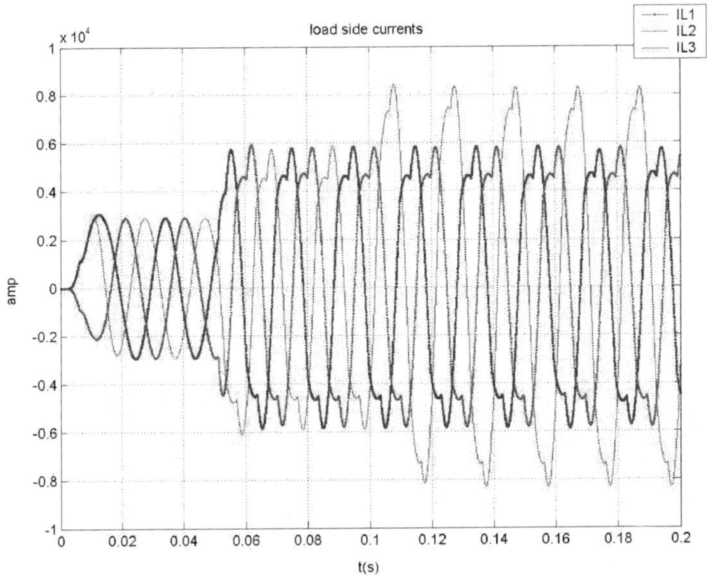

Fig. 25. Load side currents (because load is nonlinear, power system current is non sinusoidal too)

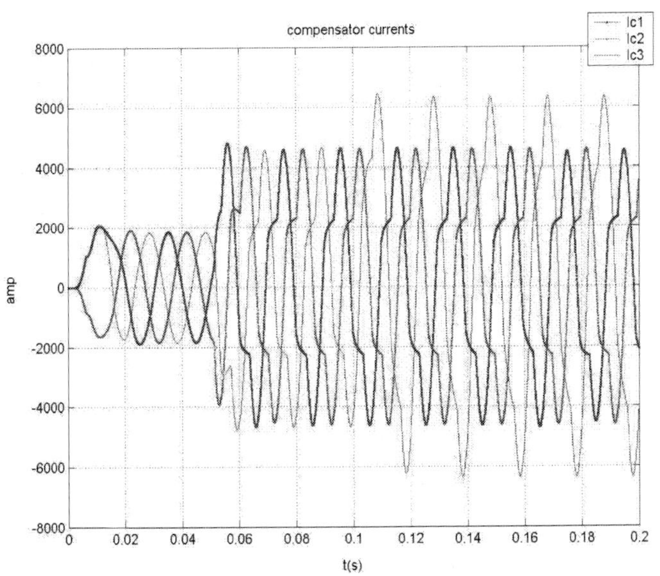

Fig. 26. Compensator currents (these currents have been injected to the power system to solve load harmonic currents) in case (1)

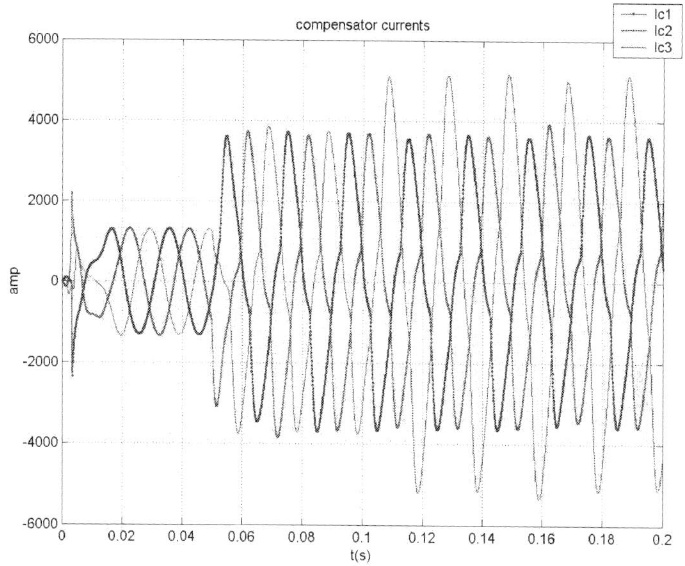

Fig. 27. Compensator currents (these currents have been injected to the power system to solve load harmonic currents) in case (2)

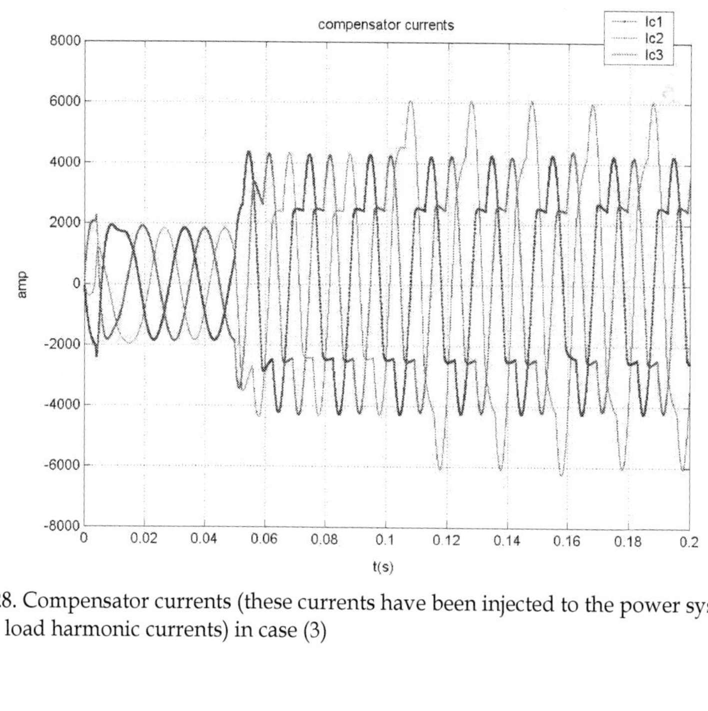

Fig. 28. Compensator currents (these currents have been injected to the power system to solve load harmonic currents) in case (3)

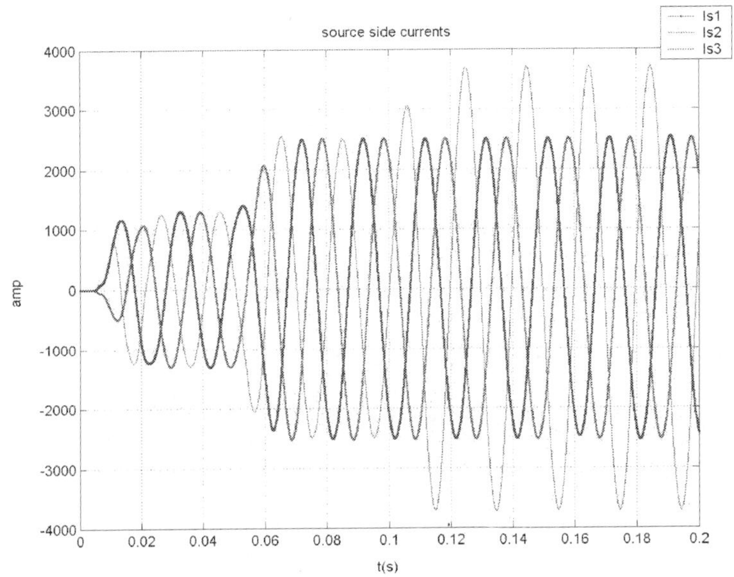

Fig. 29. Source side currents (after compensation, source side currents are sinusoidal) in case (1)

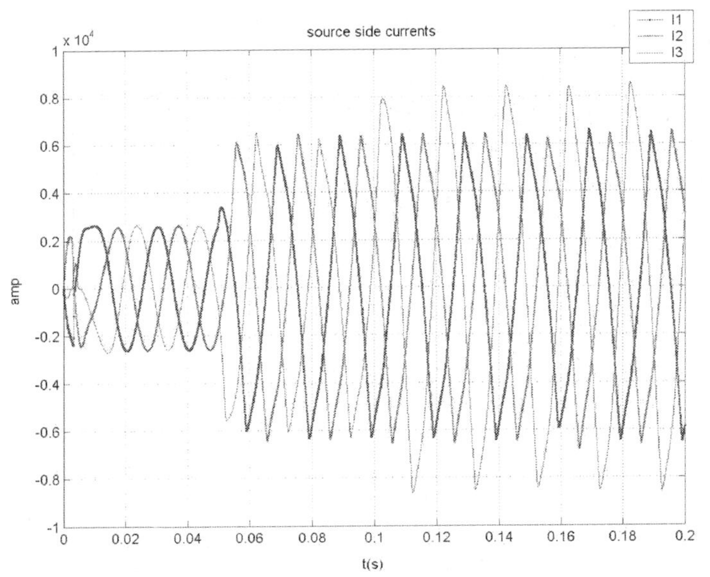

Fig. 30. Source side currents (after compensation, source side currents are sinusoidal) in case (2)

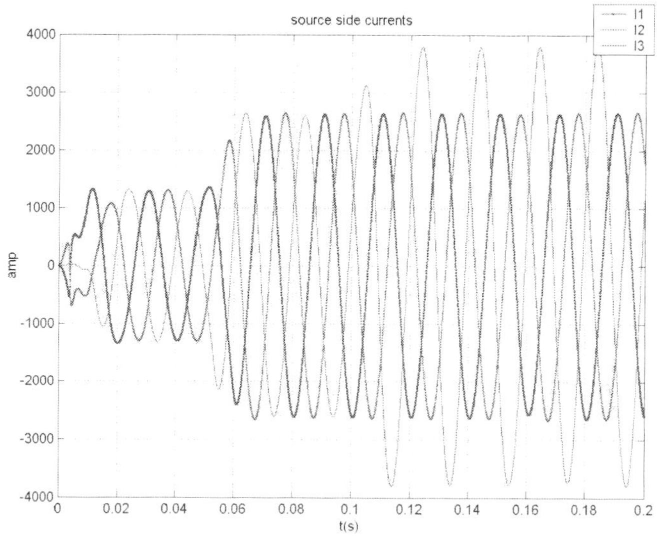

Fig. 31. Source side currents (after compensation, source side currents are sinusoidal) in case (3)

	SAF Speed Cycle	PAF Speed Cycle	Current THD	Voltage THD
Before compensation	-	-	17.6	51.2
Case (1)	0.5	0.5	0.5	0.34
Case (2)	0.3	0.3	0.27	0.29
Case (3)	0.3	0.8	11.44	0.73

Table 2. Relative speed of three control strategies

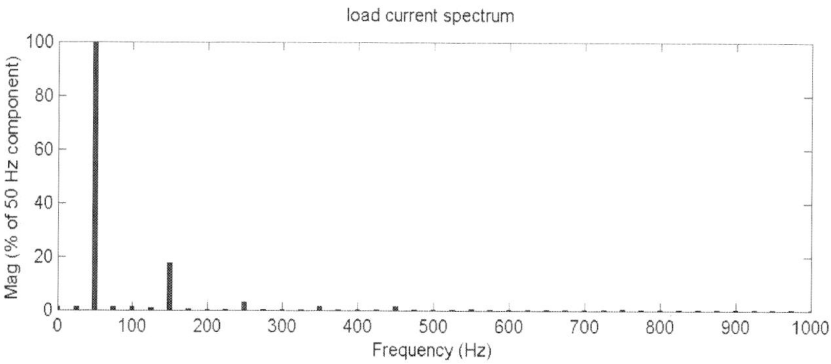

Fig. 32. Harmonic current spectrum before compensation

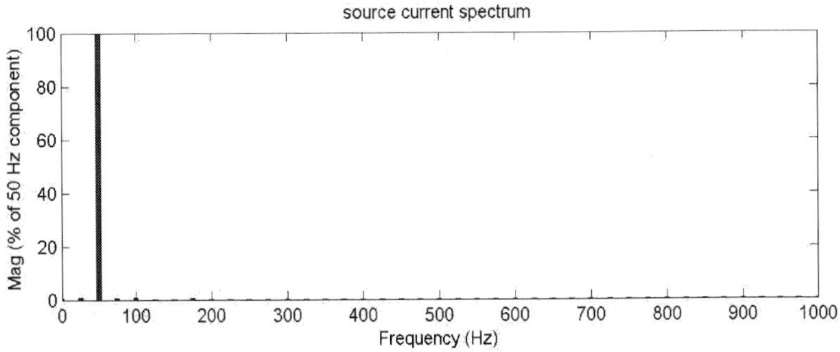

Fig. 33. Harmonic current spectrum after compensation in case (1) and case (3)

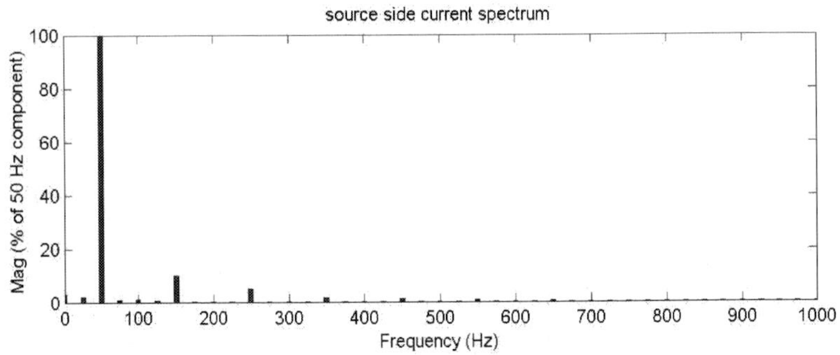

Fig. 34. Harmonic current spectrum after compensation in case (2)

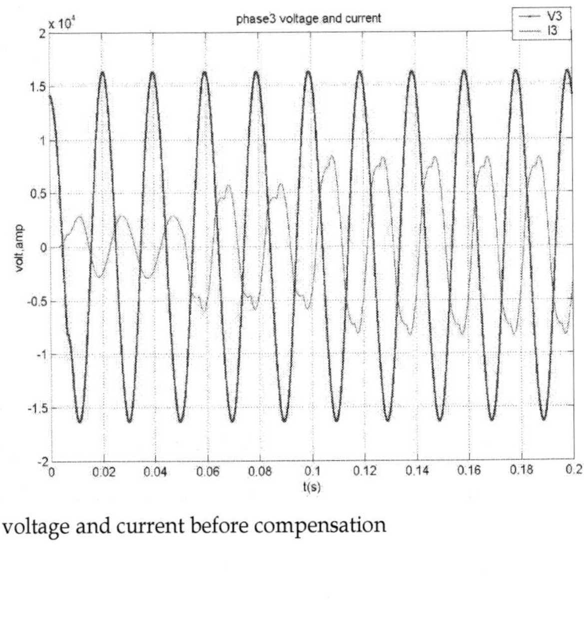

Fig. 35. Load voltage and current before compensation

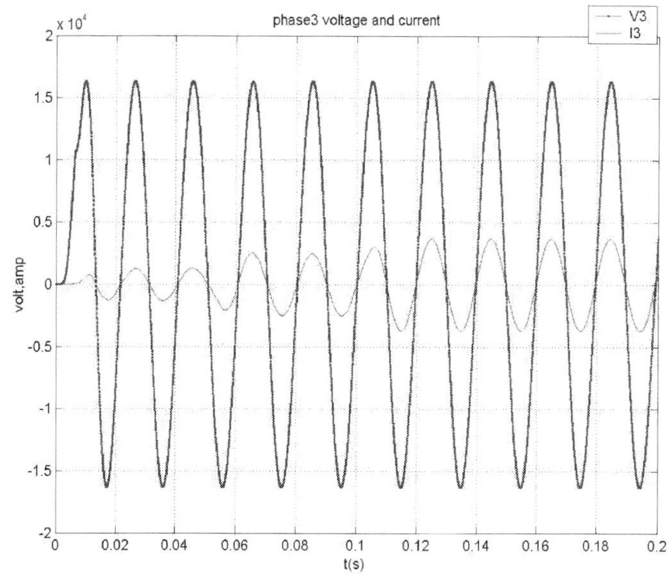

Fig. 36. Load voltage and current after compensation in case (1)

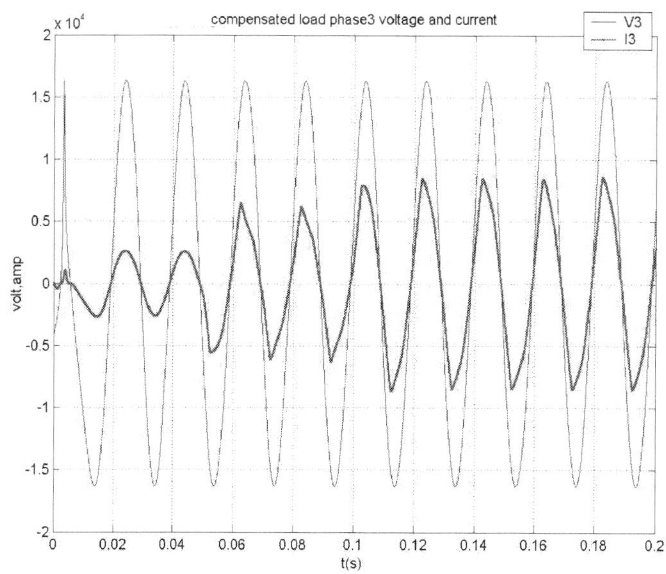

Fig. 37. Load voltage and current after compensation in case (2)

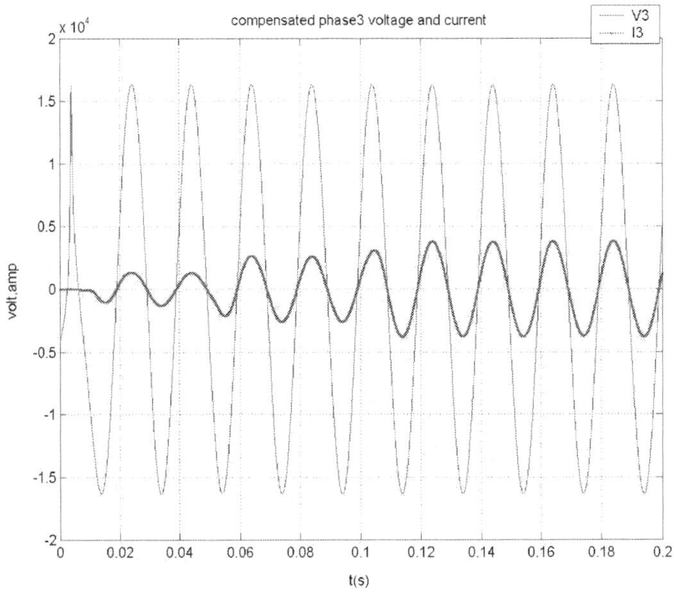

Fig. 38. Load voltage and current after compensation in case (3)

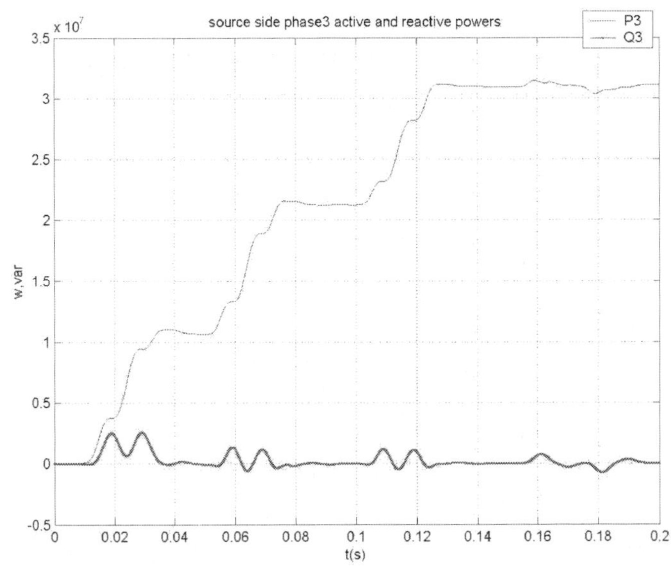

Fig. 39. Source side active and reactive powers

10. Conclusion

Power quality problems are generally divided into two parts of voltage and current problems. Voltage problems are related to the source side and current problems are related to the consumer side. Also, voltage problems can be produced from the current problems; because distribution systems have many nodes and one consumer power quality problem can affect all power quality of nodes. Voltage problems are generally voltage sag and swell, unbalance, zero sequence and harmonics. Current problems are generally zero sequence, unbalance, reactive power and harmonics. Series active filters can be used for the improvement of voltage problems and parallel active filters can be used for the improvement of current problems. Instantaneous improvement of both voltage and current problems, can be done by efficient combination of the above mentioned CP devices which is called Unified Power Quality Conditioner (UPQC).

11. References

Mulhausen, J.; Schaefer, J.; Mynam, M.; Guzman, A. & Donolo, M. (2010). Anti-Islanding Today, Successful Islanding in the Future, *Proceedings of the 63rd IEEE Annual Conferences for Protective Relay Engineers*, pp. 1-8, ISBN 978-1-4244-6073-1, Texas, U.S.A., March 29 2010-April 1 2010, IEEE, College Station

Mokhtarpour, A.; Shayanfar, H.A. & Tabatabaei, N.M. (2009). Power Quality Compensation in two Independent Distribution Feeders, *International Journal for Knowledge, Science and Technology*, Vol. 1, No. 1, (October 2009) page numbers (98-105), ISSN 1889-7762

Grainger, John J. & Stevenson, William D. (1994). *Power System Analysis*, McGraw-Hill Book Company, ISBN 0-07-061293-5, New York

Shayanfar, H.A.; Tabatabaei, N.M. & Mokhtarpour, A. (2005a). Modified Strategy for Unified Power Quality Conditioner (UPQC) Based on Intelligent Devices, *Proceedings of the Ninth World Multiconference on Systemic, Cybernetics and Informatics*, pp. 13-17, Florida, U.S.A, July 10-13 2005, Orlando

Shayanfar, H.A.; Tabatabaei, N.M. & Mokhtarpour, A. (2006a). Control of Series Active Filter Based on Modelling of ION Devices, *Proceedings of the Third International Conference on Technical and Physical Problems in Power Engineering*, pp. 915-918, Turkey, May 29-31 2006, Gazi University, Ankara

Shayanfar, H.A.; Tabatabaei, N.M. & Mokhtarpour, A. (2006b). A Novel Intelligent Control Strategy for Reactive Power and Current Harmonics Compensation, *Proceedings of the Third International Conference on Technical and Physical Problems in Power Engineering*, pp. 272-275, Turkey, May 29-31 2006, Gazi University, Ankara

Shayanfar, H.A.; Tabatabaei, N.M. & Mokhtarpour, A. (2006C). Unified Power Quality Conditioner (UPQC) Control Based on Fourier Transform, *Proceedings of the IEEE International Conference on Power System Technology (POWERCON 2006)*, pp. 1-5, ISBN 1-4244-0110-0, China, October 22-26 2006, IEEE, Chongqing

Kazemi, A.; Mokhtarpour, A. & Tarafdar Haque, M. (2006). A New Control Strategy for Unified Power Quality Conditioner (UPQC) in Distribution Systems, *Proceedings of the IEEE International Conference on Power System Technology (POWERCON 2006)*, pp. 1-5, ISBN 1-4244-0110-0, China, October 22-26 2006, IEEE, Chongqing

Shayanfar, H.A.; Tabatabaei, N.M. & Mokhtarpour, A. (2005b). Use of Intelligent Measurement and Control Devices in Electrical Energy Disturbance Compensation,

Proceedings of the Eighth Baku International Congress of Energy, Ecology, Economy, pp. 764-770, Azerbaijan, June 1-3 2005, International Ecoenergy Academy, Baku

Valtari, J.; Verho, P.; Hakala-Ranta, A. & Saarinen, J. (2009). Increasing Cost Efficiency of Substation Automation Systems by Centralized Protection Functions, *Proceedings of the 20th International Conference and Exhibition on Electricity Distribution*, pp. 1-4, ISBN 978-1-84919126-5, Czech Republic, June 8-11 2009, IEEE, Prague

Shayanfar, H.A.; Tabatabaei, N.M. & Mokhtarpour, A. (2007). Best Control Strategy for Unified Power Quality Conditioner (UPQC) Based on Simulation, *Proceedings of The IASTED International Conference on Power and Energy Systems (PES 2007)*,pp. 257-262, Florida, U.S.A., January 3-5 2007, Clearwater

8

A New, Ultra-low-cost Power Quality and Energy Measurement Technology - The Future of Power Quality Measurement

Andreas Eberhard[1]
University/organization,
USA

1. Introduction

IEC 61000-4-30 is an excellent standard that ensures that all compliant power quality instruments, regardless of manufacturer, will produce the same results when connected to the same signal. However, instruments that comply with the Class A requirements of this standard have, until now, been too expensive for common use. Now a new set of technologies developed by an American company, in cooperation with a Japanese company, demonstrate that it is possible to manufacture three-phase power quality instruments that are fully compliant with the Class A requirements of IEC 61000-4-30 at ultra-low-cost to allow putting this monitoring devices even at entry levels of individual loads.

The development uses technologies from several fields that have not previously been related to power quality, including digital cameras, power-over-ethernet, mobile phones, and submarine sonar systems.

These new technologies have been packaged in a demonstration instrument, and may be licensed to instrument manufacturers as well. This allows gathering power quality and energy consumption information throughout manufacturing facilities or commercial buildings.

Traditionally, power quality instruments have been complex and expensive - often several thousand Euro or Dollar.

The cost of power quality instruments is driven by five factors:

1. The cost of developing the instruments
2. The quantity of instruments produced - the more instruments that are produced, the lower the development cost in each instrument
3. The cost of manufacturing the instruments
4. The cost of installation, especially the cost of the communication infrastructure
5. The cost of supporting the instruments, especially the cost of supporting special-purpose software, throughout the life of the instruments.

Remarkably, in the last few years, all of these costs have been driven down simultaneously.

This paper describes the technologies in a new, ultra-low-cost power quality instrument, and explains why the costs are so low.

2. Reduction in development costs

Traditional power quality instruments have been developed, from the start, as special purpose instruments. Hardware, firmware, and software have all been developed specifically for that instrument.
However, several developments in other, unrelated industries have made that approach unnecessary.
First, the wide-spread development of digital audio (mobile phones, mobile music players, digital television, etc.) has led to rapid developments in the DSP (digital signal processor) field. New DSP chips are inexpensive, use minimal power, and have built-in analog-to-digital and digital-to-analog conversion. Best of all, they are optimized for processing multiple channels of 20 Hz to 20 kHz signals. By coincidence, power quality measurements are generally made between 50 Hz and 3 kHz - right in the middle of the optimum band. The popularity of the Ipod® means cheaper, better power quality monitors.
Second, the development of relatively complex portable devices (PDA's, mobile phone that also have computer functions, digital cameras, etc.) means that extremely tiny, high-reliability electronic devices are now readily available: connectors with large numbers of pins, tiny op amps, and passive components like resistors and capacitors. Tiny means cheap, in general, if the manufacturing is completely automatic.

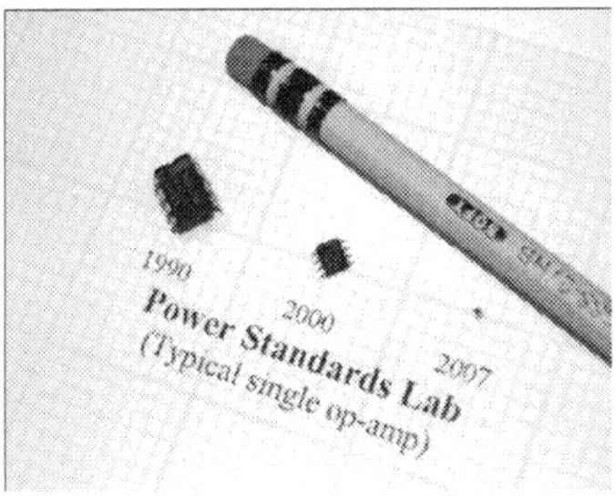

Fig. 1. Reduction in component size. Smaller components reduce costs in several ways: smaller printed wiring boards, smaller plastic packages, even smaller power supply requirements. The packages shown are, from left, through-hole DIP, surface mount, and BGA or ball-grid-array.

Third, the availability of high-voltage (1kV), low current op amps, which are generally used for driving submarine sonar transducers, means that automatic test equipment for power quality instruments can be developed far more cheaply now.
Finally, software standards for file structures mean that - just like digital cameras - a power quality instrument developers no longer need to define and support their own file structures. In fact, Windows® text files and web-based graphic file formats have become virtually universal.
Of course, some extraordinarily difficult problems must still be solved by the instrument engineer. How should one deal with a 6kV lightning impulse in such a tiny package? How can one meet the creepage and clearance requirements in the safety standards? Most important of all, by definition, a power quality instrument must work when the power is bad, and other electronic devices are failing. How can an engineer design an instrument that survives? But all of these difficult challenges can be met.

3. Increase in the quantity of power quality instruments

Traditionally, each country defined its own power quality instrument requirements. This meant that an instrument optimized for France, for example, was unlikely to find acceptance in Brazil, for example. Sometimes the situation was even worse: each electric power company would define its own requirements for power quality measurements.

As a result, the quantity for each instrument design was small, and the fraction of the development cost carried in each instrument was large. A rough example: if developing an instrument costs 1.5 million Euros, and the total expected market is 2000 instruments, each instrument must carry 750 Euros of development costs - a significant but not uncommon burden. Recent IEC standards [1][2][3][4] have solved this problem.

IEC 61000-4-30[1][2], in particular, has defined power quality measurement methods. Class A in this standard ensures that any two instruments, when connected to the same signal, will produce the same result. Figure 2 and Figure 3 give examples.

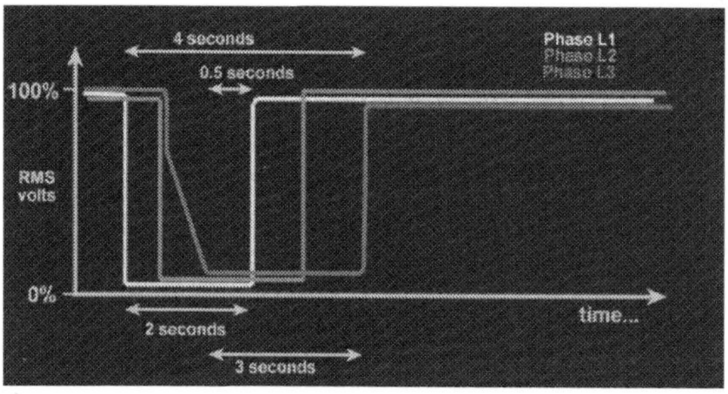

Fig. 2. Example of one of the problems solved by IEC 61000-4-30. In the graph of RMS voltages above, what is the duration of the voltage dip? The answers shown range from 0,5 seconds to 4 seconds, and <u>all of them are technically correct</u>. Simply by designating one of these answers as the requirement (4 seconds for dip duration measurement, 0.5 seconds for interruption duration measurement), 61000-4-30 reduces cost of instrumentation.

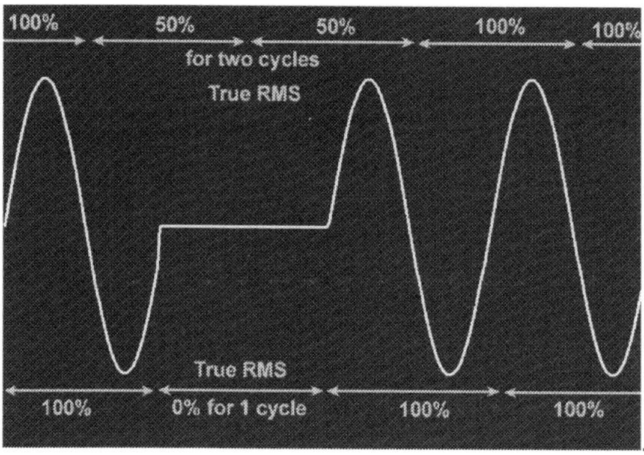

Fig. 3. Example of another problem solved by IEC 61000-4-30. Is this dip 50% for two cycles, or 0% for 1 cycle? Again, <u>both answers are technically correct</u>. 61000-4-30 makes it clear that this is a dip to 0% for 1 cycle - thus reducing cost of instrumentation.

As a result of this IEC standardization, an instrument can be designed for world-wide acceptance, and the total quantity is much higher. This means that the development cost burden is much smaller.

4. Decrease in manufacturing costs

Globalization has driven manufacturing costs down.
It is now easy to choose the best sources for parts, world-wide. For example, in the prototype shown in Figure 4, the lower case comes from Germany, the upper case comes from the United States, the display and the memory come from Japan, and the internal electronics are automatically assembled in California using parts from the U.S., Ireland, China, Japan, and other countries.
Fully-automated manufacturing (robots for placing parts, automatic testing systems for verifying that boards are working properly, automatic calibration systems that adjust internal digital constants, etc.) means that manufacturing costs can be kept very low, even in locations with high labour costs like California, without any sacrifice in quality. Indeed, the quality is generally higher than products that are produced in regions with low labour costs, due to the highly-automated production and test procedures.

Fig. 4. Prototype of the three-phase, voltage-and-current, 61000-4-30 Class A power quality monitor. The digital camera influence can be seen in the SD memory card, which holds up to 2 000 MB of data. Standard DIN-rail mounting means installation is cheap and quick.

5. Decrease in installation and communication costs

Traditional power quality instruments have been designed with their own unique packaging.
However, they are installed in locations where there are low-voltage circuit breakers. By packaging the power quality instrument in a standard 35mm DIN-rail circuit breaker package, installation is greatly simplified.
Perhaps more important, the communication cost of an installed power quality instrument, over the life of the instrument, often exceeds the cost of the instrument itself. Whether the communication is via Ethernet, or telephone modem, or short-distance radio, bringing the communication signal to the monitoring point is a significant cost.

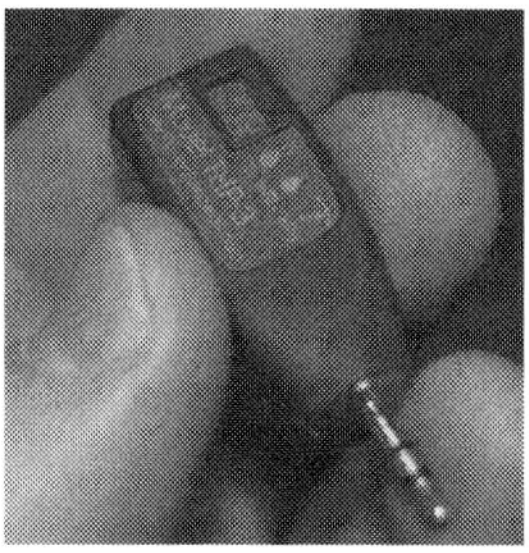

Fig. 5. Probes for parameters that may be related to power quality are included: Temperature, Humidity, Barometric Pressure, etc. A GPS satellite receiver can ensure precise timing.

(We should not forget the additional hidden cost of the damage that can be caused to the communication network, during major power disturbances. In regions with strong lightning activity, for example, telephone modems have traditionally been damaged through their connection to power quality monitors.)

Furthermore, in many power quality monitoring applications, immediate communication is not necessary.

The prototype shown in Figure 4 has Ethernet connectivity (and includes a web server, an FTP server, an e-mail client, Modbus and SNMP), wireless radio connectivity, and a G3 modem connection. However, it is optimized to function without connectivity - it can easily store a year's data on a removable SD memory card.

6. Decrease in the cost of supporting power quality instruments

Although it is often hidden from the end user, the lifetime support cost for traditional power quality instruments is significant.

This is especially true for the special-purpose software that was written for each instrument. Often, this software was written for Windows®; and the lifetime of a power quality instrument greatly exceeds the lifetime of one release of Windows®. For example, many power quality instruments are still in use that were originally issued with Windows® 3.1 software.

The lifetime costs of upgrading and supporting this software was a major cost.

However, such software is no longer necessary. By following the software model of a digital camera, the prototype power quality instrument of Figure 4 requires absolutely no software. When you connect a digital camera to your computer, you immediately see the pictures in a folder on your disk drive. The same is true for power quality data in the prototype instrument.

By eliminating the need for any software at all, we drive down the costs even further.

Again, this does not eliminate all firmware challenges. The instrument firmware must still support a wide variety of languages and character sets (Japanese, Korean, etc.), and the CSV (comma separated variable) files for spread sheets must work with European systems that use the comma symbol for other purposes. But all of these problems have been previously solved in digital cameras.

7. Conclusion

Recent changes in standards and technology have made it possible to produce a 61000-4-30 Class A power quality instrument at a very low cost. This could lead to having a Smart Grid system on the factory floor.

Biography

[1]**Andreas Eberhard** is well known in the international standards and power quality community. He is a member of various power quality standard committees around the world. Andreas holds two Master Degree in Electrical Engineering and a Master Degree in Global Technology Management from Universities in Europe and the United States. He is Vice President of Technical Services at Power Standard Labs.

8. References

[1] IEC 61000-4-30, Ed 1, "Testing and measurement techniques - Power Quality Measurement Methods". International Electrotechnical Commission. February 2003.

[2] Corrigendum 1, IEC 61000-4-30, Ed 1. August 2006.

[3] IEC 61000-4-15, Ed 1.1, "Flickermeter - Functional and design specifications". International Electrotechnical Commission. 2003.

[4] IEC 61000-4-7 Ed 2, "General guide on harmonics and interharmonics measurements and instrumentation". International Electrotechnical Commission. 2002.

Part 4

Power Quality and Mitigation

9

Active Power Filters for Harmonic Elimination and Power Quality Improvement

António Martins[1], José Ferreira[2] and Helder Azevedo[3]
[1]*University of Porto,*
[2]*Metro do Porto, S.A.,*
[3]*Efacec - Engenharia e Sistemas, S.A.*
Portugal

1. Introduction

The explosive growth in consumer electronics and domestic appliances has generated a major concern in the electricity supply industry, (Bollen, 1999). Due to its interface circuit (a diode bridge, followed by a large DC capacitor), these appliances draw current only near the peak of the mains voltage. Like this circuit other power electronics based applications draw non-sinusoidal currents, despite the applied voltage being sinusoidal. Due to the non-ideal characteristics of the voltage source, harmonic currents create voltage distortion.

Non-linear loads such as rectifiers, cycloconverters, variable speed drives and arc furnaces, large decaying DC components, asymmetrical loads and other electrical equipment can cause high disturbances in the power supply system, (Bollen, 1999).

The harmonics generated by the most common non-linear loads have the following properties:

- lower order harmonics tend to dominate in amplitude;
- if the waveform has half-wave symmetry there are no even harmonics;
- harmonic emissions from a large number of non-linear loads of the same type will be added.

The major problems caused by the mains harmonic currents are those associated with the harmonic currents themselves, and those caused by the voltage waveform distortion resulting from the harmonic currents flowing in the supply source impedance. This distortion of the voltage waveform can cause, e.g. serious effects in direct on-line induction motors, ranging from a minor increase in internal temperature through excessive noise and vibration to actual damage; electronic power supplies may fail to operate adequately; increased earth leakage current through EMI filter capacitors due to their lower reactance at the harmonic frequencies.

To minimize these effects in electricity distribution systems (non-sinusoidal voltages, harmonic currents, unbalanced conditions, power de-rating, etc) different types of compensators have been proposed to increase the electric system quality, (Bollen, 1999, Hingorani & Gyugyi, 1999). One of those compensators is the active power filter (APF), (Akagi et al., 1984).

This Chapter is organized as follows: in Section 2, it is presented a brief review of power quality and harmonic emission standards, while Section 3 addresses main active filter topologies, control methods and performance indexes. In Section 4 it is developed a prototype of an APF for demonstration purposes, including the operating principle of the current controlled filter, and Section 5 includes the design of the filter's passive elements. Simulation and experimental results in different operating conditions are presented in Section 6, including performance evaluation. Finally, Section 7 discusses conclusions related to the presented work and indicates some future research needed in this area.

2. Power quality and harmonic emissions standards

With the increased use of electrical and electronic equipment, and telecommunication and broadcasting transmissions the electromagnetic spectrum is becoming saturated. The equipment within residential, commercial or industrial installations has become increasingly sensitive to some type of electromagnetic interference (EMI) both from internal or external sources, primarily because of the use of digital technology. So there is a need for control of electromagnetic environment, namely by limiting of the harmonic emissions caused by any type of electrical or electronic equipment. In the European Union, this problem has been addressed by the Directive 2004/108/EC - the Electromagnetic Compatibility Directive; in the United States the main guideline comes from the IEEE Standard 519.

The EMC Directive incorporates standards mainly from the CISPR, the CENELEC, and the IEC organizations. The standards assist in achieving adequate power quality and in controlling it. They provide a framework within which the electricity distribution network environments, the susceptibility of equipment to low voltage quality, and the emissions from different types of equipment are all defined. Examples of standards relating to power quality characteristics and measurements are the EN 50160, the IEC 61000-4-30 and the IEEE Standard 1159.

The purpose of the EN 50160 standard is to specify the characteristics of the supply voltage with regard to the course of the curve, the voltage level, the frequency and symmetry of the three-phase network at the interconnecting point to the customer. The goal is to determine limiting values for regular operating conditions. However, facility defects may lead to major disturbances in the electricity distribution supply network. The complete breakdown of the network can no longer be described efficiently by limiting values. Thus there is no point in indicating actual limiting values. Accordingly, the standard establishes just these values as limiting values, which are not allowed to be exceeded or remained under during 95% of the controlled period. Rather than being an EMC standard the EN 50160 is a product standard giving the voltage characteristics which can be expected at the supply terminals. The EN 50160 standard is becoming a kind of reference for what should be seen as good power quality.

Like the EN 50160, the IEC 61000-4-30 and the IEEE 1159 are the first standards that define the characteristics of voltage waveform as obtained from measurements. The power quality parameters considered are grid frequency, magnitude of the supply voltage, flicker, supply voltage dips and swells, voltage interruptions, transients over voltages, supply voltage unbalance, voltage and current harmonics, voltage inter-harmonics and mains signalling on the supply voltage and rapid voltage changes. Depending on the purpose of the measurement, all of the phenomena on this list may be measured, or a subset of the phenomena on this list may be measured.

It is in this scenario of voltage/power quality issues related to the grid connection of electric/electronic loads or sources that the APF is an important player.

3. APF control methods and performance

3.1 APF topologies

In some industrial and commercial applications, electric power is distributed through three-phase four-wire systems. With incorrectly distributed or uncompensated loads such systems may suffer from excessive neutral currents caused by non-linear or unbalanced loads. In such conditions, a three-phase four-wire active filter can provide harmonic neutralisation, (Aredes et al., 1997, Montero et al., 2007). The main converter topologies for three-phase four-wire active power filters are the conventional three-leg converter with neutral point connection in the DC bus and the four-leg converter; the fundamental difference between them is the number of power semiconductor devices. In some conditions, even in three-phase installations, single-phase compensation can be advantageous. In such cases, the single-phase shunt active filter is often used, (Komurcugil & Kruker, 2006).

However, three-phase systems without neutral conductor are more general and will be the object of the present work.

3.2 Control methods and strategies

Different approaches such as notch filter, (Newman et al., 2002), scalar control, (Chandra et al., 2000), instantaneous reactive power theory, (Furuhashi et al., 1990, Akagi et al., 2007), synchronous detection method, (Chen et al., 1993), synchronous d-q frame method, (Mendalek et al., 2003), flux-based control, (Bhattacharya et al., 1996), and closed loop PI, (Bhattacharya et al., 1996), internal model control, (Marconi et al., 2007), and sliding mode control, (Saetieo et al., 1995), can be used to improve the active filter performance. Also, the direct power control method has found application in active filters, (Chen & Joós, 2008). Specific harmonics can be cancelled out in the grid using the selective harmonic elimination method (Lascu et al., 2007). In all cases, the goal is to design a simple but robust control system for the filter.

Usually, the voltage-source is preferred over the current-source to implement the parallel active power filter since it has some advantages, (Routimo et al., 2007). Using higher voltages in the DC bus is desirable and can be achieved with a multilevel inverter (Lin & Yang, 2004). In this Chapter it is used the voltage-source parallel topology, schematically shown in Fig. 1.

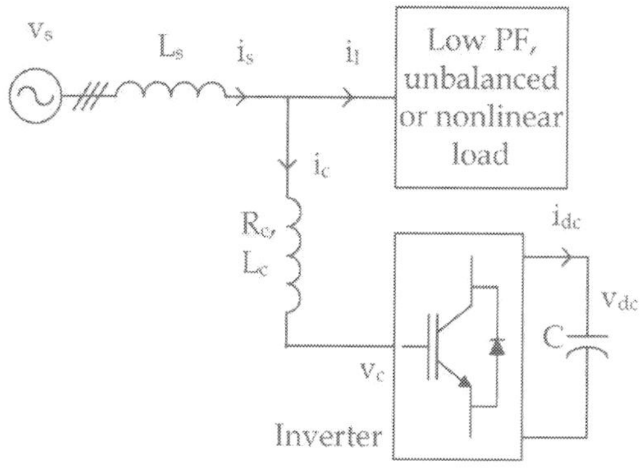

Fig. 1. Connection diagram of a voltage source active power filter.

The filter generates currents in the connection point in order to: 1- cancel/minimize the harmonic content in the AC system, 2- correct the power factor at fundamental frequency, 3- regulate the voltage magnitude, and 4- balance loads. So, the AC distribution system only carries the active fundamental component of the load current.
Very different current control algorithms can be applied to the active filter, (Akagi, 2005). The current reference for the active filter connection node usually satisfies one of the two following strategies: 1- power factor correction, harmonic elimination, and load unbalance compensation or, 2- voltage regulation, harmonic elimination, and load unbalance compensation.
The voltage regulation strategy is a concurrent objective faced to the power factor compensation because the two depend on the reactive current. However, any control algorithm has enough flexibility to be configured, in real-time, to either objectives or for the two, in a weighted form.
Even under the same compensation strategy, the filter can be controlled with different control algorithms. Two main approaches are common: voltage control, and current control. Both methods have advantages and weaknesses. In Fig. 2 it is represented a block diagram, with the variables shown in Fig. 1, of a voltage-based control algorithm of an active power filter implementing vector control.

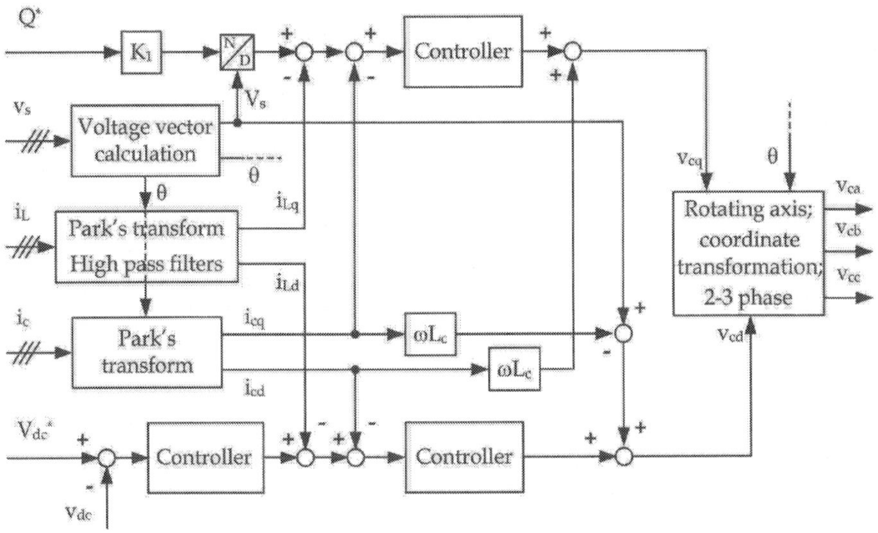

Fig. 2. Voltage control of the active power filter.

In the presence of the power quality concern the mains current is the most important variable to be controlled. This method indirectly controls the mains current through the filter output voltage.
In Fig. 3 it is shown a block diagram of a current-based control algorithm of an active power filter also capable of implementing different compensation strategies. This control method directly controls the mains current through the i_{sRef} signals. However, it has slower dynamics than the voltage-based method (Chandra et al., 2000).

A static power converter, like the one shown in Fig. 1, capable of doing (almost) all the above referred functions is necessarily very complex. This complexity arises from the following considerations:

- the converter dynamic behaviour must be very fast in order to be capable of compensate currents in a large spectra,
- the control algorithm must deal with a large number of variables such as mains voltages and currents, load currents, DC voltage and current, and
- high dynamic performance and better active and reactive power decoupling can demand direct and inverse coordinate transformation and a large amount of signal processing.

So, fast power electronics semiconductors, with high switching frequencies, and powerful control platforms are needed to build this type of power electronics systems.

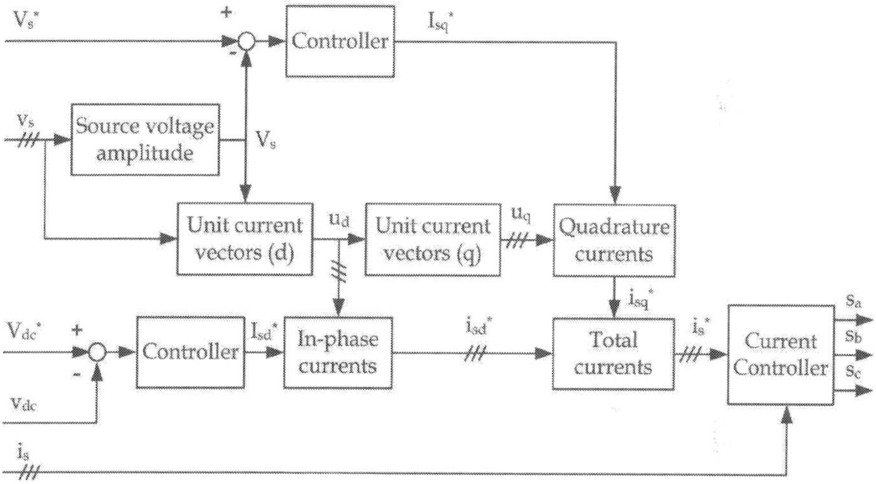

Fig. 3. Current control of the active power filter.

3.3 Performance evaluation of APFs

In order to analyze and evaluate the performance of an active power filter, different aspects must be considered. Two different performance types can be considered: the qualitative ones and the quantitative measures.

Qualitative evaluation

The qualitative value of an active power filter is a consequence of different technical merits. The most important ones are briefly described:

- power semiconductor characteristics, namely of diodes, GTOs and IGBTs, and converter topology,
- type of control system, scalar or vector control, and operating modes,
- converter supervision, diagnostics and remote control.

The active filter must accomplish one, or more than one, specific objective. So, its working conditions must be in agreement with the established purposes: reactive power

compensation; voltage regulation; load balancing or harmonic compensation. These operating modes should be programmable, remotely or on a local basis.
The filter is connected to the mains in a specified point of connection, thus guaranteeing a maximum level in the mains reactive current and/or in the total harmonic distortion flowing through the network.

Quantitative performance measures

The filter performance should be evaluated in a typical distribution system with different loads, linear and non-linear. The relevant performance indexes will be characterized by the total harmonic distortion (THD) of the mains current, with and without filter, in the following two indexes basis: filter effectiveness index and filter capacity index.
The Filtering Effectiveness index (FE) is the relation between the total harmonic distortion of the current supplied by the mains with and without filter in a pre-defined frequency range, according to some standard, e.g. EN 61000-3-2 or IEEE 519, (total harmonic cancellation results in a null factor):

$$FE = \frac{THD_{APF}}{THD} \tag{1}$$

The Filtering Capacity index (FC) is the relation between the total apparent power supplied by the filter and the total mains or load apparent power:

$$FC = \frac{S_{APF}}{S_{Load}} \tag{2}$$

These two indexes are the basis for evaluating the filter performance in static operation. In transient operation, only special conditions can be evaluated and they usually are not under the restrictions of power quality measurements, like those referred by the EN 50160 and the IEC 61000-4-30.

4. An APF example

Among the different alternatives, the direct current control strategy that generates the reference waveform for the AC source current was chosen for demonstration purposes (Chandra et al., 2000). It does not need to measure the load current or power, requires a low processing time and allows a fast calculation of the reference currents. The current-controlled pulse width modulation (PWM) with a medium frequency fixed carrier ensures enough bandwidth to implement the different active filtering goals.
To demonstrate these advantages, a three-phase 5 kVA prototype of an active filter is designed, and tested in dynamic and stationary operation with different load types.
The main block diagram of the system operation is shown in Fig. 3. It handles the two referred control strategies (Section 3.2) being capable of dealing with load harmonics elimination, power factor correction, load unbalancing compensation, and/or voltage regulation. The lower side generates the active current reference and the upper side the reactive component.
In this Chapter, it is implemented the first operating mode: power factor correction, harmonic elimination, and load unbalance compensation. So, the AC source power factor will be approximately one, and there is no need to control the voltage at the connection point.

The control algorithm needs the measurement of several variables like the three-phase AC source voltages and currents and the DC-link voltage. In a voltage distorted grid one of two compensating strategies can be chosen: 1- imposing sinusoidal currents in the grid or, 2- imposing unity power factor, (Cavallini & Montanari, 1994). However, the last strategy implies the circulation of harmonic currents in the grid; since medium voltage grids are usually very little distorted, it is not very used.
In a distorted grid and in order to impose sinusoidal currents in the source it is needed to low-pass filter the AC voltage, so obtaining the fundamental component. In an unbalanced grid it is needed to estimate the symmetrical components from the measured voltage signals. In this condition, the instantaneous positive sequence components can be obtained in the time domain with simple algebraic manipulation, (Hsu, 1998).
In steady-state, and neglecting losses in the active power filter, the active power supplied from the AC source should be equal to the demanded load active power, since no active power flows into the DC capacitor.
However, once the source voltage varies or the load power changes, the active power balance between the AC source and the load will not be maintained. This transient drives the average voltage of DC capacitor away from the reference voltage. So, in order to keep the active power filter operation, the amplitude of the grid current must be adjusted. The active power supplied from the source is then changed proportionally in order to compensate the active power supplied/received by the DC capacitor and match the active power consumed by the load. So, the AC source current amplitude can be obtained by regulating the DC capacitor voltage.
The active power balance in the DC-link determines the reference current of the AC source and the use of a PI controller allows a smooth control of the filter current and improves the system dynamic response. In this case, the schematic in Fig. 3 represents the essential block diagram of the current reference. The error in the DC voltage is transformed in active power to be controlled in the AC source.

4.1 Grid synchronization

With a three-phase balanced system or with the positive sequence component, the RMS voltage source amplitude, V_s, is calculated at the sampling frequency, f_s, from the source phase voltages, v_{sa}, v_{sb}, v_{sc}. At each sampling instant, it is expressed as in (3).

$$V_s = \sqrt{\frac{2}{3}(v_{sa}^2 + v_{sb}^2 + v_{sc}^2)} \tag{3}$$

The direct (or in-phase) unit current vectors are obtained from the AC source phase voltages and the RMS amplitude of the source voltage, V_s.

$$u_{si} = v_{si} / V_s; \qquad i = a,b,c \tag{4}$$

The unit current vectors implement one important function in the grid connection of a power electronics converter, the synchronization. This method is simple and robust and compares favourably with other methods like the decomposition of single-phase into orthogonal components method or the linear estimation of phase method, (Thomas & Woolfson, 2001).

4.2 Voltage controller

The AC source current has two active components: 1- the filter current, which maintains the DC bus voltage at a constant value and 2- the load current. So, all the filter losses, AC, DC and switching, are automatically compensated.
The reference stored energy on the DC-link capacitor is given by

$$E_{dcn} = \frac{1}{2} C V_{dcn}^2, \tag{5}$$

where V_{dcn} is the reference/nominal voltage across the capacitor, C.
When the capacitor is charged with a V_{dc} voltage the energy unbalance in the DC-link capacitor is

$$\Delta E_{dc} = \frac{1}{2} C (V_{dcn}^2 - V_{dc}^2) . \tag{6}$$

This energy unbalance must be supplied by the three-phase AC grid. Imposing a sinusoidal input current, the change in the capacitor energy must satisfy (7),

$$\Delta E_{dc} = \int_0^T \left[\sum_{i=0}^{2} V_m \sin(\omega_g t - \frac{2\pi}{3} i) I_{m1} \sin(\omega_g t - \frac{2\pi}{3} i) dt \right], \tag{7}$$

where I_{m1} is the active current supplied to the DC-link capacitor and ω_g is the grid angular frequency. So, the reference current to maintain the DC voltage is given by

$$I_{m1}^* = \frac{2}{3} \frac{\Delta E_{dc}}{T V_m} = C \frac{V_{dcn}^2 - V_{dc}^2}{3 T V_m}, \tag{8}$$

where T is the time interval where the averaging is calculated.
The total active component of the AC source current, I_s, is the sum of the filter current and the load current, I_l,

$$I_s = I_{m1} + I_l \tag{9}$$

However, the inclusion of the load current in the control algorithm implies additional current sensors and more signal processing to estimate the load active power, thus increasing the cost and decreasing the dynamic response, (Newman et al., 2002). A direct controller is then used; the voltage controller must directly provide the total active component of the AC source current.
The active component amplitude of the reference currents in the AC source, i_{sad}^*, i_{sbd}^*, i_{scd}^*, is calculated through a PI controller with anti-windup as shown in Fig. 4. The proportional and integral gains determine the controller behaviour in dynamic and static operation, (Newman et al., 2002, Marconi et al., 2007, Saetieo et al., 1995, Buso et al., 1998). As stated before, the correct active component, I_{sd}^*, of the AC source current is determined from the power balance in the DC-link. The current reference is then scaled with the unit current vectors in phase with the source voltage:

$$i_{si}^* = I_{sd}^* \cdot u_{si}; \quad i = a,b,c \tag{10}$$

4.3 Current controller

Hysteresis is the easiest control method to implement current control. One disadvantage is that it is difficult to limit the minimum and maximum switching frequencies in order to have a good tracking of the reference current.

Several solutions are known to overcome this limitation but they increase the controller complexity, (Zeng et al., 2004). Additionally, other features should be referred: 1- there is a lack of intercommunication between the individual hysteresis controllers and therefore no strategy to generate zero voltage vectors; 2- it has a tendency to lock into limit cycles of high frequency switching; and 3- it causes the generation of sub-harmonic current components, (Holmes & Lipo, 2003).

The direct current control method, in combination with a carrier-based pulse width modulation, gives a good performance at medium switching frequencies (in the range of 5 to 7 kHz). Its performance can only be degraded if used at low switching frequencies; this degradation is caused by the response time of the current controlled inverter, which leads to a phase shift between the reference currents and the output currents. So, the PWM-based current controller was selected for implementation.

In the modulation stage, shown in Fig. 5, the total reference currents are subtracted from the source current, thus obtaining a current error adapted according to the amplitude of the triangular carrier. Since the current error signal is always kept within the negative and positive peaks of the triangular waveform, the system has an inherent over current protection. The PWM output is completed with the introduction of an appropriate dead time in the control signals of the inverter transistors.

The filter output current is defined by the AC source voltage, the filter output voltage and the AC-link inductance.

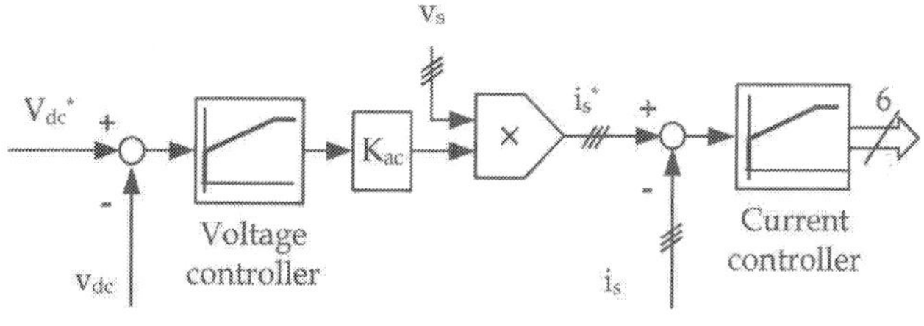

Fig. 4. Voltage and current controllers operation.

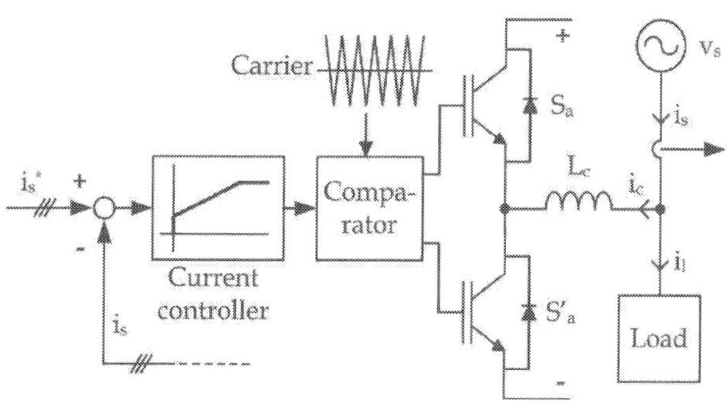

Fig. 5. Current control and PWM stage.

The current controller is designed according to the diagram in Fig. 6, where K_p and T_i are the PI controller gains, K_{inv} is the inverter gain and T_{inv} is the inverter time delay, equal to half of the switching period. Additionally, $T_L=L_c/R_c$ and $K_L=1/R_c$ and the measured current is filtered with a time constant T_f.

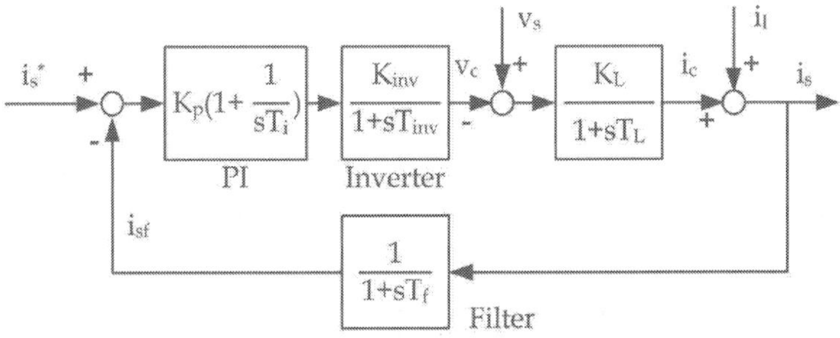

Fig. 6. Detailed model of the current control loop.

It is worthwhile to note that the closed loop transfer function is the same that is obtained with vector control; the only difference in the diagram of Fig. 6 is the absence of the cross term $\omega L_c i_{cq}$ in the voltage adding node, which is considered a disturbance.
The current controller open loop transfer function is

$$G_{oi}(s) = \frac{K_p K_{inv} K_L (1 + sT_i)}{sT_i(1 + sT_{inv})(1 + sT_L)(1 + sT_f)}. \tag{11}$$

Assuming that $T_i=T_L=L_c/R_c$, and that the two time constants, T_{inv} and T_f, can be approximated by one because they are quite small, the closed loop transfer function can be expressed as:

$$\frac{I_s}{I_s^*} = \frac{\omega_n^2}{s^2 + 2\xi\omega_n s + \omega_n^2}, \tag{12}$$

with

$$\xi = \frac{1}{2}\sqrt{\frac{1}{KT_{sf}}}; \qquad \omega_n = \sqrt{\frac{K}{T_{sf}}} \tag{13}$$

In (13) the two constants are given by:

$$T_{sf} = T_{inv} + T_f; \qquad K = \frac{K_p K_{inv} K_L}{T_i}. \tag{14}$$

According to the assumption made, $T_i=T_L$, the parameter K_p determines the damping factor of the control loop and, simultaneously, the speed response. Thus, knowing the filter parameters and imposing the dynamic behaviour the current controller gains can be obtained.

5. Inductive and capacitive components

The selection of the AC-link inductance and the DC-link capacitor values affects directly the performance of the active power filter. Static VAR compensators and active filters implemented with voltage-source inverters present the same power topology, but the criteria used to select the values of L and C are different. The active filter can implement the two compensation modes simultaneously, so it is presented the two main design criteria to accomplish these objectives: fundamental component reactive power compensation and harmonics compensation.

5.1 Inductive filter design

One possibility to design the inductance is to consider the maximum current, I_{max}, that the filter must supply in order to compensate a totally inductive load. In this condition the inductance and the inverter must be dimensioned taking into consideration the apparent power to be compensated. A simple phasor diagram analysis, shown in Fig. 7, gives the inductance value as in (15):

$$L_{\min} = \frac{\Delta V_{\min}}{\omega_g I_{\max}} \tag{15}$$

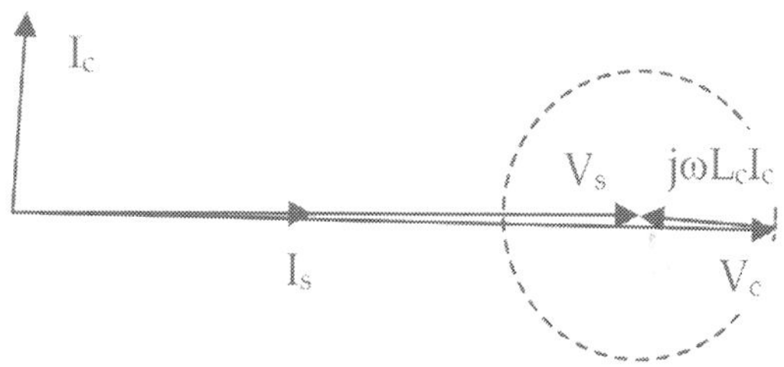

Fig. 7. Maximum output filter voltage with reactive power compensation.

ΔV_{min} is the difference between the RMS source voltage and the fundamental component RMS inverter voltage, which depends on the values of the DC-link voltage and the modulation index, and ω_g is the value of the grid frequency. If the inductive voltage drop, ΔV, is made small the inductance will be small and there is a better utilization of the DC voltage. However, a very small inductance implies a high voltage gain and introduces a higher complexity in the controller design. Additionally, the ripple current will increase. In order to keep the ripple current at a reduced level, the inductive voltage drop should be kept at a minimum value.

As an active filter, the inductance value should be decreased to be capable of higher surge currents and harmonic currents. In this condition, another criterion can be evaluated, like the one referred in Moran et al. (1995). Imposing a fixed switching frequency, f_s, and the maximum di/dt of the current to be compensated, the inductance value is now given by:

$$L_{\max} = \frac{\Delta V_{\max}}{4(di/dt)_{\max} \cdot f_s} \tag{16}$$

ΔV_{max} is the maximum difference between the instantaneous AC source voltage and the instantaneous inverter voltage. From this point of view a higher DC-link voltage allows a higher di/dt, so increasing the active filter frequency response.

If the ripple current can be reduced through an increasing in the switching frequency or an increasing in the controller dynamics the inductance value would not be so influent in the global filter performance. The connecting inductances also decouple the output inverter voltage from the AC source voltage.

Considering, designing, and ranking all the referred factors, including the filter compensation modes, an indicative value for the inductance can be obtained.

5.2 DC-link capacitor design

The DC-link capacitor can be designed according to distinct objectives, the most important one being keeping the DC-link voltage fluctuation limited. The reactive power supplied by the filter corresponds to a stored energy in the AC inductors. When a reactive power compensation change is demanded, there occurs a variation in the stored energy associated with the final and initial values of the filter reactive current, I_1 and I_0, respectively:

$$\Delta E = \frac{3}{2} L (I_0^2 - I_1^2)\,. \tag{17}$$

In the worst case of a totally inductive to totally capacitive compensation change, it can be assumed that the energy change in the inductors has to be supplied by the DC-link capacitor. So, the DC voltage fluctuates in transient operation, according to supplying or absorbing the energy; the DC-link capacitor has to be designed to moderate the DC voltage fluctuation.

When the DC voltage changes from V_{dc0} to V_{dc1}, the energy released from the DC-link capacitor, is given by (18) if it is assumed a small change in V_{dc}.

$$\Delta E_{dc} = \frac{1}{2} C (V_{dc0}^2 - V_{dc1}^2) \cong C V_{dc0} (V_{dc0} - V_{dc1})\,, \tag{18}$$

Introducing the ratio of the DC voltage change, ε, defined by

$$\varepsilon = \frac{V_{dc0} - V_{dc1}}{V_{dc0}} = \frac{\Delta V_{dc}}{V_{dc0}}\,, \tag{19}$$

the required DC capacitor is given by

$$C = \frac{3L(I_0^2 - I_1^2)}{2\varepsilon V_{dc0}^2}\,. \tag{20}$$

Analysis of (20) shows that the required capacitance of the DC capacitor is proportional to the line inductance and inversely proportional to the specified DC voltage fluctuation. The value of the DC-link capacitor can also be designed in order to supply active power to the load during a pre-defined time interval in case of AC source absence.

Thus, knowing the AC connecting inductance, the nominal DC voltage and the allowed voltage fluctuation, the DC capacitor value can be obtained.

6. Simulation and experimental results

The Saber Designer software package was used to design, simulate, and test the active filter control algorithm. In order to implement the algorithm in digital hardware platform it was built a simulation model of a microcontroller. So, the C code written and validated in the Saber environment can be easily transferred to any hardware target.

6.1 Simulation results

Different active filtering conditions, power factor compensation, and load balancing tests have been done, in transient and static operation and for linear and non-linear loads. The transient operation tests were made putting the active filter into operation when the load was already connected to the AC source voltage and also with simultaneous connection of both, load and filter.

In Fig. 8 the non-linear load is a three-phase diode rectifier, with a DC current of 7 A and a DC voltage of 250 V. [The other filter parameters are given in Table 1.] The transient operation test was made first connecting the filter and then connecting the load under test.

Nominal power	5 kVA
Supply voltage	110/190 V, 50 Hz
DC voltage	350 V, DC capacitor: 1.6 mF
Switching frequency	5 kHz, dead time: 4 μs
Filter inductance	L_c, R_c: 2 mH, 0.2 Ω
DC voltage controller	K_{pv}, T_{iv}: 0.01, 100 ms
Current controller	K_{ip}, T_{ii}: 0.007, 10 ms

Table 1. System parameters (simulation and experimental).

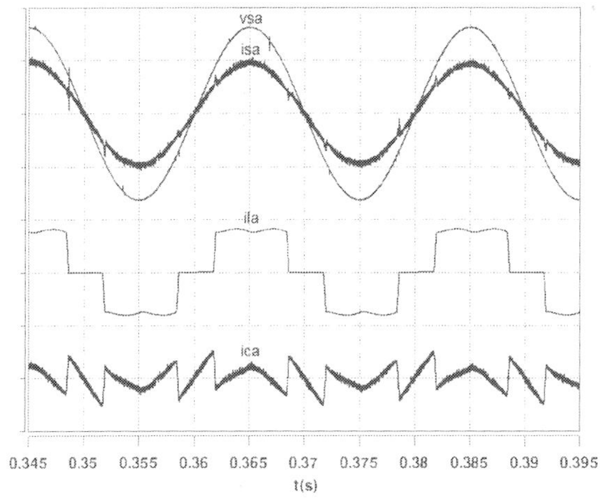

Fig. 8. Steady-state compensation of a three-phase diode rectifier. Traces from top to bottom (phase a): source voltage, v_{sa}, (100 V/div); source current, i_{sa}, (10 A/div); load current, i_{la}, (10 A/div); filter current, i_{ca}, (10 A/div).

As can be seen, except for the high di/dt intervals, the active filter completely compensates the rectifier harmonic currents. A smaller inductance would produce a better low-frequency compensation but with high ripple current. The APF performance is demonstrated in Fig. 9, showing the harmonic distortion of the source current with and without filter.

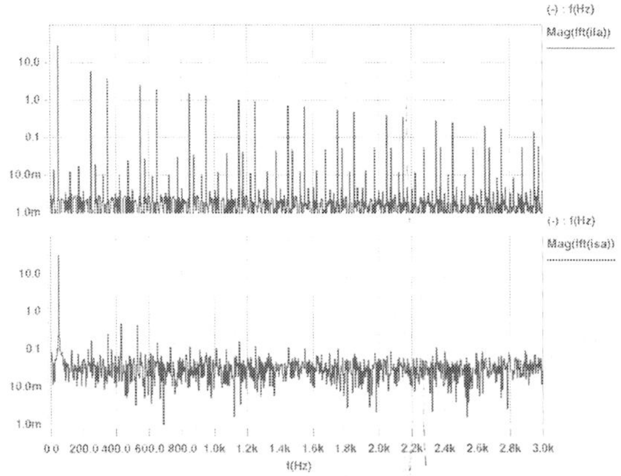

Fig. 9. Harmonic spectrum of phase a non-linear load current (top) and AC source current (bottom), with the APF connected, corresponding to Fig. 8.

Transient operation with the same load is shown in Fig. 10. It can be noticed a short interval where the AC source current is out of phase with the voltage but it quickly gets in phase.

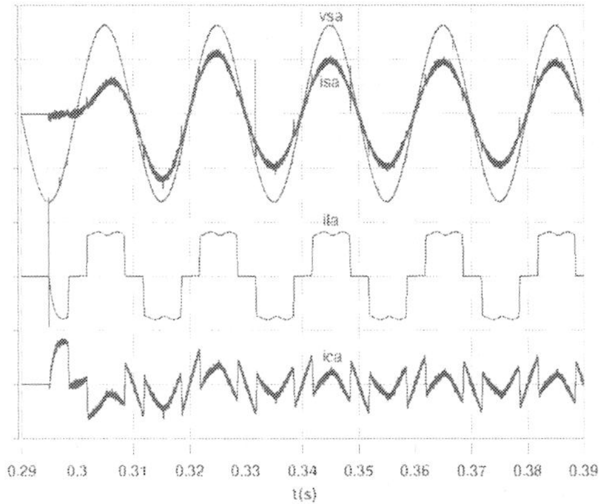

Fig. 10. Transient compensation of a three-phase diode rectifier; filter connected before the load. Traces from top to bottom (phase a): source voltage, v_{sa}, (100 V/div); source current, i_{sa}, (10 A/div); load current, i_{la}, (10 A/div); filter current, i_{ca}, (10 A/div).

The same non-linear load type, but now a three-phase thyristor bridge was used to perform the test shown in Fig. 11. The bridge firing angle is 30°, thus generating not only low-frequency harmonics but also a reactive current component at fundamental frequency. Fig.

11 shows the same waveforms already presented in Fig. 8, and Fig. 12 shows the source current spectrum without and with compensation. Differently from Fig. 10, Fig. 13 shows the transient response when the filter is connected to the system with the non-linear load already supplied by the source voltage.

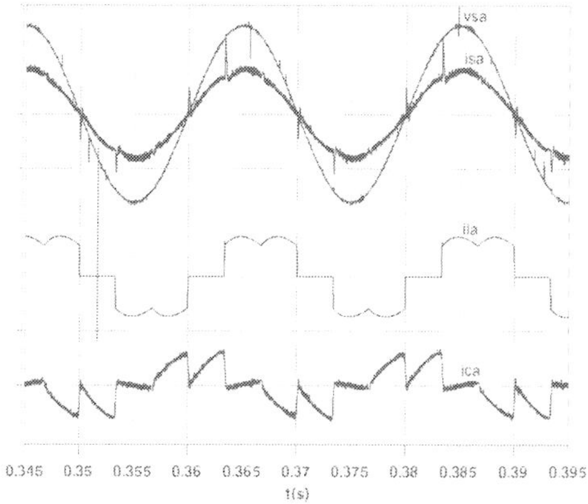

Fig. 11. Steady-state compensation of a three-phase thyristor rectifier, with a firing angle of 30°. Traces from top to bottom (phase a): source voltage, v_{sa}, (100 V/div); source current, i_{sa}, (10 A/div); load current, i_{la}, (10 A/div); filter current, i_{ca}, (10 A/div).

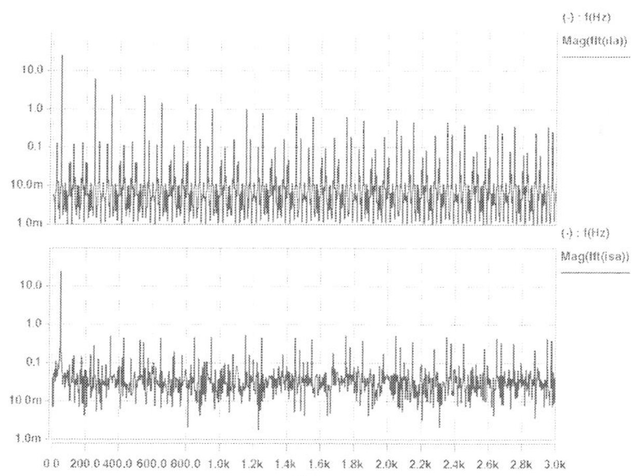

Fig. 12. Harmonic spectrum of phase a non-linear load current (top) and AC source current (bottom), with the APF connected, corresponding to Fig. 11.

The simulation of the filter operation in load unbalance compensation is shown in two conditions, with linear (Fig. 14) and non-linear load (Fig. 15). In Fig. 14, a linear load is connected between phases a and b, leaving phase c in open circuit.

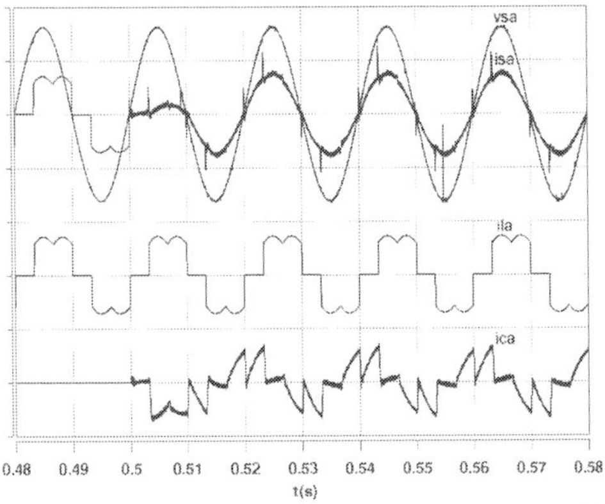

Fig. 13. Transient compensation of a three-phase thyristor rectifier, with a firing angle of 30°; filter connected after the load. Traces from top to bottom (phase a): source voltage, v_{sa}, (100 V/div); source current, i_{sa}, (10 A/div); load current, i_{la}, (10 A/div); filter current, i_{ca}, (10/div).

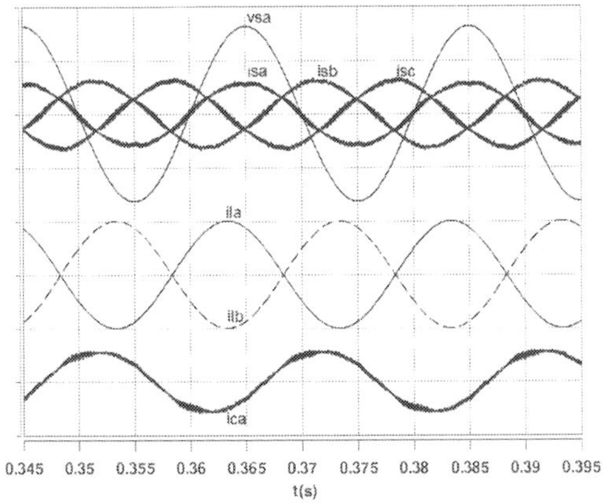

Fig. 14. Load unbalance compensation with linear load. Traces from top to bottom: source voltage (phase a), v_{sa}, (100 V/div); source currents, i_{sa}, i_{sb}, i_{sc}, (10 A/div); load currents, i_{la}, i_{lb}, (10 A/div); filter current, i_{ca}, (10 A/div).

In Fig. 15, the load is a single-phase diode rectifier bridge, connected between two phases (a and b); the third phase is opened. In either condition the unbalance compensation is not perfect, as can be also concluded from Fig. 16, which shows the spectrum of phase a source current. It should be referred the effect of the small DC voltage fluctuation and the effect of the relatively high value of the AC inductance.

If the unbalanced load is the condition that would be the most important in the filter operation then the DC-link capacitor value could be increased. Also, the AC inductance value can be decreased but the switching frequency must rise to keep the high order harmonics at low levels. Alternatively, small passive filters can be used at the point of common coupling.

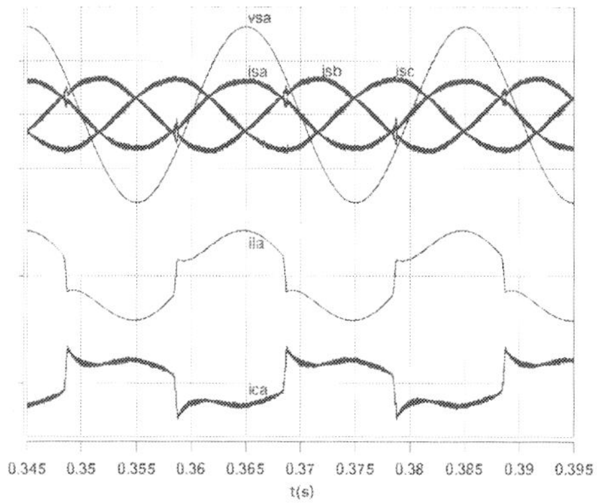

Fig. 15. Load unbalance compensation (non-linear load). Traces from top to bottom: phase a source voltage, v_{sa}, (100 V/div); source currents, i_{sa}, i_{sb}, i_{sc}, (10 A/div); load current, i_{la}, (10 A/div); filter current, i_{ca}, (10 A/div).

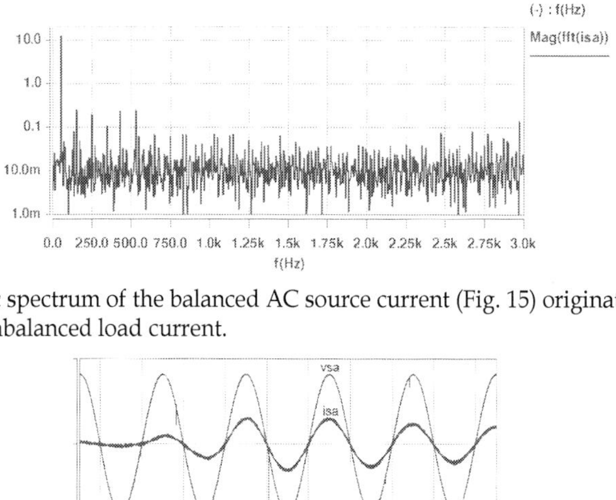

Fig. 16. Harmonic spectrum of the balanced AC source current (Fig. 15) originated from a non-linear and unbalanced load current.

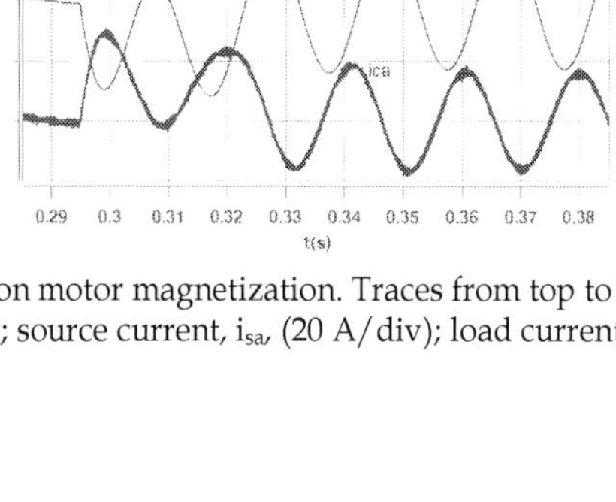

Fig. 17. No-load induction motor magnetization. Traces from top to bottom (phase a): source voltage, v_{sa}, (100 V/div); source current, i_{sa}, (20 A/div); load current, i_{la}, (10 A/div); filter current, i_{ca}, (10 A/div).

Finally, Fig. 17 shows the simulation of the filter operation when an induction motor is directly connected to the source in free acceleration mode and no mechanical load. In this case, it can be observed a DC transient in the motor current, quickly compensated by the filter as can be verified in the source current.

The conclusion from the presented simulation results is that the control algorithm succeeded in the most important functions of the active power filter, harmonics mitigation, reactive power compensation and load unbalance compensation.

6.2 Experimental results

An experimental three-phase active filter prototype has been designed for a nominal power of 5 kVA. It is based on a voltage source inverter with MOSFET transistors and a switching frequency of 5 kHz. The control platform is build over a TERN 586-Engine, a module with a C++ programmable microprocessor, with a 32-bit CPU operating at 100 MHz that includes a math coprocessor for floating point operations. Several I/O interfaces are available, including 8 channel, 12-bit A/D converter; series D/A converter; 32 I/O lines; 15 external interrupts, and seven 16-bit timers.

Similar experimental tests have been conducted and the results can be viewed in two perspectives: validation of the global simulation model, and measure of the active filter performance. The experimental tests were conducted in a laboratory with an important distortion in the mains voltage, due to the proximity of a large number of non-linear loads. However, the mains current waveform after compensation is almost a sine wave due to the extraction of the fundamental component of the mains voltage through low-pass filters. This filter action allows the imposition of sine wave references for the source current; it could be possible to impose non-sine wave references "in phase" with the mains voltage, (Cavallini & Montanari, 1994).

The non-linear load present in Fig. 18 and Fig. 19 is a three-phase diode rectifier. In Fig. 18 it can be seen the steady-state compensation, where the AC source current gets a sine wave form, being only slightly distorted in the instants of the diode switching. This is due to the high di/dt occurring in these points, which is impossible to compensate unless with a very high DC voltage or a very low AC inductance. There should be a compromise between the active filter dimensioning and the characteristics of the load current to be compensated. Fig. 19 shows the filter transient operation with the same load condition.

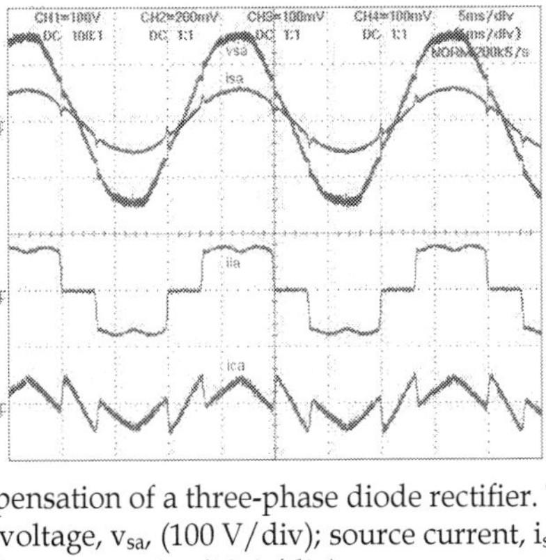

Fig. 18. Steady-state compensation of a three-phase diode rectifier. Traces from top to bottom (phase a): source voltage, v_{sa}, (100 V/div); source current, i_{sa}, (20 A/div); load current, i_{la}, (10 A/div); filter current, i_{ca}, (10 A/div).

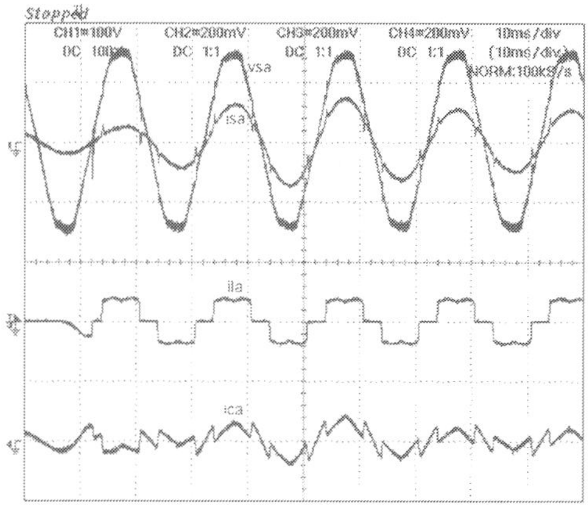

Fig. 19. Transient compensation of a three-phase diode rectifier. Traces from top to bottom (phase a): source voltage, v_{sa}, (100 V/div); source current, i_{sa}, (20 A/div); load current, i_{la}, (20 A/div); filter current, i_{ca}, (20 A/div).

In Fig. 20 it is presented the balancing action of the active filter compensating an unbalanced load (a single-phase diode rectifier bridge, connected between two phases; the third phase is an open circuit).

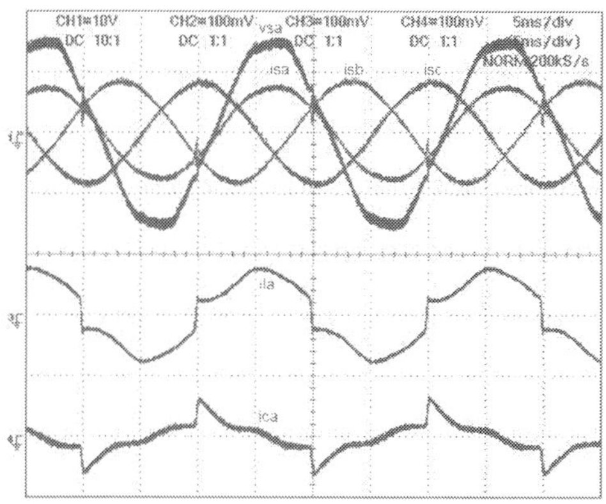

Fig. 20. Load unbalance compensation. Traces from top to bottom: source voltage (phase a), v_{sa}, (100 V/div); source currents, i_{sa}, i_{sb}, i_{sc}, (10 A/div); load current, i_{la}, (10 A/div); filter current, i_{ca}, (10 A/div).

In Fig. 21 it can be seen the active filter behaviour in a no-load induction motor magnetization. The filter quickly compensates the load power factor and the decaying DC component. In this case, the transient response is also dependent on the time interval where the load current has a non-zero DC value.

As expected, the experimental results are in close agreement with the simulation ones thus validating the control algorithm as in its operating principle as in its digital implementation.

6.3 Discussion

Four simulated results can be compared with the experimental ones. Figures 8 and 18 show the steady-state operation with the filter compensating a typical non-linear load, and close agreement between the two can be noticed even with a distorted grid in the lab.

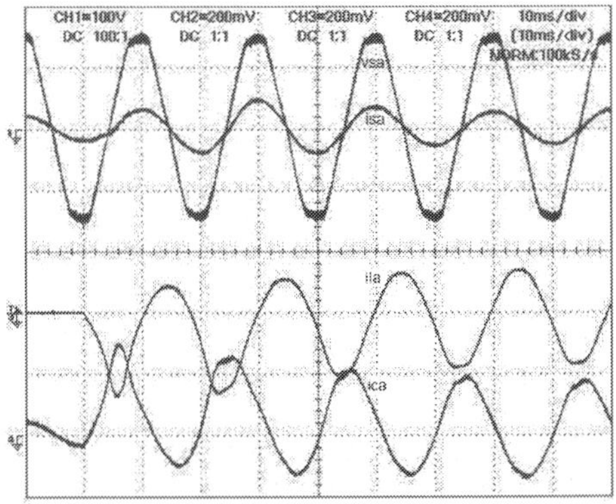

Fig. 21. No-load induction motor magnetization. Traces from top to bottom (phase a): source voltage, v_{sa}, (100 V/div); source current, i_{sa}, (20 A/div); load current, i_{la}, (10 A/div); filter current, i_{ca}, (10 A/div).

Figures 10 and 19, related to the transient operation with the same non-linear load, show very similar results; the filter operating principle, based on energy balance between AC inductance and DC capacitor, is validated.

Load balancing, analyzed in Figures 15 and 20, is an easy task with this control method. It is automatically achieved; it does not need decomposition of the load current into symmetrical components.

Figures 17 and 21 demonstrate an additional benefit of the active power filter: minimization of DC transients. This is an important feature especially in weak grids where long DC current transients cause also DC transients in the AC voltage, particularly harmful for transformers and electric machines.

7. Conclusions and further research

7.1 Conclusions

An active power filter is a high performance power electronics converter and can operate in different modes: harmonics elimination, power factor correction, voltage regulation and load unbalance compensation. Different control approaches are possible but they all share a common objective: imposing sinusoidal currents in the grid, eventually with unity power factor, even in the case of highly distorted mains voltage. This Chapter analyzes and compares different approaches to be used in the control of the APF.

As a demonstration of the capabilities of the APF, one control approach has been selected. It uses a simple and robust power circuit interface without load current sensors and an efficient signal processing without heavy or complex computations. Simulation results under different operating conditions demonstrate the overall possibilities of the control method and of the active power filter globally.

Also, the presented experimental implementation has been validated and can be applied in different operating modes: harmonics active filtering, power factor correction, and balancing of linear or non-linear loads, single or grouped, which cause great perturbation and performance degradation in the power quality of an electrical distribution system.

7.2 Further research

Active power filters are now well established in the market. However, some issues still need further research. The filter dynamics depends on the switching frequency; higher frequencies given better results but also higher losses. Specific modulation strategies and control algorithms must be improved. In particular selective harmonic elimination methods can bring additional performance. Also, multilevel based topologies allow the APF to reach higher voltages and power and so give the filter the possibility of being applied in the power systems domain.

8. References

Akagi, H., Kanazawa, Y. & Nabac, A. (1984). Instantaneous reactive power compensators comprising switching devices without energy storage components. *IEEE Transactions on Industry Applications*, vol. 20, n° 3, (May/June 1984), pp. 625-630, ISSN 0093-9994.

Akagi, H. (2005). Active harmonic filters. *Proceedings of the IEEE*, vol. 93, n° 12, (Dec. 2005), pp. 2128-2141, ISSN 0018-9219.

Akagi, H., Watanabe, E.H. & Aredes, M. (2007). *Instantaneous Power Theory and Applications to Power Conditioning*. IEEE Press, ISBN 978-0-470-10761-4, Piscataway, New Jersey.

Aredes, M., Hafner, J. & Heumann, K. (1997). Three-phase four-wire shunt active filter control strategies. *IEEE Transactions on Power Electronics*, vol. 12, n° 2, (March 1997), pp. 311-318, ISSN 0885-8993.

Bhattacharya, S., Veltman, A., Divan, D.M. & Lorenz, R.D. (1996). Flux-based active filter controller. *IEEE Transactions on Industry Applications*, vol. 32, n° 3, (May/June 1996), pp. 491-502, ISSN 0093-9994.

Bollen, M.H. (1999). *Understanding Power Quality Problems: Voltage Sags and Interruptions*. Wiley-IEEE Press, ISBN 0-7803-4713-7, Piscataway, New Jersey.

Buso, S., Malesani, L. & Mattavelli P. (1998). Comparison of current control techniques for active filter applications. *IEEE Transactions on Industrial Electronics*, vol. 45, n° 5, (Oct. 1998), pp. 722-729, ISSN 0278-0046.

Cavallini, A. & Montanari, G.R. (1994). Compensation strategies for shunt active filter control. *IEEE Transactions on Power Electronics*, vol. 9, n° 6, (Nov. 1994), pp. 587-593, ISSN 0885-8993.

Chandra, A., Singh, B., Singh, B.N., Al-Haddad, K. (2000). An improved control algorithm of shunt active filter for voltage regulation, harmonic elimination, power factor correction and balancing of nonlinear loads. *IEEE Transactions on Power Electronics*, vol. 15, n° 3, (May/Jun. 2000), pp. 495-507, ISSN 0885-8993.

Chen, B.S., & Joós, G. (2008). Direct power control of active filters with averaged switching frequency regulation. *IEEE Transactions on Power Electronics*, vol. 23, n° 6, (Nov. 2008), pp. 2729-2737, ISSN 0885-8993.

Chen, C.L., Lin, C.E. & Huang, C.L. (1993). Reactive and harmonic current compensation for unbalanced three-phase systems using the synchronous detection method. *Electric Power Systems Research*, vol. 26, n° 3, (Apr. 1993), pp. 163-170, ISSN 0378-7796.

Furuhashi, T., Okuma, S. & Uchikawa, Y. (1990). A study on the theory of instantaneous reactive power. *IEEE Transactions on Industrial Electronics*, vol. 37, n° 1, (Jan./Feb. 1990), pp. 86-90, ISSN 0278-0046.

Hingorani, N.G. & Gyugyi, L. (1999). *Understanding Facts: Concepts and Technology of Flexible AC Transmission Systems*. Wiley-IEEE Press, ISBN 0-7803-3455-8, Piscataway, New Jersey.

Holmes, D.G. & Lipo, T.A. (2003). *Pulse Width Modulation for Power Converters. Principles and Practice*. IEEE Press, ISBN 0-471-20814-0, Piscataway, New Jersey.

Hsu, J.S. (1998). Instantaneous phasor method for obtaining instantaneous balanced fundamental components for power quality control and continuous diagnostics. *IEEE Transactions on Power Delivery*, vol. 13, n° 4, (Oct. 1998), pp. 1494-1500, ISSN 0885-8977.

Komurcugil, H. & Kukrer, O. (2006). A new control strategy for single-phase shunt active power filters using a Lyapunov function. *IEEE Transactions on Industrial Electronics*, vol. 53, n° 1, (Feb. 2006), pp. 305-312, ISSN 0278-0046.

Lascu, C., Asiminoaei, L., Boldea, I. & Blaabjerg, F. (2007). High performance current controller for selective harmonic compensation in active power filters. *IEEE Transactions on Power Electronics*, vol. 22, n° 5, (Sept. 2007), pp. 1826-1835, ISSN 0885-8993.

Lin, B.-R. & Yang, T.-Y. (2004). Three-level voltage-source inverter for shunt active filter. *IEE Proceedings Electric Power Applications*, vol. 151, n° 6, (Nov. 2004), pp. 744-751, ISSN 1350-2352.

Marconi, L., Ronchi, F. & Tilli, A. (2007). Robust nonlinear control of shunt active filters for harmonic current compensation. *Automatica*, vol. 43, n°2, (Feb. 2007), pp. 252-263, ISSN 0005-1098.

Mendalek, N., Al-Haddad, K., Fnaiech, F. & Dessaint, L.A. (2003). Nonlinear control technique to enhance dynamic performance of a shunt active power filter. *IEE Proceedings Electric Power Applications*, vol. 150, n° 4, (July 2003), pp. 373-379, ISSN 1350-2352.

Montero, M.I.M., Cadaval, E.R. & González, F.B. (2007). Comparison of control strategies for shunt active power filters in three-phase four-wire systems. *IEEE Transactions on Power Electronics*, vol. 22, n° 1, (Jan. 2007), pp. 229-236, ISSN 0885-8993.

Morán, L.A., Dixon, J.W. & Wallace, R.R. (1995). A three-phase active power filter operating with fixed switching frequency for reactive power and current harmonic compensation. *IEEE Transactions on Industrial Electronics*, vol. 42, n° 4, (Aug. 1995), pp. 402-408, ISSN 0278-0046.

Newman, M.J., Zmood, D.N. & Holmes, D.G. (2002). Stationary frame harmonic reference generation for active filter systems. *IEEE Transactions on Industry Applications*. vol. 38, n° 6, (Nov./Dec. 2002), pp. 1591-1599, ISSN 0093-9994.

Routimo, M., Salo, M. & Tuusa, H. (2007). Comparison of voltage-source and current-source shunt active power filters. *IEEE Transactions on Power Electronics*, vol. 22, n° 2, (March 2007), pp. 636-643, ISSN 0885-8993.

Saetieo, S., Devaraj, R. & Torrey, D.A. (1995). The design and implementation of a three-phase active power filter based on sliding mode control. *IEEE Transactions on Industry Applications*, vol. 31, n° 5, (Sept./Oct. 1995), pp. 993-1000, ISSN 0093-9994.

Thomas, D.W.P. & Woolfson, M.S., (2001). Evaluation of frequency tracking methods. *IEEE Transactions on Power Delivery*, vol. 16, n° 3, (July 2001), pp. 367-371, ISSN 0885-8977.

Zeng, J., Yu, C., Qi, Q., Yan, Z., Ni, Y., Zhang, B.L., Chen, S. & Wu, F.F. (2004). A novel hysteresis current control for active power filter with constant frequency. *Electric Power Systems Research*, vol. 68, (2004), pp. 75-82, ISSN 0378-7796.

10

Power Quality Enhancement using Predictive Controlled Multilevel Converters

João Dionísio Simões Barros[1] and José Fernando Alves da Silva[2]
[1]*University of Madeira,*
[2]*Instituto Superior Técnico, Technical University of Lisbon*
Portugal

1. Introduction

Electrical power is one of the factors that most influences the economic development of our society. Since the beginning of the use of electricity the continuous improvement of generation, distribution and use of electricity tries to satisfy the ever increasing quality and performance needs of most sectors of human activity.

The power quality of the electrical network is mainly related to the technical characteristics of the voltage waveform at a given point of the electrical power network. The quality of the AC voltage can be evaluated, considering that electrical power networks should provide constant frequency constant amplitude pure sinusoidal voltages, with 120° phase shift between phases. Voltage deviations from these characteristics of the sinusoidal waveforms, imply a loss of power quality (Moreno-Muñoz, 2007), (Tan et al., 2005).

To mitigate some power quality issues, dynamic adjustable solutions based on electronic power converters have been developed. These power converters must use fully controlled (turn-on and turn-off) power semiconductor switching devices to reduce the harmonic content of currents in the electrical power network. These converters can perform as active power filters (APF) (Singh et al., 1999), to cancel disturbing current or voltage harmonics, or Unity Power Factor Rectifiers (UPFR) to obtain almost sinusoidal and in phase AC currents.

The dynamic voltage restorer (DVR) is an advantageous solution to mitigate power quality problems such as sags, ripple, flicker, transients or swells, and to reduce the susceptibility of loads (or people) sensitive to them.

This chapter presents the optimal predictive control of power converters applied to power quality enhancement procedures. A three-level, three-phase neutral point clamped multilevel (NPC) converter, useful in medium voltage applications, is considered. The predictive optimal controllers and fundamental frequency synchronizers are designed, and applied to control the NPC converter AC currents and to balance the DC capacitor voltages in applications such as:

1. Active Power Filters (APF)
2. Unity Power Factor Rectifiers (UPFR)
3. High Quality Dynamic Voltage Restorers (DVR)

Simulation and experimental results show that the AC currents and voltages are almost sinusoidal (THD less than 1%) in steady state operation, but also when facing balanced and unbalanced sags, distorted network voltages and short interruptions with unbalanced loads, or high level of non-linear distorted currents.

2. Optimal predictive control for power converters

Multilevel converters are designed to inherently share their usually DC high voltage power supply evenly between cascaded power semiconductors. Therefore, they are suitable for high voltage, high power applications despite DC voltage balancing problems and complex modulation, compared to two-level converters (Holmes & Lipo, 2003).
There are some available multilevel converter topologies, being the Neutral Point Clamped (NPC) (Nabae & Takahashi, 1981), Flying Capacitor (Meynard & Foch, 1992), and Cascaded H-bridge (Marchesoni et al., 1988) the most reported in literature (Holmes & Lipo, 2003).
Pulse-width modulation (PWM) and space vector modulation (SVM) are the most common control techniques used in multilevel power converters (Holmes & Lipo, 2003), even though the controlled outputs are often affected by power semiconductor switching times. Correctly designed control methods for multilevel converters based on hysteresis comparators, or sliding mode approaches, are robust to component mismatches, semiconductor switching times, present zero steady-state error, but need variable switching frequency, and DC voltages both usually higher than SVM and PWM to obtain better performance (Kazmierkowski & Malesani, 1998).
Optimal predictive controllers are based on the linear optimal control systems theory. Their aim is to solve the minimization problem of a cost functional. Therefore, predictive controllers can be tailored to minimize system output errors, DC voltage, and/or the switching frequency, at a given sampling frequency, being suitable to simultaneously control currents and voltages with non-linear and coupled dynamics (Kwakernaak, & Sivan, 1972).

2.1 Dynamic modeling of NPC multilevel converters

The dynamic models of the NPC multilevel converters are here derived and used to design the AC currents and DC voltages optimal predictive controllers. From the three-phase NPC 27 output voltage vectors are available. The optimal predictive controllers choose the best output voltage vector that minimizes the AC line alfa-beta (αβ) current errors, the total DC voltage regulation error and the unbalancing of partial DC capacitor voltages, U_{C1} and U_{C2}.
The DC side of a Neutral-Point-Clamped multilevel converter (Fig. 1) includes a DC source U_{dc} and DC voltage dividing capacitors C_1 and C_2. The switching variables γ_k represent the state of the multilevel converter active switches, S_{kj}, with $k \in \{1, 2, 3\}$ and $j \in \{1, 2, 3, 4\}$. Assuming ideal semiconductors, the three combinations of the binary states of the switches S_{kj} of each leg k can be defined as:

$$\gamma_k = \begin{cases} 1 & (S_{k1}=1 \wedge S_{k2}=1) \wedge (S_{k3}=0 \wedge S_{k4}=0) \\ 0 & (S_{k1}=0 \wedge S_{k2}=1) \wedge (S_{k3}=1 \wedge S_{k4}=0) \\ -1 & (S_{k1}=0 \wedge S_{k2}=0) \wedge (S_{k3}=1 \wedge S_{k4}=1) \end{cases}. \quad (1)$$

Assuming that the optimum predictive controller balances the capacitor voltages U_{C1} and U_{C2}, so that $U_{C1} \approx U_{C2} \approx U_{dc}/2$, the voltage u_{mk} from each leg to the neutral point of the multilevel converter is:

$$u_{mk} = \gamma_k \frac{U_{dc}}{2}. \quad (2)$$

The NPC converter AC side is connected to the electrical power network voltages U_{L1}, U_{L2}, U_{L3} trought three-phase inductors, L, with losses represented by resistor, R.

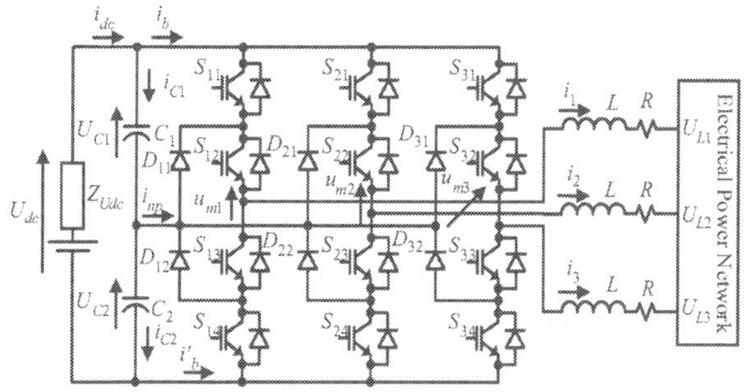

Fig. 1. Neutral-point-clamped multilevel converter circuit

2.1.1 Switched state-space multilevel converter model

Applying Kirchhoff laws to the NPC converter circuit (Fig. 1) and doing some algebraic manipulations, the dynamic equations of the AC currents, i_1, i_2, and i_3, and the capacitor voltages, U_{C1} and U_{C2}, are obtained as functions of the circuit parameters and switching variables (3),

$$\begin{bmatrix} \frac{di_1}{dt} \\ \frac{di_2}{dt} \\ \frac{di_3}{dt} \\ \frac{dU_{C1}}{dt} \\ \frac{dU_{C2}}{dt} \end{bmatrix} = \begin{bmatrix} -\frac{R}{L} & 0 & 0 & \frac{\Xi_{11}}{L} & \frac{\Xi_{12}}{L} \\ 0 & -\frac{R}{L} & 0 & \frac{\Xi_{21}}{L} & \frac{\Xi_{22}}{L} \\ 0 & 0 & -\frac{R}{L} & \frac{\Xi_{31}}{L} & \frac{\Xi_{32}}{L} \\ -\frac{\Gamma_{11}}{C_1} & -\frac{\Gamma_{12}}{C_1} & -\frac{\Gamma_{13}}{C_1} & 0 & 0 \\ -\frac{\Gamma_{21}}{C_2} & -\frac{\Gamma_{22}}{C_2} & -\frac{\Gamma_{23}}{C_2} & 0 & 0 \end{bmatrix} \begin{bmatrix} i_1 \\ i_2 \\ i_3 \\ U_{C1} \\ U_{C2} \end{bmatrix} + \begin{bmatrix} -\frac{1}{L} & 0 & 0 & 0 \\ 0 & -\frac{1}{L} & 0 & 0 \\ 0 & 0 & -\frac{1}{L} & 0 \\ 0 & 0 & 0 & \frac{1}{C_1} \\ 0 & 0 & 0 & \frac{1}{C_2} \end{bmatrix} \begin{bmatrix} U_{L1} \\ U_{L2} \\ U_{L3} \\ i_{dc} \end{bmatrix} \tag{3}$$

where

$$\Gamma_{1k} = \frac{\gamma_k(\gamma_k+1)}{2};\ \Gamma_{2k} = \frac{\gamma_k(1-\gamma_k)}{2};\ \Xi_{ki} = \frac{1}{3}\left[2\Gamma_{ik} - \sum_{\substack{j=1 \\ j\neq k}}^{3}\Gamma_{ij}\right]. \tag{4}$$

For AC networks with isolated neutral (zero homopolar current), these equations can be simplified using αβ coordinates.

2.1.2 Switched state-space multilevel system model in αβ coordinates

The relationship between the variables $\mathbf{X}_{123}$ represented in system coordinates and in αβ coordinates (Jones, 1967), $\mathbf{X}_{\alpha\beta}$, is ($\mathbf{C}$ is the Clarke-Concordia transformation matrix)

$$\mathbf{X}_{123} = \mathbf{C}\mathbf{X}_{\alpha\beta 0}\ ;\ \mathbf{C} = \sqrt{\frac{2}{3}} \begin{bmatrix} 1 & 0 & \frac{\sqrt{2}}{2} \\ -\frac{1}{2} & \frac{\sqrt{3}}{2} & \frac{\sqrt{2}}{2} \\ -\frac{1}{2} & -\frac{\sqrt{3}}{2} & \frac{\sqrt{2}}{2} \end{bmatrix}. \tag{5}$$

Applying (5) to the model (3), and assuming zero homopolar current the multilevel model (6) in $\alpha\beta$ coordinates is derived.

$$\begin{bmatrix} \frac{di_\alpha}{dt} \\ \frac{di_\beta}{dt} \\ \frac{dU_{C1}}{dt} \\ \frac{dU_{C2}}{dt} \end{bmatrix} = \begin{bmatrix} -\frac{R}{L} & 0 & \frac{\Gamma_{1\alpha}}{L} & \frac{\Gamma_{2\alpha}}{L} \\ 0 & -\frac{R}{L} & \frac{\Gamma_{1\beta}}{L} & \frac{\Gamma_{2\beta}}{L} \\ -\frac{\Gamma_{1\alpha}}{C_1} & -\frac{\Gamma_{1\beta}}{C_1} & 0 & 0 \\ -\frac{\Gamma_{2\alpha}}{C_2} & -\frac{\Gamma_{2\beta}}{C_2} & 0 & 0 \end{bmatrix} \begin{bmatrix} i_\alpha \\ i_\beta \\ U_{C1} \\ U_{C2} \end{bmatrix} + \begin{bmatrix} -\frac{1}{L} & 0 & 0 \\ 0 & -\frac{1}{L} & 0 \\ 0 & 0 & \frac{1}{C_1} \\ 0 & 0 & \frac{1}{C_2} \end{bmatrix} \begin{bmatrix} U_{L\alpha} \\ U_{L\beta} \\ i_{dc} \end{bmatrix} \tag{6}$$

Where

$$\Gamma_{i\alpha} = \sqrt{\frac{2}{3}}\left(\Gamma_{i1} - \frac{\Gamma_{i2}}{2} - \frac{\Gamma_{i3}}{2}\right);\ \Gamma_{i\beta} = \sqrt{\frac{2}{3}}\left(\frac{\sqrt{3}}{2}\Gamma_{i2} - \frac{\sqrt{3}}{2}\Gamma_{i3}\right). \tag{7}$$

This model is suitable to design the optimal predictive current controller, and the 27 possible combinations of $\Gamma_{i\alpha}$ and $\Gamma_{i\beta}$ values can be used to plot the 27 vectors of Fig. 2. (Values of γ_1, γ_2, and γ_3 are inside brackets).

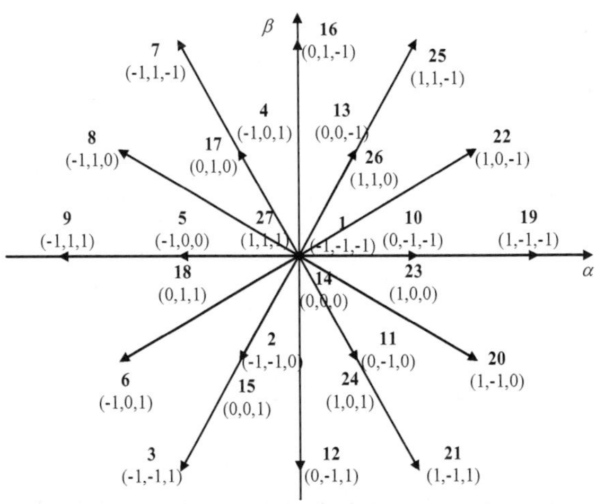

Fig. 2. Available voltage vectors at the output of the multilevel converter

2.1.3 Switched state-space multilevel system model in dq coordinates

Variables $\mathbf{X}_{\alpha\beta}$ in αβ coordinates are expressed in dq coordinates, $\mathbf{X}_{dq}$, using the Park transformation (Jones, 1967):

$$\mathbf{X}_{\alpha\beta} = \mathbf{D}\mathbf{X}_{dq}\,;\, \mathbf{D} = \begin{bmatrix} \cos\theta & -\sin\theta \\ \sin\theta & \cos\theta \end{bmatrix}. \tag{8}$$

The argument $\theta=\omega t+\theta_i$ is the angular phase of the electrical network AC line voltage with angular frequency ω and initial phase θ_i. Applying the Park transformation (8) to (6), the NPC converter model (9) in dq coordinates is obtained.

$$\begin{bmatrix} \frac{di_d}{dt} \\ \frac{di_q}{dt} \\ \frac{dU_{C1}}{dt} \\ \frac{dU_{C2}}{dt} \end{bmatrix} = \begin{bmatrix} -\frac{R}{L} & \omega & \frac{\Gamma_{1d}}{L} & \frac{\Gamma_{2d}}{L} \\ -\omega & -\frac{R}{L} & \frac{\Gamma_{1q}}{L} & \frac{\Gamma_{2q}}{L} \\ -\frac{\Gamma_{1d}}{C_1} & -\frac{\Gamma_{1q}}{C_1} & 0 & 0 \\ -\frac{\Gamma_{2d}}{C_2} & -\frac{\Gamma_{2q}}{C_2} & 0 & 0 \end{bmatrix} \begin{bmatrix} i_d \\ i_q \\ U_{C1} \\ U_{C2} \end{bmatrix} + \begin{bmatrix} -\frac{1}{L} & 0 & 0 \\ 0 & -\frac{1}{L} & 0 \\ 0 & 0 & \frac{1}{C_1} \\ 0 & 0 & \frac{1}{C_2} \end{bmatrix} \begin{bmatrix} U_{Ld} \\ U_{Lq} \\ i_{dc} \end{bmatrix} \tag{9}$$

Where

$$\begin{bmatrix} \Gamma_{id} \\ \Gamma_{iq} \end{bmatrix} = \begin{bmatrix} \cos\theta & \sin\theta \\ -\sin\theta & \cos\theta \end{bmatrix} \begin{bmatrix} \Gamma_{i\alpha} \\ \Gamma_{i\beta} \end{bmatrix};\ \text{with } i \in \{1, 2\}. \tag{10}$$

This quasi-linear model of the state space variables is suitable to design non-optimum but simple linear controllers to control the multilevel converter.

2.2 Optimal predictive controller design

The obtained αβ NPC multilevel converter dynamic model (6) must be solved to estimate (predict) the state variables values at the next sampling period, for all the 27 available vectors of the NPC converter.

Using the predicted values and sampling the AC output currents and DC capacitor voltages, control errors are calculated for the next sampling time, using linearized models and considering the application of every one of the 27 output voltage vectors. Then, a suitable quadratic weighed cost functional is defined and evaluated to choose the voltage vector that minimizes the AC current tracking errors, the DC voltage steady-state error and the input DC capacitor voltage unbalancing.

2.2.1 Predictive equations for AC currents and DC capacitor voltages

From the decoupled model (6) the solution of the AC currents, i_x, with $x \in \{\alpha,\beta\}$, are found to be (Kwakernaak & Sivan, 1972)

$$i_x(t_{s+1}) \approx i_x(t_s) - \frac{R}{L}\Delta T i_x(t_s) + \frac{\Gamma_{1x}}{L}\Delta T U_{C1}(t_s) + \frac{\Gamma_{2x}}{L}\Delta T U_{C2}(t_s) - \frac{\Delta T}{L} U_{Lx}(t_s). \tag{11}$$

Where $i_x(t_s)$ are the currents $i_x(t)$ at the sampling instant $t_s = k\Delta T$ and $i_x(t_{s+1}) = i_x(t = (k+1)\,\Delta T)$ are the currents to be predicted for the next $(k+1)$ sampling interval ΔT. Assuming a sampling time ΔT small enough, $U_{C1}(t_s)$, $U_{C2}(t_s)$, and $U_{Lx}(t_s)$ can all be considered nearly constant during ΔT (or its change can be further estimated). In order to predict the capacitor voltage difference, $U_{C1}(t)$-$U_{C2}(t)$, from the dynamic equations (6), the solution of the capacitor voltages are (Kwakernaak & Sivan, 1972)

$$U_{Ci}\left(t_{s+1}\right) \approx U_{Ci}\left(t_s\right) - \frac{\Gamma_{i\alpha}}{C_i}\Delta T i_\alpha\left(t_s\right) - \frac{\Gamma_{i\beta}}{C_i}\Delta T i_\beta\left(t_s\right) + \frac{\Delta T}{C_i} i_{dc}\left(t_s\right);\ \text{with } i \in \{1, 2\}. \quad (12)$$

Where $U_{Ci}(t_s)$ are the sampled capacitor voltages at $t = t_s = k\Delta T$. Considering $C_1 \approx C_2 \approx C$ and ΔT small enough so that the DC current, $i_{dc}(t)$, is nearly constant and assuming that the AC line currents follow their references, the capacitors voltage error is:

$$U_{C1}\left(t_{s+1}\right) - U_{C2}\left(t_{s+1}\right) \approx U_{C1}\left(t_s\right) - U_{C2}\left(t_s\right) + \left(\Gamma_{2\alpha} - \Gamma_{1\alpha}\right)\frac{\Delta T}{C} i_\alpha\left(t_s\right) + \left(\Gamma_{2\beta} - \Gamma_{1\beta}\right)\frac{\Delta T}{C} i_{L\beta}\left(t_s\right). \quad (13)$$

2.2.2 Quadratic cost functional definition

A three-phase NPC converter operated as a current source has three variables to control, $i_\alpha(t)$, $i_\beta(t)$, and $U_{C1}(t)$-$U_{C2}(t)$ to ensure $U_{C1} \approx U_{C2} \approx U_{dc}/2$. The main objective of the optimizing controller is the minimization of both the AC currents errors and the capacitor voltage difference. Therefore, the quadratic cost functional contains weighted current errors and DC voltage difference (14).

$$C\left(t_{s+1}\right) = \sqrt{\frac{e_\alpha^2\left(t_{s+1}\right)}{\rho_\alpha} + \frac{e_\beta^2\left(t_{s+1}\right)}{\rho_\beta} + \frac{e_{UC}^2\left(t_{s+1}\right)}{\rho_{UC}}} \quad (14)$$

Where

$$\begin{aligned} e_\alpha\left(t_{s+1}\right) &= i_{\alpha Ref}\left(t_{s+1}\right) - i_\alpha\left(t_{s+1}\right); \\ e_\beta\left(t_{s+1}\right) &= i_{\beta Ref}\left(t_{s+1}\right) - i_\beta\left(t_{s+1}\right); \\ e_{UC}\left(t_{s+1}\right) &= U_{C1}\left(t_{s+1}\right) - U_{C2}\left(t_{s+1}\right). \end{aligned} \quad (15)$$

In the cost functional (14) the errors are weighted with the weights ρ_α, ρ_β, and ρ_{UC}. The weights have two purposes: they normalize the distinct errors, which have different units and ranges, with a maximum error; and they define the priority level of each error variable. In (15), $i_{\alpha ref}(t_{s+1})$ and $i_{\beta ref}(t_{s+1})$, are the AC current references one sample time forward, t_s+1, to compensate for the processor calculation delay. To assure adjacent level voltage transition in the multilevel output voltage, thus minimizing the number of semiconductor commutations, only vectors adjacent to the vector in use at t=t_s are analysed, since they need just the switching of two semiconductors per leg. The adjacent vectors verify $|\gamma_k(t_{s+1}) - \gamma_k(t_s)| < 2$, with $k \in \{1, 2, 3\}$, in the 3 legs of the NPC converter.

2.3 Predictive fundamental frequency synchronization

The measurement of the fundamental frequency and phase of the network voltage is crucial in power quality applications. The AC voltage phase detector must be fast to ensure a synchronization frame always in phase with the fundamental frequency of the network AC voltage. To achieve this a robust and fast phase detection/synchronization method together with fundamental frequency generation (for short interruptions) must be implemented. The phase and frequency recognition is hard to obtain from the AC network voltages when sags or interruptions occur. The synchronization must use the network AC waveforms, which can have amplitude and frequency variations, voltage sags/swells, short interruptions and distortion. The synchronization method must rapidly capture the phase of the network AC voltage after a short interruption, it must be robust to amplitude variations, and must handle AC voltages with unbalanced amplitudes.

To provide an optimal solution to the phase detection and fundamental frequency generation, an optimal predictive phase detection and fundamental frequency generation based on the dq Park transformation, is introduced.

The application of the Clarke-Concordia and the Park transformations to ideally balanced network AC voltages with synchronous frame, θ, can provide a zero quadrature, U_q, component. The synchronous frame, θ, can be obtained by finding the optimal value that results in a zero quadrature, U_q, AC voltage component.

To handle temporary total failures on the network AC voltage, the synchronization frame must be generated in a self-running mode at a constant frequency (50 Hz).

Applying the Clarke-Concordia to the network constant amplitude U AC voltages and Park transformations with phase error $\Delta\theta(t_s)$, the direct component, $U_d(t_s)$, and the quadrature component, $U_q(t_s)$, calculated at the present sampling time, $t_s = k\Delta T$, are (Barros & Silva, 2010):

$$U_d(t_s) = \sqrt{\frac{3}{2}} U \cos\Delta\theta(t_s); \; U_q(t_s) = -\sqrt{\frac{3}{2}} U \sin\Delta\theta(t_s). \tag{16}$$

The phase error, $\Delta\theta(t_s)$, can be derived from the measured direct $U_d(t_s)$, and quadrature, $U_q(t_s)$, AC voltage components. At the next sampling time $t_{s+1} = (k+1)\Delta T$ the fundamental frequency has a phase variation of $\omega\Delta T$. Thus, the prediction of the phase error $\Delta\theta(t_{s+1})$ is

$$\Delta\theta(t_{s+1}) = \Delta\theta(t_s) - \omega\Delta T. \tag{17}$$

The prediction of the synchronous frame at the next sapling time, $\theta(t_{s+1})$, is obtained by subtracting the predicted phase error $\Delta\theta(t_{s+1})$ to the actual synchronous frame, $\theta(t_s)$,

$$\theta(t_{s+1}) = \theta(t_s) - \Delta\theta(t_{s+1}). \tag{18}$$

The Fig. 3 shows the predicted vector $\theta(t_{s+1})$ obtained from the phase error, $\Delta\theta(t_s)$, and the actual synchronous frame, $\theta(t_s)$, vectors.

The feedback control loop to obtain a synchronous frame, $\theta(t_{s+1})$, (Fig. 4) calculates the Clarke-Concordia and Park transformations to obtain the quadrature voltage component, $U_q(t_s)$, which is compared to its reference value, $U_{q\mathrm{Ref}} = 0$. The $U_q(t_s)$ and $U_d(t_s)$ AC voltage components values feed an optimal predictive phase quadrature, (Barros & Silva, 2010), compensator that computes the phase error $\Delta\theta(t_s)$, using equations (16), generating the synchronous phase $\theta(t_{s+1})$, to obtain a quadrature voltage component, U_q, that ensures $U_{q\mathrm{Ref}} = 0$.

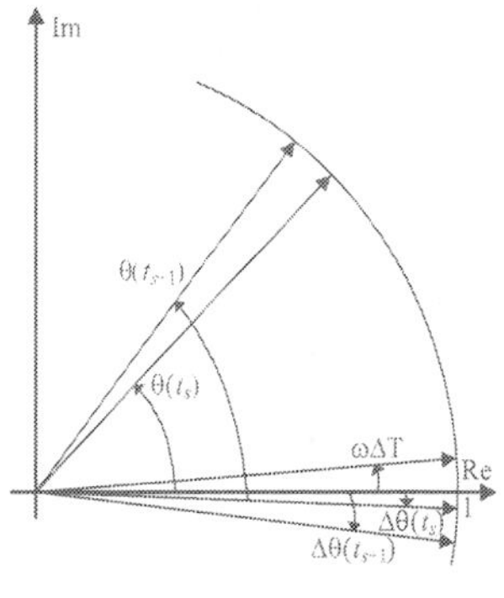

Fig. 3. Predictive synchronous frame vectors

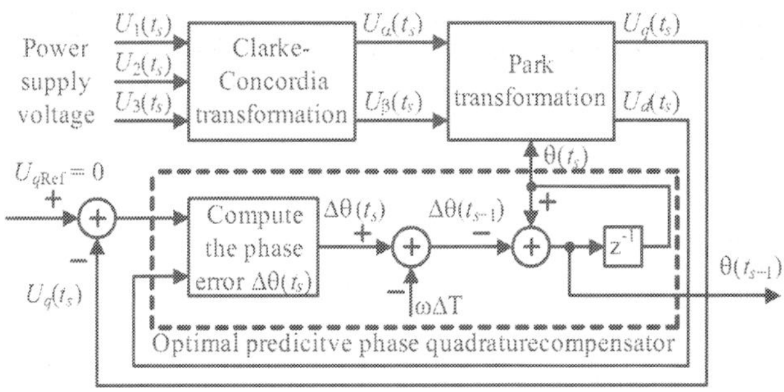

Fig. 4. Principle of optimal predictive phase and frequency detection based on dq Park transformation

Using the values of the voltage components $U_d(t_s)$ and $U_q(t_s)$ the optimal predictive phase quadrature compensator (Barros & Silva, 2010) predicts the optimal value of $\theta(t_{s+1})$ ($\cos[\theta(t_{s+1})]$ and $\sin[\theta(t_{s+1})]$) that minimizes the quadrature voltage component, U_q, of the network AC voltage, at the next sampling time $t_{s+1} = (k+1)\Delta T$,

$$\begin{aligned} \cos\left[\theta(t_{s+1})\right] &= \cos\left[\theta(t_s)\right]\cos\left[\Delta\theta(t_{s+1})\right] + \sin\left[\theta(t_s)\right]\sin\left[\Delta\theta(t_{s+1})\right], \\ \sin\left[\theta(t_{s+1})\right] &= \sin\left[\theta(t_s)\right]\cos\left[\Delta\theta(t_{s+1})\right] - \cos\left[\theta(t_s)\right]\sin\left[\Delta\theta(t_{s+1})\right]. \end{aligned} \tag{19}$$

Where the phase error $\Delta\theta(t_{s+1})$, (17), at the next sampling time (and $\cos[\Delta\theta(t_{s+1})]$, $\sin[\Delta\theta(t_{s+1})]$) depends on the actual phase shift error $\Delta\theta(t_s)$ (as well as $\cos[\Delta\theta(t_s)]$, $\sin[\Delta\theta(t_s)]$). The actual phase shift error $\Delta\theta(t_s)$ can be related to the direct voltage component $U_d(t_s)$ (16), the quadrature voltage component, $U_q(t_s)$ (16), and the fundamental frequency phase variation $\omega\Delta T$ (Barros & Silva, 2010), as follows

$$\begin{aligned} \cos\left[\Delta\theta(t_{s+1})\right] &= \cos\Delta\theta(t_s)\cos(\omega\Delta T) + \sin\Delta\theta(t_s)\sin(\omega\Delta T), \\ \sin\left[\Delta\theta(t_{s+1})\right] &= \sin\Delta\theta(t_s)\cos(\omega\Delta T) - \cos\Delta\theta(t_s)\sin(\omega\Delta T). \end{aligned} \tag{20}$$

To handle the network AC voltage short interruptions, the synchronization frame must be generated in a self-running mode at a constant frequency (50 Hz) (Barros & Silva, 2010). The prediction of the self-running synchronous frame at the next sapling time, $\theta(t_{s+1})$, is

$$\theta(t_{s+1}) = \theta(t_s) + \omega\Delta T \,. \tag{21}$$

The optimal predictive phase quadrature compensator can be designed for unbalanced AC voltages, which is especially useful to detect the phase and frequency during unbalanced sags (Barros & Silva, 2010). Considering U_1, U_2, and U_3 the magnitudes of the unbalanced network AC voltage and applying the Clarke-Concordia and the Park transformations with synchronous frame, θ, the quadrature, $U_q(t_s)$, component (Barros & Silva, 2010), is

$$U_q(t_s) = \sqrt{\frac{1}{6}}\left(-\frac{\sqrt{3}}{2}U_2 + \frac{\sqrt{3}}{2}U_3\right)\cos 2\theta(t_s) + \sqrt{\frac{1}{6}}\left(-U_1 + \frac{U_2}{2} + \frac{U_3}{2}\right)\sin 2\theta(t_s)\,. \tag{22}$$

The quadrature, $U_q(t_s)$, component has second harmonic terms, $U_{2q}(t_s)$. The DC term, $U_{0q}(t_s)$ should be zero and can be obtained by subtracting the second harmonics term, $U_{2q}(t_s)$, to the quadrature component,

$$\begin{aligned} U_{0q}(t_s) &= U_q(t_s) - U_{2q}(t_s) = U_q(t_s) - \\ &-\sqrt{\frac{1}{6}}\left(-\frac{\sqrt{3}}{2}U_2 + \frac{\sqrt{3}}{2}U_3\right)\cos 2\theta(t_s) - \\ &-\sqrt{\frac{1}{6}}\left(-U_1 + \frac{U_2}{2} + \frac{U_3}{2}\right)\sin 2\theta(t_s) \\ &= 0 \end{aligned} \tag{23}$$

The optimal predictive phase quadrature compensator for unbalanced AC network voltages is designed to obtain the synchronous frame, θ, which results in zero DC quadrature component, U_{0q}, in a way similarly to the balanced AC network voltage fundamental frequency synchronization.

2.4 Evaluation of NPC multilevel converter predictive current controllers

To evaluate the performance of the optimal predictive AC current control algorithms and the capacitor voltage balancing in NPC converters, a DC source, U_{dc} (120 V) with internal resistance 0.2 Ω, is connected in the DC side of the NPC converter (Fig. 1), the currents references, $i_{\alpha Ref}$ and $i_{\beta Ref}$, are obtained from DSP based lookup tables (i_{ac} = 4 A), U_L is obtained from 230/400V through a 400/230V transformer, and the following parameters were used: C_1 = 4.4 mF, 200 V, C_2 = 4.4 mF, 200 V, L = 15.1 mH, R = 0.1 Ω, f_{ac} = 50 Hz (T = 0.02 s and ω = 314.2 rad/s), $\rho_\alpha = 0.09$ A^2, $\rho_\beta = 0.09$ A^2, and $\rho_{UC} = 0.04$ V^2.
Experimental results of the AC currents, i_1, i_2, and i_3, with ΔT=28 μs and steady state operation (Fig. 5), show that they are nearly sinusoidal with very small ripple factor (less than 3%), and a total harmonic distortion (THD) lower than 1%. Applying a positive step (1 A to 4 A) in the AC current references (Fig. 5, at 5 ms) and a negative step (4 A to 1 A), at 45 ms the results show a fast dynamic response, with no overshoot, and no stationary error.

The optimal predictive controller also chooses the optimal vector that minimizes the capacitor voltage errors and these voltages are balanced (Fig. 6), with a mean error lower than 1%. Near perfect capacitor DC voltage balancing is obtained, while reducing AC current harmonic distortion.

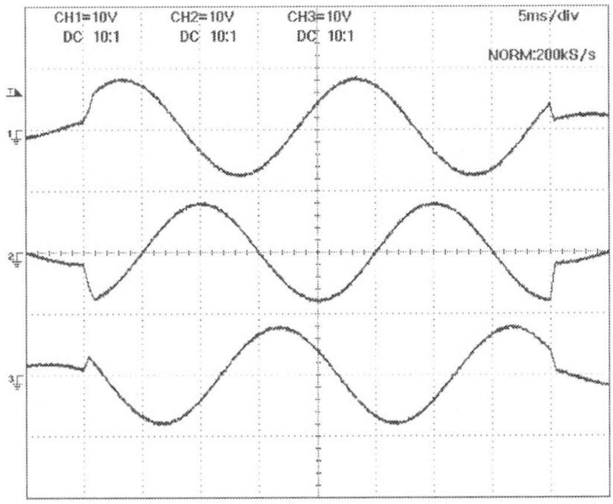

Fig. 5. Experimental results of sinusoidal AC currents, i_1, i_2, and i_3, with an amplitude step. i_1 has a displacement of 2 divisions and i_3 a displacement of -2 divisions (vertical - 5 A/Div and horizontal - 5 ms/Div)

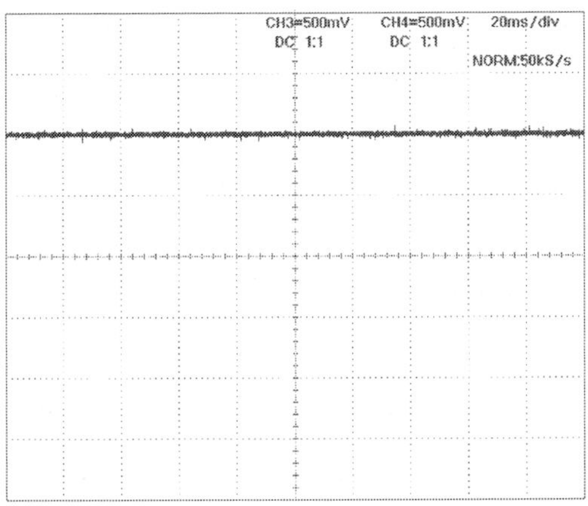

Fig. 6. Capacitor voltages, U_{C1} and U_{C2} (vertical - 10 V/Div and horizontal - 20 ms/Div)

The optimal predictive phase quadrature synchronization controller is tested during a short electrical network voltage interruption. The experimental results (Fig. 7a) show the electrical network voltage short interruption, U_{L1}, U_{L2}, and U_{L3}. The Fig. 7b show the phase θ of the optimal predictive synchronizer and a binary signal signalling that the electrical network voltages level voltage is lower than the minimum limit, $U_{dq\min}$. The results show that the optimal predictive synchronizer is able to internally generate without interruptions a phase signal, at the fundamental frequency (50 Hz).

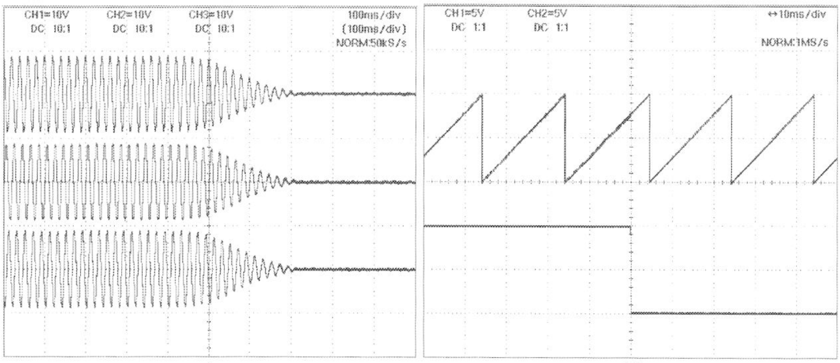

a) b)

Fig. 7. a) The electrical power network voltages, U_{L1} + 740 V, U_{L2} and U_{L3} - 740 V. b) The phase θ and the signal showing the electrical power network voltages interruption (vertical 370 V/Div; π/Div and horizontal 100 ms/Div; 10 ms/Div)

3. Power quality applications of predictive controlled multilevel converters

Using the optimal predictive controllers and multilevel converters, power quality enhancing technologies are proposed, such as active power filters (APF) with reactive power compensation, unity power factor rectifiers (UPFR), and dynamic voltage restorers (DVR), which are tested using both MATLAB/SIMULINK simulations and a digital signal processor based laboratory prototype.

3.1 Active power filter

The NPC multilevel, connected as a shunt APF (Fig. 8), can be controlled to compensate the power factor and the AC line i_{Lk} current harmonics introduced by non-linear loads such as three-phase bridge diode rectifier (Fig. 8). The NPC multilevel AC side is shunt connected to the electrical network voltages U_{L1}, U_{L2}, and U_{L3} using three inductors, L, with resistive R losses. The APF injects currents to cancel the harmonic components of the non-linear load currents. The DC voltage, U_{dc}, must be regulated and the capacitor voltages, U_{C1} and U_{C2}, balanced.

The NPC converter APF model is obtained and the optimal predictive control of the electrical power network AC currents is derived from the principles outlined in section 2.1. Proportional integral (PI) type regulators, optimal predictive, sliding-mode and μ-synthesis are control methodologies proposed to regulate the DC capacitor voltage U_{dc}.

3.1.1 Switched state-space model of the multilevel APF

Applying Kirchhoff laws to the multilevel converter APF (Fig. 8), the Clarke-Concordia transformation (5) and the Park transformation (8) the dynamic equations (24) of the AC line currents, i_{Ld}, and i_{Lq}, and the capacitor voltages, U_{C1} and U_{C2}, can be written in function of the circuit parameters, as in 2.1.

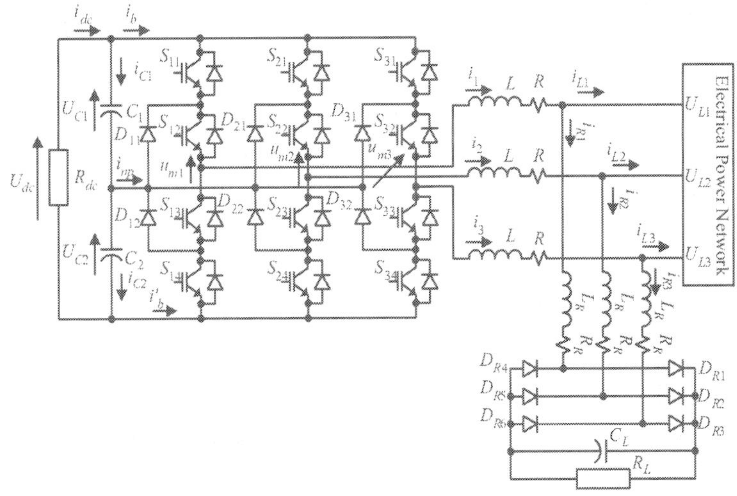

Fig. 8. NPC multilevel converter circuit, working as an active power filter

$$
\begin{bmatrix} \frac{di_{Ld}}{dt} \\ \frac{di_{Lq}}{dt} \\ \frac{dU_{C1}}{dt} \\ \frac{dU_{C2}}{dt} \end{bmatrix} = \begin{bmatrix} -\frac{R}{L} & \omega & \frac{\Gamma_{1d}}{L} & \frac{\Gamma_{2d}}{L} \\ -\omega & -\frac{R}{L} & \frac{\Gamma_{1q}}{L} & \frac{\Gamma_{2q}}{L} \\ -\frac{\Gamma_{1d}}{C_1} & -\frac{\Gamma_{1q}}{C_1} & 0 & 0 \\ -\frac{\Gamma_{2d}}{C_2} & -\frac{\Gamma_{2q}}{C_2} & 0 & 0 \end{bmatrix} \begin{bmatrix} i_{Ld} \\ i_{Lq} \\ U_{C1} \\ U_{C2} \end{bmatrix} + \begin{bmatrix} -\frac{1}{L} & 0 & 0 \\ 0 & -\frac{1}{L} & 0 \\ 0 & 0 & \frac{1}{C_1} \\ 0 & 0 & \frac{1}{C_2} \end{bmatrix} \begin{bmatrix} U_{Ld} \\ U_{Lq} \\ i_{dc} \end{bmatrix} + \\
+ \begin{bmatrix} -\frac{R}{L} & \omega & -1 & 0 \\ -\omega & -\frac{R}{L} & 0 & -1 \\ -\frac{\Gamma_{1d}}{C_1} & -\frac{\Gamma_{1q}}{C_1} & 0 & 0 \\ -\frac{\Gamma_{2d}}{C_2} & -\frac{\Gamma_{2q}}{C_2} & 0 & 0 \end{bmatrix} \begin{bmatrix} i_{Rd} \\ i_{Rq} \\ \frac{di_{Rd}}{dt} \\ \frac{di_{Rq}}{dt} \end{bmatrix} \tag{24}
$$

The DC voltage, U_{dc}, dynamic equation is obtained directly from (24) since $U_{dc} = U_{C1} + U_{C2}$,

$$
\frac{dU_{dc}}{dt} = -\left(\frac{\Gamma_{1d}}{C_1} + \frac{\Gamma_{2d}}{C_2}\right) i_{Ld} - \left(\frac{\Gamma_{1q}}{C_1} + \frac{\Gamma_{2q}}{C_2}\right) i_{Lq} - \\
-\left(\frac{\Gamma_{1d}}{C_1} + \frac{\Gamma_{2d}}{C_2}\right) i_{Rd} - \left(\frac{\Gamma_{1q}}{C_1} + \frac{\Gamma_{2q}}{C_2}\right) i_{Rq} + \left(\frac{1}{C_1} + \frac{1}{C_2}\right) i_{dc} . \tag{25}
$$

This DC voltage quasi-linear model, U_{dc}, is suitable to design a linear predictive controller to regulate the DC voltage of the APF.

3.1.2 Predictive equations for AC line currents and DC capacitor voltages

The AC current and DC voltage predictive optimal controllers are designed to choose the best output voltage vector able to minimize the AC line current, $i_{L\alpha}$ and $i_{L\beta}$ errors, U_{dc} voltage regulation error and the unbalancing of DC capacitor voltages, U_{C1} and U_{C2}. The obtained αβ converter model is solved (like in section 2.2.1) to predict the state variable values at the next sampling period, for all the 27 available vectors.

$$\begin{aligned} i_{Lx}(t_{s+1}) \approx\ & i_{Lx}(t_s) + i_{Rx}(t_s) - i_{Rx}(t_{s+1}) - \frac{R}{L}\Delta T i_{Lx}(t_s) - \frac{R}{L}\Delta T i_{Rx}(t_s) + \\ & + \frac{\Gamma_{1x}}{L}\Delta T U_{C1}(t_s) + \frac{\Gamma_{2x}}{L}\Delta T U_{C2}(t_s) - \frac{\Delta T}{L} U_{Lx}(t_s) \end{aligned} \tag{26}$$

$$\begin{aligned} U_{C1}(t_{s+1}) - U_{C2}(t_{s+1}) \approx\ & U_{C1}(t_s) - U_{C2}(t_s) + (\Gamma_{2\alpha} - \Gamma_{1\alpha})\frac{\Delta T}{C}\left[i_{L\alpha}(t_s) + i_{R\alpha}(t_s)\right] + \\ & + (\Gamma_{2\beta} - \Gamma_{1\beta})\frac{\Delta T}{C}\left[i_{L\beta}(t_s) + i_{R\beta}(t_s)\right] \end{aligned} \tag{27}$$

3.1.3 Quadratic cost functional definition

In the APF the NPC multilevel converter is operating as a current source, controlling the two AC line currents $i_{L\alpha}(t)$, $i_{L\beta}(t)$, and the capacitive unbalance $U_{C1}(t)$-$U_{C2}(t)$, thus the quadratic cost functional (14) is rewritten to minimize the AC line currents errors, $e_\alpha(t_{s+1}) = i_{L\alpha Ref}(t_{s+1}) - i_{L\alpha}(t_{s+1})$ and $e_\beta(t_{s+1}) = i_{L\beta Ref}(t_{s+1}) - i_{L\beta}(t_{s+1})$, and the capacitor voltage difference, $e_{UC}(t_{s+1}) = U_{C1}(t_{s+1}) - U_{C2}(t_{s+1})$. The $i_{L\alpha Ref}(t_{s+1})$ and $i_{L\beta Ref}(t_{s+1})$, are the AC line current references one sample time forward, t_{s+1}. These current references will be computed to ensure near unity power factor and to regulate the U_{dc} voltage.

3.1.4 Multilevel converter APF optimal predictive controller

The block diagram of the multilevel converter APF with the optimal controller (Fig. 9a) includes as inputs, the states of the multilevel converter switches (all the vectors to be tested), $\gamma_1(t_s)$, $\gamma_2(t_s)$, and $\gamma_3(t_s)$, the sampled capacitor voltages, $U_{C1}(t_s)$ and $U_{C2}(t_s)$, the AC line current references, one sampling time forward, $i_{L\alpha Ref}(t_{s+1})$ and $i_{L\beta Ref}(t_{s+1})$, and the AC currents, $i_{L1}(t_s)$, $i_{L2}(t_s)$, $i_{L3}(t_s)$, $i_{R1}(t_s)$, $i_{R2}(t_s)$, and $i_{R3}(t_s)$, sampled and transformed to αβ coordinates, $i_{L\alpha}(t_s)$, $i_{L\beta}(t_s)$, $i_{R\alpha}(t_s)$, and $i_{R\beta}(t_s)$. The controller makes use of these inputs to compute the optimal vector, in order to apply it to the NPC converter, at the next sampling interval. Adjacent level voltage transition is also ensured.

3.1.5 Active power filter current reference generation and power flow control

The generation of the AC line current references, i_{LdRef} and i_{LqRef}, for the active power filter must assure that the AC line currents are sinusoidal, the power factor is near unity and the DC voltage is regulated.

A linear optimal predictive regulator is here designed to generate references, i_{LdRef} and i_{LqRef}, able to cancel the mean value of the DC voltage error, $e_{Udc}(t) = U_{dcRef}(t) - U_{dc}(t)$, during the time interval T_{Udc} (28).

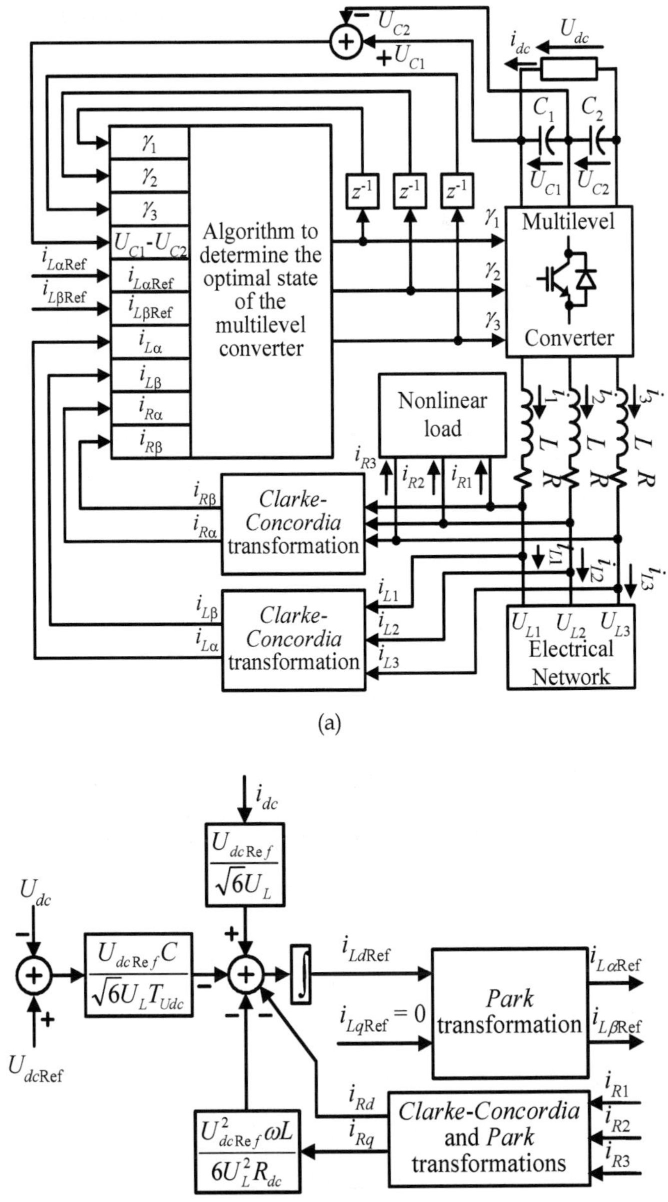

Fig. 9. Feedback system of the multilevel converter APF with predictive optimal controller. (a) NPC multilevel optimal predictive controller. (b) APF current reference generation

$$\overline{e_{Udc}}(t)=\frac{1}{T_{Udc}}\int_{T_{Udc}} e_{Udc}(t)dt=\frac{1}{T_{Udc}}\int_{T_{Udc}} U_{dcRef}(t)-U_{dc}(t)dt=0 \tag{28}$$

To obtain a quasi-unity power factor the AC line current must be in phase with the AC line voltage, forcing i_{LqRef} = 0. Using the dynamic equation of the DC voltage (25) in (28), assuming that the optimal predictive controller forces the i_{Ld}=i_{LdRef} and i_{LqRef} = 0, the optimal predictive control law of i_{LdRef} is:

$$i_{LdRef}(t)\approx-\overline{i_{Rd}}(t)-\frac{U_{dcRef}C}{\sqrt{6}U_L}\frac{\overline{U_{dcRef}}(t)-\overline{U_{dc}}(t)}{T_{Udc}}-\frac{U_{dcRef}^2\omega L}{6U_L^2R_{dc}}\overline{i_{Rq}}(t)+\frac{U_{dcRef}\omega L}{\sqrt{6}U_L}\overline{i_{dc}}(t). \tag{29}$$

Where T_{Udc} is the time interval used to compute the mean value of the DC voltages, U_{dc} and U_{dcRef}, the DC current, i_{dc}, and the non-linear currents, i_{Rd} and i_{Rq}. U_L is the AC line voltage amplitude and ω is the fundamental angular frequency. The time interval T_{Udc} must be much larger than the period, T, of AC line currents (T_{udc}>> T) to maintain the AC line currents sinusoidal, (the time constant $R_{dc}C$ is of no concern since R_{dc} represents the safety discharge resistors of the capacitors, $TU_{dc} << R_{dc}C$). Therefore, the active power drawn from the AC network just compensates the converter losses. In (29) the switching variables, $\Gamma_{1d}+\Gamma_{2d}$ and $\Gamma_{1q}+\Gamma_{2q}$, were obtained similarly as in (Silva, 1999) and are given in (30).

$$\Gamma_{1d}+\Gamma_{2d}\approx\frac{\sqrt{6}U_L}{U_{dcRef}};\ \Gamma_{1q}+\Gamma_{2q}\approx\frac{2\omega LU_{dcRef}}{\sqrt{6}U_LR_{dc}} \tag{30}$$

In the control of the APF, the DC voltages (U_{dc}) and currents (i_{dc}, i_{R1}, i_{R2}, and i_{R3}) are sampled to generate the AC line current references (Fig. 9b), which are applied to the optimal predictive controller of the multilevel converter (Fig. 9a) to force the AC line currents to follow their references.

The regulation of DC capacitor voltage can also be done using other regulators such as (Barros & Silva, 2008): PI-type regulators (31), sliding-mode (32), and μ-synthesis (33).

The DC PI regulator parameters, K_{pUdc} and K_{iUdc}, (31), depend on two system parameters, ω_n and ζ, which characterize the second-order system closed loop behaviour. The damping factor, ζ, is usually chosen with values close to 0.707, to avoid severe overshoots in the step response to the voltage DC reference, U_{dcRef} (or to minimize the integral of time multiplied by absolute error - the ITAE criterion). The natural frequency ω_n should be a much lower frequency than the AC voltage frequency, $\omega_n << \omega$ ($\omega=2\pi50$ rad/s), so that the amplitude of the currents of the electrical power network shows much slower variations than the AC currents frequency (50 Hz).

$$\begin{aligned}
&i_{LdRef}=\left(k_{pUdc}+\frac{k_{iUdc}}{s}\right)\left(U_{dcRef}-U_{dc}\right);\\
&K_{pUdc}=\frac{U_{dcRef}-U_{dcRef}\zeta\omega_nR_{dc}C}{\sqrt{6}U_LR_{dc}};\\
&K_{iUdc}=-\frac{U_{dcRef}\omega_n^2C}{2\sqrt{6}U_L}.
\end{aligned} \tag{31}$$

In the sliding mode control law (32), the time constant β_{eUdc} defines the role of the instantaneous error in the DC voltage control variable, i_{LdRef}. In the APF, the control variables, that regulate the DC voltage, are the network AC current amplitudes, which must be almost sinusoidal and therefore can not have abrupt variations. The time constant β_{eUdc} should be much lower than the integration period ($\beta_{eUdc} << T_{Udc}$) so that AC current amplitudes are slowly time variant.

$$
\begin{aligned}
i_{Ld\mathrm{Re}f}(t) \approx & -\overline{i_{Rd}}(t) - \frac{\beta_{eUdc}}{T_{Udc}} \frac{U_{dc\mathrm{Re}f}C}{\sqrt{6}U_L} \frac{U_{dc\mathrm{Re}f}(t) - U_{dc}(t)}{T_{Udc}} - \\
& - \frac{U_{dc\mathrm{Re}f}C}{\sqrt{6}U_L} \frac{\overline{U_{dc\mathrm{Re}f}}(t) - \overline{U_{dc}}(t)}{T_{Udc}} - \\
& - \frac{U_{dc\mathrm{Re}f}^2 \omega L}{6U_L^2 R_{dc}} \overline{i_{Rq}}(t) + \frac{U_{dc\mathrm{Re}f}}{\sqrt{6}U_L} \overline{i_{dc}}(t)
\end{aligned}
\tag{32}
$$

The overall structure to synthesis the μ-synthesis compensator (33), with the methodology based on structured singular values, having a general model, $P(s)$ and a dynamics of uncertainty, $\Delta(s)$, was implemented in MATLAB with the μ–Synthesis tool. A compensator (33) was obtained to ensure robust stability and robust performance for the APF (Barros et al., 2005). The values of the parameters, components, and uncertainties used to design the compensator are (Barros & Silva, 2008): $C_1 = C_2 = 4.4$ mF, $C_L = 1$ mF, $L = 15$ mH, $L_R = 1$ mH, $R = 0.7\ \Omega$, $R_{dc} = 1\ \mathrm{k}\Omega$, $R_L = 10\ \Omega$, and $R_R = 0.1\ \Omega$.

$$
i_{Ld\mathrm{Re}f} = \frac{-0.097\left(s-1.4\times10^4\right)\left(s+3269\right)\left(s^2+36.5s+548.6\right)}{\left(s+306.8\right)\left(s+2.4\right)\left(s+0.33\right)\left(s^2+275.3s+4.7\times10^4\right)}\left(U_{dc\mathrm{Re}f} - U_{dc}\right)
\tag{33}
$$

3.1.6 Simulation and experimental results

The shunt connected multilevel APF (Fig. 8) is controlled to compensate the load current harmonics and to regulate the power factor at unity. The AC current references of the APF are generated to regulate the mean value of the DC voltage, $U_{dcRef} = 240$ V, the power factor and the AC line currents.

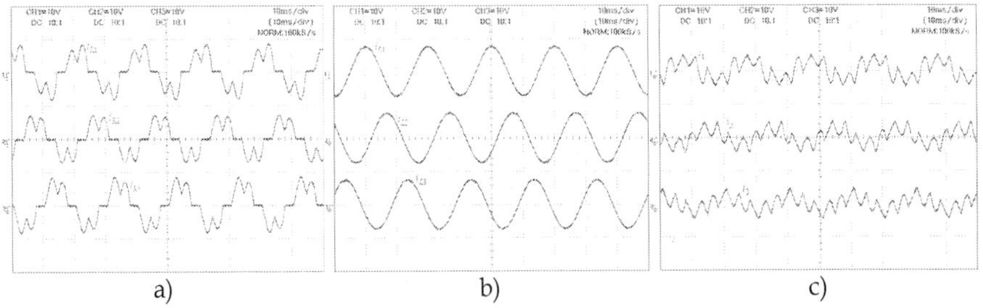

Fig. 10. Experimental results of: a) the non-linear currents, i_{R1}, i_{R2}, and i_{R3}; b) the AC line currents, i_{L1}, i_{L2}, and i_{L3}; c) the output currents, i_1, i_2, and i_3 of the active power filter. i_{R1} , i_{L1}, and i_1 have a displacement of 2 divisions and i_{R3}, i_{L3}, and i_3 a displacement of –2 divisions (vertical - 12 A/Div and horizontal - 10 ms/Div).

The experimental results of the rectifier non-linear currents, i_{R1}, i_{R2}, and i_{R3} (Fig. 10a), show that the THD is about 36% and the power factor is 0.91. The multilevel APF injects currents (Fig. 10c) to cancel the current harmonic components of this non-linear load. The THD of AC line currents is reduced from 36% to nearly 1% (Fig. 10b).
The DC voltage, U_{dc} (Fig. 11a) shows that the DC voltage follows it references (U_{dcRef} = 240 V) with almost no steady state error (0.3%).
Figure 11b shows the experimental results of the AC line currents, $-i_{L1}$ and $-i_{L2}$, simultaneously with the AC line voltages, U_{L1} and U_{L2}. The results indicate that the AC line voltage and the AC line currents are sinusoidal and in phase, showing that the power factor is near unity. The power factor was improved from 0.910 (without APF) to 0.997 (with APF). The optimization of the multilevel current control improves the performance of the active power filter performance, by reducing the total harmonic distortion of AC line currents.

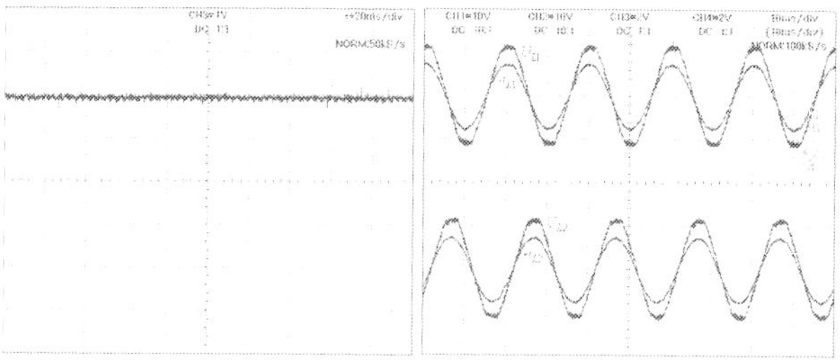

a) b)

Fig. 11. a) Experimental results of the DC voltage, U_{dc}, of the active power filter (vertical - 40 V/Div and horizontal - 20 ms/Div). b) The AC line currents, $-i_{L1}$, and $-i_{L2}$ (vertical - 12 A/Div), and the AC line voltages, U_{L1} and U_{L2} (vertical - 40 V/Div), of the active power filter. i_{L1} and U_{L1} have a displacement of 2 divisions and i_{L2} and U_{L2} a displacement of -2 divisions (horizontal - 10 ms/Div)

3.2 Unity power factor rectifier

The UPFR is based on the three-phase NPC multilevel converter. The optimal predictive controller enforces the AC currents to be almost sinusoidal and in phase with the electrical power network. Therefore, the electrical power network sees the rectifier load almost like a pure resistance avoiding current (and voltage) distortion, which improves power quality.
The multilevel rectifier has two control feedback loops: a slow and external for the control of DC voltage (a PI controller is used) and the internal current control loop (predictive controller) that also balances DC capacitors voltages. The predictive optimum controller minimizes the ripple factor and total harmonic distortion of input currents of UPFR. The amplitudes of the AC current references are obtained using the PI controller of the rectifier DC voltage.
The predictive methodology is compared with a fast non-linear method (sliding-mode) to evaluate the performance improvement resulting from the use of optimal predictive control of multilevel converter, ensuring that experiments are done under the same conditions and with the same PI controller parameters to control the DC voltage.

3.2.1 Switched state-space multilevel UPFR model

To operate as rectifier, the three-phase NPC multilevel converter has the three legs connected to the electrical power network, through coupling inductors, L (with loss resistor, R), and the load is connected to the DC side seeing the sum of the voltages of the capacitors C_1 and C_2 (Fig. 12). The main power flow direction is from electrical power network to the DC load. Yet, the UPFR is bidirectional and can transfer power from the DC side to the AC side.

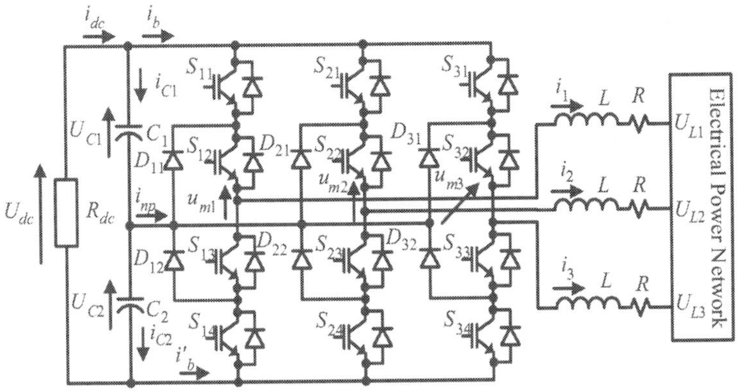

Fig. 12. Multilevel NPC rectifier with three legs

The dynamic equations of the AC currents and the capacitors voltage, U_{C1} and U_{C2} are similar to the NPC converter in section 2.1. Since $U_{dc} = U_{C1} + U_{C2}$, the dynamic equation of the rectifier DC voltage, U_{dc}, is obtained directly from (9),

$$\frac{dU_{dc}}{dt} = -\left(\frac{\Gamma_{1d}}{C_1} + \frac{\Gamma_{2d}}{C_2}\right)i_d - \left(\frac{\Gamma_{1q}}{C_1} + \frac{\Gamma_{2q}}{C_2}\right)i_q + \left(\frac{1}{C_1} + \frac{1}{C_2}\right)i_{dc} \,. \tag{34}$$

The design of the PI linear controller for the rectifier DC voltage is based on the dq coordinates dynamic equation (34), since it is time invariant at steady state operation.

3.2.2 Predictive control of AC line currents and DC capacitor voltages

The dynamic equations of the state space variables in multilevel rectifier are similar to the dynamic equations of the multilevel inverter (6). The design of the optimal predictive controller for AC line currents is done as in Section 2.2 (multilevel current inverter). In the multilevel converter, working as inverter, the current references, $i_{\alpha Ref}$ and $i_{\beta Ref}$, were obtained from lookup tables. However, in the rectifier mode, the current references must be derived from the voltage regulator, which in each moment computes the current references to ensure that the voltage DC, U_{DC}, follow its reference, U_{dcRef}.

3.2.3 UPFR current reference generation and DC voltage control

To ensure near unity power factor, the AC currents at the input of the multilevel rectifier must be sinusoidal and in phase with the voltages. The DC voltage PI controller generates the AC references current, i_{dRef} and i_{qRef}, of the multilevel converter for the DC voltage, U_{dc}, to follow its reference, U_{dcRef} without steady-state errors. The dynamics of the DC voltage rectifier (34) is similar to the dynamic DC voltage of the APF (25). Therefore, the design of the DC voltage regulator rectifier was made with a PI compensator with the same parameters of the APF (31).

3.2.4 Simulation and experimental results

In UPFR simulations and experiments the parameters have the following values: C_1 = 20 mF, C_2 = 18.6 mF, R_{dc} = 100 Ω, R = 0.1 Ω, L = 15.5 mH, U_{dcRef} = 100 V, U_{LRMS} = 24 V, f_{ac} = 50 Hz, ΔT = 28 μs, ζ = 0.71, ω_n = 4 rad/s, ρ_α = 0.09 A^2, ρ_β = 0.09 A^2, and ρ_{UC} = 0.04 V^2.

In steady-state operation the experimental and simulation results of the rectified DC voltage, U_{dc}, (Fig. 13) show that this voltage follows the reference, U_{dcRef}, without stationary error.

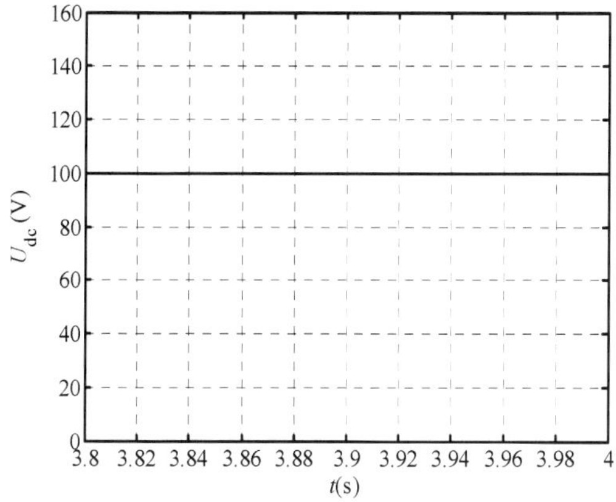

(a) Simulation results

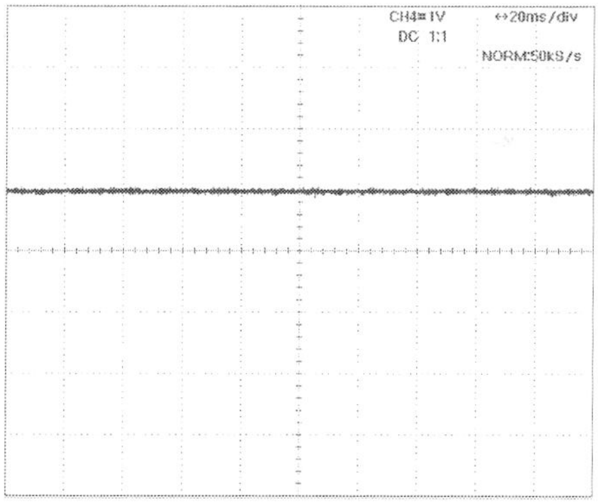

b) Experimental results

Fig. 13. U_{dc} voltage in steady state operation. Each vertical division is 20 V/Div

The line AC currents, at the input of the multilevel rectifier, (Fig. 14b) are quasi sinusoidal, as is needed, with an rms ripple error near to 1% (0.02 A) having an expressive improvement comparatively to the sliding mode controller (Fig. 14a), which have an rms ripple error of approximately 9% (0.1 A).

The rms error of the electric network voltage, U_{L1}, U_{L2}, and U_{L3}, is about 7% (3.5 V). The AC line currents are in phase with the electric network voltage (Fig. 14), showing that both controllers have a near unity power factor.

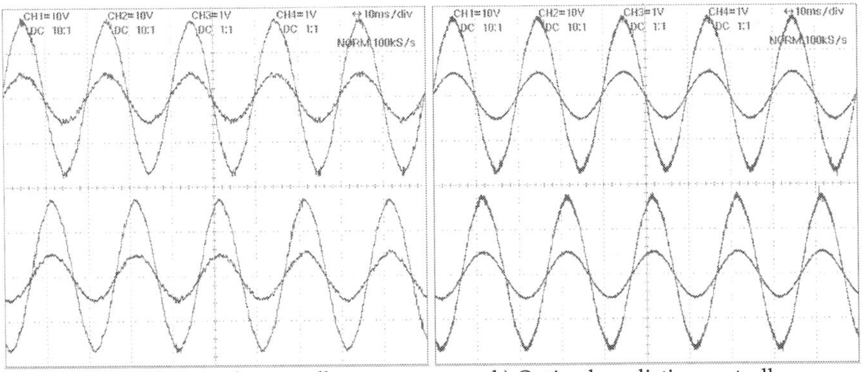

a) Sliding mode controller b) Optimal predictive controller

Fig. 14. AC currents, $-i_1$ and $-i_2$, and AC line voltages, U_{L1} and U_{L2}, in steady state operation. The AC current vertical division is 5 A/Div and the AC voltage vertical division is 20 V/Div

3.3 Dynamic voltage restorer

Dynamic voltage restorers are mainly used to protect sensitive loads from the electrical network voltage disturbances, such as sags or swells, and could be used to reduce flicker and harmonic distortion of AC voltages (acting as series active filters). Voltage sags are abrupt reductions (between 10% and 99%) in the AC voltage root mean square value, lasting less than 60 seconds.

The DVR is here designed using optimal predictive output voltage controllers to ensure sensitive load AC sinusoidal voltages with constant amplitude, without interruptions and unbalances, while reducing the AC voltage THD, even during the presence of sags, short interruptions and high values of THD. The pre-sag compensation is used to restore the existing amplitude voltage and phase prior to the voltage short interruption, and the in-phase compensation is used to mitigate sags, minimizing the DVR voltage amplitude. The sensitive load AC voltage waveform quality is improved using the designed optimal predictive NPC based DVR. The synchronous dq frame predictive controller of the output AC voltage, able to mitigate AC voltage disturbances is designed and it is compared to a synchronous dq frame PI controller and asynchronous (P+resonant) proportional integral controller in simulation and experimental results. Using the state-space model of the three-phase NPC converter, the obtained optimal predictive controller in addiction reduces the THD of the AC voltages. The optimal predictive phase quadrature synchronizer (section 2.3) detects the phase of the AC network voltages and generates the fundamental frequency.

3.3.1 Optimal predictive controller of the sensitive load AC voltage

The control system of DVR uses an outer loop for the AC load voltage control, which generates the current references, $i_{\alpha Ref}$ and $i_{\beta Ref}$, for the inner predictive AC current loop.

The structure of the optimal predictive multilevel DVR showing the interconnection of sub-systems and feedback control loops is represented in Fig. 15. The optimal predictive outer loop controller of the AC sensitive load voltages generates, through a limiter, the current references, $i_{\alpha Ref}$ and $i_{\beta Ref}$, for the inner AC current loop optimal predictive controllers (reported in section 3.3.4). An equivalent DVR model connected to an AC network is derived next to design the outer loop controllers, including the optimal predictive controller for the load sensitive AC voltages.

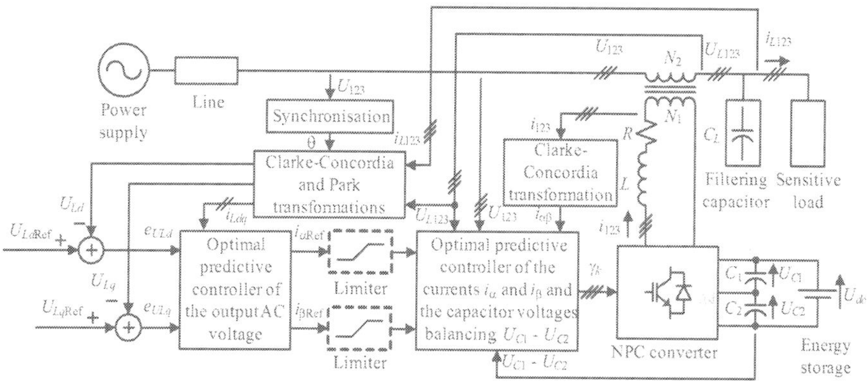

Fig. 15. Optimized feedback system of the DVR to improve the power quality of the output AC voltage of critical sensitive loads

The AC network voltages (U_1, U_2, and U_3) are represented using the equivalent network model with line inductances (L_{Line}) and equivalent loss resistors (R_{Line}) (Fig. 16). One three phase transformer with 3 separated secondary windings, each one series connected with each line voltage, is used to series inject the compensation voltage for the critical load. The filtering capacitors C_L (Fig. 16) are placed on the line-side to obtain high control bandwidth while reducing the NPC converter high frequency switching harmonics.

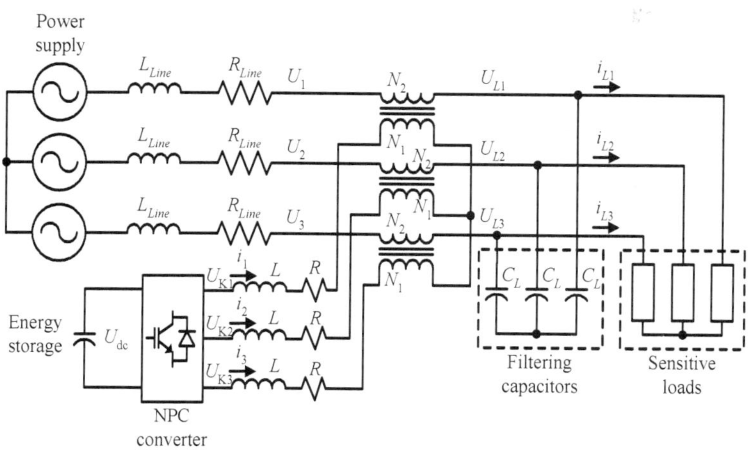

Fig. 16. Three-phase DVR simplified power circuit

Assuming balanced operation and using Kirchhoff laws in the DVR circuit (Fig. 16), the dynamic equations (35) of the sensitive load AC voltages (U_{L1}, U_{L2}, and U_{L3}), can be written as functions of the circuit parameters (filter capacitor C_L and transformer ratio $n=N_2/N_1$).

$$\frac{dU_{Lj}}{dt}=\frac{1}{nC_L}i_j-\frac{1}{C_L}i_{Lj};\quad j=1,2,3 \tag{35}$$

These equations (35) are suitable to design the outer loop controllers for U_{L1}, U_{L2}, and U_{L3} voltages. Their control inputs are the NPC converter output AC currents i_1, i_2, and i_3. From the control viewpoint, the load currents, i_{L1}, i_{L2}, and i_{L3}, are considered as disturbances.
For control purposes, the isolated neutral output AC voltages, can be advantageously written in αβ and then in dq coordinates. Using the Clarke-Concordia transformation the αβ coordinates model of the output AC voltage (35), is (36).

$$\begin{bmatrix}\dfrac{dU_{L\alpha}}{dt}\\ \dfrac{dU_{L\beta}}{dt}\end{bmatrix}=\begin{bmatrix}\dfrac{1}{nC_L} & 0\\ 0 & \dfrac{1}{nC_L}\end{bmatrix}\begin{bmatrix}i_\alpha\\ i_\beta\end{bmatrix}+\begin{bmatrix}-\dfrac{1}{C_L} & 0\\ 0 & -\dfrac{1}{C_L}\end{bmatrix}\begin{bmatrix}i_{L\alpha}\\ i_{L\beta}\end{bmatrix} \tag{36}$$

Using the Park transformation in (36), the model of the output AC voltage (37) in dq coordinates is obtained.

$$\begin{bmatrix}\dfrac{dU_{Ld}}{dt}\\ \dfrac{dU_{Lq}}{dt}\end{bmatrix}=\begin{bmatrix}0 & \omega\\ -\omega & 0\end{bmatrix}\begin{bmatrix}U_{Ld}\\ U_{Lq}\end{bmatrix}+\begin{bmatrix}\dfrac{1}{nC_L} & 0\\ 0 & \dfrac{1}{nC_L}\end{bmatrix}\begin{bmatrix}i_d\\ i_q\end{bmatrix}-\begin{bmatrix}\dfrac{1}{C_L} & 0\\ 0 & \dfrac{1}{C_L}\end{bmatrix}\begin{bmatrix}i_{Ld}\\ i_{Lq}\end{bmatrix} \tag{37}$$

An optimal predictive controller of the sensitive load AC voltages intends to predict the value of the multilevel current references needed to minimize the errors of load AC voltages (relatively to their references). At time step $t_s = k\Delta T_{ULdq}$ (ΔT_{ULdq} is the voltage loop sampling step) the prediction of the multilevel current references for the next sampling step $t_{s+1} = (k+1)\Delta T_{ULdq}$, $i_{dRef}(t_{s+1})$ and $i_{qRef}(t_{s+1})$ (38), is obtained trough inversion of the dynamic equations of the sensitive load AC voltages (37), considering that the actual ($t_s = k\Delta T_{ULdq}$) sampled load voltages $U_{Ld}(t_s)$ and $U_{Lq}(t_s)$, must follow their references $U_{LdRef}(t_{s+1})$ and $U_{LqRef}(t_{s+1})$ within one voltage loop sampling time ΔT_{ULdq} (the next sampling step).

$$\begin{aligned}i_{d\mathrm{Re}f(t_{s+1})}&=nC_L\frac{U_{Ld\mathrm{Re}f(t_{s+1})}-U_{Ld(t_s)}}{\Delta T_{ULdq}}-nC_L\omega U_{Lq(t_s)}+ni_{Ld(t_s)};\\ i_{q\mathrm{Re}f(t_s+1)}&=nC_L\frac{U_{Lq\mathrm{Re}f(t_s+1)}-U_{Lq(t_s)}}{\Delta T_{ULdq}}+nC_L\omega U_{Ld(t_s)}+ni_{Lq(t_s)}\end{aligned} \tag{38}$$

These predictive equations, (38), are discrete feedback control laws (sampling time $\Delta T_{ULdq} = 360$ μs) for the sensitive load AC voltages in dq coordinates, giving the reference currents $i_{d\text{Ref}}(t_{s+1})$ and $i_{q\text{Ref}}(t_{s+1})$ needed to track the critical load reference voltages, $U_{Ld\text{Ref}}(t_{s+1})$ and $U_{Lq\text{Ref}}(t_{s+1})$.

3.3.2 Synchronous dq frame PI controller of the output AC voltage

The design of the PI controller can be advantageously made by decoupling the cross coupling terms on the dynamic equations of the sensitive load AC voltage in (37). The PI control laws of $i_{d\text{Ref}}$ and $i_{q\text{Ref}}$ (39), are functions of the output AC voltage errors and the K_p and K_I parameters:

$$\begin{aligned} i_{d\text{Re}f} &= K_P\left(U_{LdRef} - U_{Ld}\right) + K_I \int\left(U_{LdRef} - U_{Ld}\right)dt - nC_L\omega U_{Lq}; \\ i_{q\text{Re}f} &= K_P\left(U_{LqRef} - U_{Lq}\right) + K_I \int\left(U_{LqRef} - U_{Lq}\right)dt + nC_L\omega U_{Ld}. \end{aligned} \tag{39}$$

The parameters, $K_P = 2n\zeta\omega_n C_L$ and $K_I = C_L n\omega^2{}_n$, of the PI are obtained as functions of the desired damping ratio, ζ, and undamped natural frequency, ω_n, of the closed-loop second order system ($s^2 + 2\zeta\omega_n + \omega^2{}_n$) (Ogata, 2002).

3.3.3 Asynchronous proportional integral (P+resonant) controller of the output AC voltage

The dynamic equations of the output AC voltage (36), in $\alpha\beta$ coordinates, do not present cross coupling terms. Therefore a proportional integral controller, with ω_0 frequency shift, can be used to generate the current references of the DVR (Zmood et al., 2001),

$$H_{PI}(s) = K_P + \frac{K_I\omega_{cut}s}{s^2 + 2\omega_{cut}s + \omega_0^2}. \tag{40}$$

These asynchronous PI controllers, called P+resonant (Li et al., 2007), (Zmood et al., 2001), are tuned to the fundamental frequency ω_0 ($2\pi 50$ rad/s). The practical controller (40) is obtained by approximating the ideal ω_0 integrator using a high-gain low-pass filter with cut-off frequency ω_{cut} (Zmood et al., 2001). An analysis of the P+resonant controller, applied to DVRs, can be found in (Li et al., 2007).

3.3.4 Optimal predictive multilevel DVR current control system

The global performance of the DVR depends significantly on its feedback system, which can be improved using an optimal synchronizer to the fundamental frequency and an optimal predictive controller to track the needed $i_{d\text{Ref}}$ and $i_{q\text{Ref}}$ current references.

The NPC multilevel converter generates the AC currents i_1, i_2, and i_3, tracking references (from $i_{d\text{Ref}}$ and $i_{q\text{Ref}}$) obtained from the output AC voltage controllers (sections 3.3.1 to 3.3.3). These currents are needed to enforce the output AC voltages, and must present low ripple and zero tracking error, to ensure high quality voltage waveforms.

Applying the Kirchhoff laws to the converter circuit (Fig. 16), and the Clarke-Concordia transformation, the dynamic equations of the NPC converter AC output currents, i_1, i_2, and i_3, and the capacitor voltages, U_{C1} and U_{C2}, are advantageously represented in $\alpha\beta$ coordinates, i_α and i_β (41).

$$\begin{bmatrix} \frac{di_\alpha}{dt} \\ \frac{di_\beta}{dt} \\ \frac{dU_{C1}}{dt} \\ \frac{dU_{C2}}{dt} \end{bmatrix} = \begin{bmatrix} -\frac{R}{L} & 0 & \frac{\Gamma_{1\alpha}}{L} & \frac{\Gamma_{2\alpha}}{L} \\ 0 & -\frac{R}{L} & \frac{\Gamma_{1\beta}}{L} & \frac{\Gamma_{2\beta}}{L} \\ -\frac{\Gamma_{1\alpha}}{C_1} & -\frac{\Gamma_{1\beta}}{C_1} & 0 & 0 \\ -\frac{\Gamma_{2\alpha}}{C_2} & -\frac{\Gamma_{2\beta}}{C_2} & 0 & 0 \end{bmatrix} \begin{bmatrix} i_\alpha \\ i_\beta \\ U_{C1} \\ U_{C2} \end{bmatrix} + \begin{bmatrix} -\frac{1}{L} & 0 & 0 \\ 0 & -\frac{1}{L} & 0 \\ 0 & 0 & \frac{1}{C_1} \\ 0 & 0 & \frac{1}{C_2} \end{bmatrix} \begin{bmatrix} \frac{U_{L\alpha} - U_\alpha}{n} \\ \frac{U_{L\beta} - U_\beta}{n} \\ i_{dc} \end{bmatrix} \quad (41)$$

Several control methods for the currents, i_α and i_β, were proposed (Barros & Silva, 2008), (Kazmierkowski & Malesani, 1998), and (Nabae & Takahashi, 1981). However, the optimal predictive control method (Barros & Silva, 2008) is advantageous, as it both minimizes AC current, i_α and i_β, errors, and the DC unbalance of capacitor voltages, U_{C1} and U_{C2}. Results show comparably better performances in the current waveforms (lower ripple and THD) and on the balancing of capacitor DC voltages. The optimal predictive current controller has the potential to improve the overall performance of the DVR (Barros & Silva, 2010).
The current optimum controller minimizes both the AC currents errors and the capacitor voltage difference using the quadratic cost functional (14) of the tracking errors (15).

3.3.5 Simulation and experimental results

Experimental results of the AC voltages at the sensitive loads, U_{L1}, U_{L2}, and U_{L3}, in steady state operation (Fig. 17) show that predictive, PI controllers and P+ressonant controllers ensure tracking of their references showing a nearly perfect sinusoidal waveform.

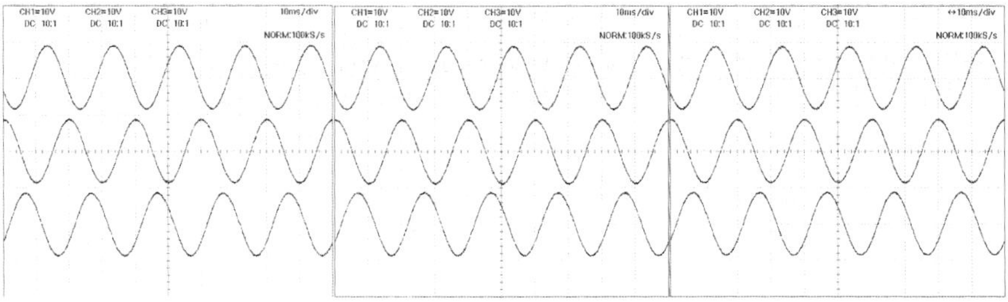

a) Predictive controller. b) Classical PI controller. c) P+resonant controller

Fig. 17. The sensitive load AC voltage at sensitive loads, U_{L1} + 740 V, U_{L2}, and U_{L3} - 740 V, in steady state operation (vertical: 370 V/Div and horizontal: 10 ms/Div)

To analyse the influence of load disturbances on the output AC voltage, the load resistive component was altered from the nominal values, R_L, to unbalanced values of $R_L/2$, R_L, and $2R_L$, respectively in phases 1, 2, and 3, (changes occur at the fifth horizontal division of Fig. 18). The results show (Fig. 18a) that using the optimal predictive output AC voltage controller the sensitive load voltages track their references, being almost insensitive to the unbalanced load step variation. The AC voltage PI controller (Fig. 18b) is not able to control the individual voltage amplitude of the three phases supplying unbalanced loads. The output

AC voltages controlled via the P+resonant (Fig. 18c) are balanced, but present a slowly recovering distortion.

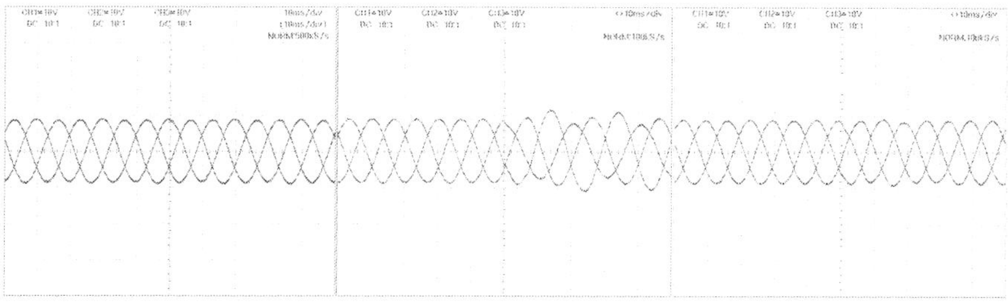

a) Predictive controller. b) Classical PI controller. c) P+resonant controller

Fig. 18. The sensitive load AC voltage, U_{L1}, U_{L2}, and U_{L3}, when is connected an unbalanced load (vertical: 370 V/Div and horizontal: 10 ms/Div)

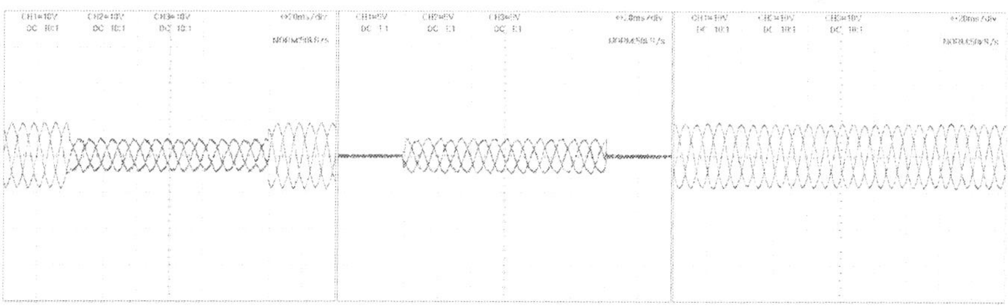

a) AC voltage b) Injected voltages c) Output AC voltage

Fig. 19. Experimental results of the AC voltage sag, U_1, U_2, and U_3, the injected voltages of the DVR, U_{L1}-U_1, U_{L2}-U_2, and U_{L3}-U_3, and the load AC voltage at the sensitive load, U_{L1}, U_{L2}, and U_{L3}, using the optimal predictive controller (370V/Div and 20 ms/Div).

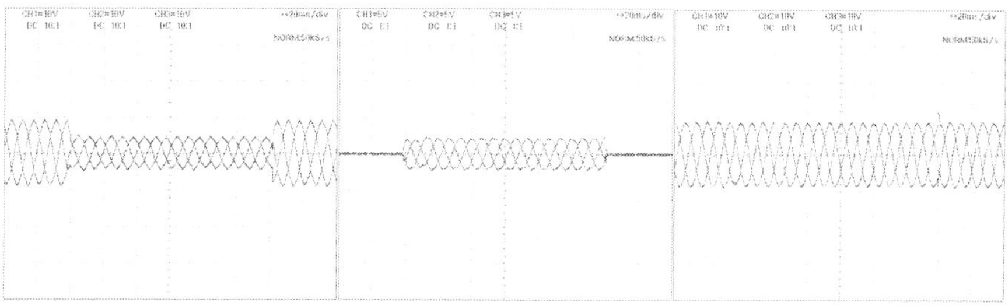

a) AC voltage b) Injected voltages c) Output AC voltage

Fig. 20. Experimental results of the AC voltage sag, U_1, U_2, and U_3, the injected voltages of the DVR, U_{L1}-U_1, U_{L2}-U_2, and U_{L3}-U_3, and the load AC voltage at the sensitive load, U_{L1}, U_{L2}, and U_{L3}, with the filtering capacitor on the NPC converter-side (370V/Div and 20 ms/Div).

During a sag occurrence, (Fig. 19a), reducing to 50% the AC voltages during 120 ms, the NPC converter injects voltages, (Fig. 19b), to mitigate the sag, whose references are calculated using the output AC voltage optimal predictive controller, showing a near perfect critical load voltage waveform.
The NPC converter with the filtering capacitor on the NPC converter-side causes sensitive load transient notches or overshoots at beginning or at the end of the sag, respectively (Fig. 20). Oppositely, the line-side capacitor filtering topology enables sensitive load voltages with constant amplitude and no notches or overshoots (Fig. 19c). As expected this topology improves the power quality and protects sensitive loads from network AC voltage sags, without notches or overshoots. Results obtained when using the PI and the P+resonant controllers show lower performances.

4. Conclusion

This chapter presented an optimized control method to the multilevel NPC converter for applications in electrical power quality. The experimental results confirmed these improvements through three applications: filtering the currents of non-linear loads, with an APF, unity power factor AC/DC conversion, with a multilevel UPFR, and mitigation of sags, over-voltages, short interruptions, harmonic distortion, and unbalances in the AC electrical power network with a multilevel DVR.
The proposed optimal predictive controller predicts in real time the state space voltages and currents of the multilevel converter and computes a quadratic cost functional to choose the optimal vector.
Obtained simulation and experimental results show that the NPC converter AC side currents track their references showing comparatively smaller ripple, total harmonic distortion less than 1%, and almost no steady state error (0.3%). The capacitor voltages are balanced with an error lower than 1%, and the balancing are assured even when NPC redundant vectors are not chosen. Near perfect capacitor DC voltage balancing is obtained while reducing current ripples, harmonic distortion and switching frequency.
In the first application to improve power quality, the NPC multilevel converter was connected as APF in parallel to non-linear loads to reduce the injection of harmonic components in the electrical power network, working as APF filter and reactive power compensator so that the power factor was nearly one.
Results confirm that the predictive optimal control of the NPC multilevel converter operating as APF can regulate the DC voltage and reduce the THD of AC line current from 36% to 1% with near-unity power factor (0.997), significantly improving power quality of line voltages.
In section 3.2, the synthesis of UPFR optimum controllers was presented.
The optimal predictive controller of the multilevel converter was used to control the AC currents and to balance the DC capacitors voltage. Comparisons were done to other controllers such as sliding mode, aiming to study the influence of state space control methods in the multilevel converter performance in applications to improve the electrical power quality.
Results show that the multilevel UPFR DC voltage follows the reference value without steady error and the input AC currents are sinusoidal in phase with the AC voltages. The

optimal predictive controller reduces the AC current ripple factor from 9% to 1%, and reduces THD of the AC current from 8% to 1%, compared to the sliding mode controller.
The optimized multilevel DVR here proposed vastly improves power quality at sensitive critical loads. The performance of the optimal predictive controller generating the reference voltages to inject was compared to the synchronous dq frame PI controller and to the P+resonant controller. Results show that predictive controllers present the lowest THD levels in the load AC voltages, are able to regulate unbalanced loads AC voltages, allow plug-in and out of extra loads without causing swells, sags, notches or overshoots, and reduce the distortion of AC voltages in non-linear loads. Load AC voltages are almost sinusoidal, when facing balanced and unbalanced sags, and short interruptions with unbalanced loads. Voltage THD is reduced to values lower than 1%, the DVR behaving also as a series active power filter for the AC voltages strongly improving the power quality of sensitive loads.

5. References

Barros, J. D.; Fekri, S. & Athans, M. (2005). Robust mixed-mu synthesis performance for mass-spring system with stiffness uncertainty, *Proceedings of IEEE 13th Mediterranean Conference on Control and Automation MED05*, pp. 743-748, ISBN: 0-7803-8936-0, Cyprus, June 2005, IEEE, Limassol

Barros, J. D. & Silva, J. F. (2008). Optimal predictive control of three-phase NPC multilevel converter for power quality applications. *IEEE Transactions on Industrial Electronics*, Vol. 55, No. 10, (October 2008) pp. 3670-3681, ISSN: 0278-0046

Barros, J. D. & Silva, J. F. (2010). Multilevel optimal predictive dynamic voltage restorer. *IEEE Transactions on Industrial Electronics*, Vol. 57, No. 8, (August 2010) pp. 2747-2760, ISSN: 0278-0046

Holmes, D. G. & Lipo, T. A. (2003). *Pulse Width Modulation for Power Converters*, Wiley-Interscience, John Wiley & Sons, Canada

Jones, C. V. (1967). *The Unified Theory of Electrical Machines*, Plenum, New York

Kazmierkowski, M. P. & Malesani, L. (1998). Current control techniques for three-phase voltage-source PWM converters: a survey. *IEEE Trans. Ind. Electron.*, Vol. 45, No. 5, (October 1998) pp. 691-703, ISSN: 0278-0046

Kwakernaak, H. & Sivan, R. (1972). *Linear Optimal Control Systems*, Wiley-Interscience, John Wiley & Sons, ISBN: 0-471-51110-2, Canada

Li, Y. W.; Blaabjerg, F.; Vilathgamuwa, D. M. & Loh, P. C. (2007). Design and comparison of high performance stationary-frame controllers for DVR implementation. *IEEE Trans. Power Electron.*, Vol. 22, (March 2007) pp. 602-612, ISSN: 0885-8993

Marchesoni, M.; Mazzucchelli, M. & Tenconi, S. (1988). A non conventional power converter for plasma stabilization, *Proceedings of Power Electronics Specialists Conference PESC88*, pp. 122-129, Japan, April 1988, IEEE, Kyoto

Meynard, A. & Foch, H. (1992). Multi-level choppers for high voltage applications. *EPE Journal*, Vol. 2, (March 1992) pp. 45-50

Moreno-Muñoz, A. (2007). *Power Quality: Mitigation Technologies in a Distributed Environment*, Springer, ISBN: 1612-1287, London

Nabae, A. & Takahashi, I. (1981). A new neutral-point-clamped PWM inverter. *IEEE Trans. Ind. Applicat.,* Vol. IA-17, (Sep./Oct. 1981) pp. 518-523, ISSN: 0093-9994

Ogata, K. (2002). *Modern Control Engineering,* Prentice Hall, 4th edition, USA

Silva, J. (1999). Sliding-mode control of boost-type unity-power-factor PWM rectifiers. *IEEE Trans. on Ind. Electron.,* Vol. 46, No. 3, (June 1999) pp. 594-603, ISSN: 0278-0046

Singh ,B.; Al-Haddad, K. & Chandra, A. (1999). A review of active filters for power quality improvement. *IEEE Transactions on Industrial Electronics,* Vol. 46, No. 5, (October 1999) pp. 960-971, ISSN: 0278-0046

Tan, T. L.; Chen, S. & Choi, S. S. (2005). An overview of power quality state estimation, Proceedings of the 7th International Power Engineering Conference IPEC 2005, pp. 1-5, ISBN: 981-05-5702-7, Singapore, November/December 2005, IEEE

Zmood, D. N.; Holmes, D. G. & Bode, G. H. (2001). Frequency-domain analysis of three-phase linear current regulators. *IEEE Trans. Ind. Applicat.,* Vol. 37, No. 2, (March/April 2001) pp. 601-610, ISSN: 0093-9994

Part 5

Power Quality and Critical Components

11

Numerical Relay: Influenced by and Accessing the Power Quality

Ruth P.S. Leao, Giovanni C. Barroso, Nelber X. Melo, Raimundo F. Sampaio, Janaína A. Barbosa and Fernando L.M. Antunes
Federal University of Ceara - UFC
Brazil

1. Introduction

With the latest technological advances in the area of monitoring, protection and control systems, together with the availability of digital communication technologies, the power utilities have been requested to deliver a greater amount of specific information to a broad range of users such as customers, grid operators, regulation agencies and market operators.

The power systems make use of permanently installed, integrated, measuring devices that collect, measure, and evaluate one or more characteristic parameters with the intention of automatically determining and reporting the status of the monitored system (Mackrell et al, 2010). Information provided by the real-time monitoring system enables the right decisions to be made for improving security, reliability and performance of the networks.

Relay devices form a major part of all monitoring, protection and control systems. These protection relays are multifunctional devices, primarily employed to provide related equipment and system protection, and more recently, they have been adapted to assist the power quality evaluation of power systems.

Power quality defines an acceptable set of limits that allow power systems and ultimately the equipment-user to operate within their intended manner with efficiency and expected life span. Power quality deviation refers to a wide variety of electromagnetic phenomena that characterizes the voltage and current at a given time and at a given location on the power system (IEEE Std. 1159, 1995). An electromagnetic phenomenon is a disturbance that may cause variation on voltage and current rms, frequency and waveform that degrade the performance of a device, equipment, process or system.

Sometimes power quality can be affected by transient phenomena such as system fault occurrence such as voltage interruption, voltage sag and voltage swell. The protective relays are designed to operate reliably when there are faults in the power system. The relay device must trip on detection of any fault condition and must not trip during normal operating conditions. When the fault is cleared by a relay device, quite often the supply voltage will be restored automatically to its normal operating level. The relay can therefore contribute to restoring the voltage quality of the power system. Protection relaying is primary concerned with clearing faults while power quality is concerned with the delivery of reliable power to defined quality standards (Cease and Kunsman, 2003).

The modern multifunction digital relays are able to measure and register parameter values (currents and voltages) that are useful to the power quality assessment. The disturbances registered by the digital relays can be assessed in order to characterize the events and their severity, in terms of magnitude, duration, and the frequency of occurrence. The digital relay, equipped with advanced digital processing and high density memory chips are able therefore to concurrent performance of protection and quality analysis functions.

The protective relays, however, are among the devices that may suffer from the disturbances in the power system. When the quality of power is poor, protective relays can have problems making the correct decisions on whether to trip or not. Some power quality related phenomena may cause incorrect protection system operations. This is because the poor power quality conditions cause the relay to register erroneous input values, and for that reason, it acts erroneously. The opposite situation is also possible whereby the relay does not trip when it should, due to the poor power quality (Zamora et al, 2004). The sensitivity and the immunity of the relays are evaluated through tests performed according to international standards like IEC 60255-11 and IEC 61000-4-11.

This chapter deals with the relationship between power quality and protection relays. As the protection relays are used to assess power quality, they may suffer influence from the quality of supply. The objective of this chapter is to evaluate the impact of power disturbances upon protection relay devices. The responses of the relay to specific disturbances are evaluated and the immunity of the device is presented.

To access the power quality of a site, the protection relays can provide useful information that can be quantified by using appropriate power quality indices. This chapter aims also to assess the power quality from information provided by the protection relays.

The chapter is organized as follows section two provides an overview about the common disturbances accountable for degrading the power quality, the importance of power quality monitoring and some indices to quantify the severity of the disturbances; section three discusses issues concerning numerical relays, the influence of the power quality on their performance, and the relay responses to tests performed according to international standards. The application of numerical relays for evaluating the power quality is also included within this section; section four presents the latest advances on wide area monitoring protection and control which are revolutionizing the way state estimation is being performed; finally, the conclusions are given in the fifth section.

2. Overview of power quality

Power Quality is a technical term that has practical implications on the performance and productivity of end-user equipment. It can be considered as power that does not impair the operation of a customer's equipment. A power quality problem is any occurrence that manifests itself in voltage, current, or frequency deviation that results in failure or misoperation of customer equipment. If the quality of power is inadequate for those needs, then the quality is lacking.

Power quality disturbances that are common in a power system include: voltage sags, voltage swells, short-term interruptions, transients, voltage unbalance, harmonics, and voltage fluctuations.

Voltage Sag is defined as a rapid short-term decrease in voltage. As with sags, swells are usually associated with system fault conditions, but they are not as common as voltage sags. One way that a swell can occur is from the temporary voltage rise on the unfaulted phases during a single line-to-ground fault.

Voltage Swell is defined as a rapid short-term increase in voltage typically caused by the switching off bank of capacitors and large block of loads.

Voltage Interruption is the disappearance of the supply voltage on one or more phases. When the interruption is longer than 1 min, it is considered as sustained interruption. Interruptions can be caused by faults, control malfunctions, or equipment failures.

Transient is a sudden non-power frequency change in the steady-state condition of voltage or current, or both. When it is unidirectional in polarity (either positive or negative) it is defined as an impulsive transient, otherwise, when both polarities are included, it is considered as an oscillatory transient.

Voltage Unbalance is characterized by any differences that exist in the three phase voltage magnitude or any phase shift in the 120 degree separation between the three phases. Causes of voltage unbalance include unequal impedances of three-phase transmission and distribution system lines, large and/or unequal distribution of single-phase loads, phase-to-phase loads and unbalanced three-phase loads.

Harmonics are sinusoidal voltages or currents having frequencies that are integer multiples of the frequency at which the supply system is designed to operate. Harmonics combine with the fundamental voltage or current, and produce waveform distortion (IEEE 1159, 1997).

Voltage Fluctuation is defined as a series of rms voltage changes or a cyclical variation of the voltage waveform envelope. The primary cause of voltage fluctuations is the time variability of the reactive power component of variable loads. Such loads include arc furnaces, rolling mill drives, and main winders – all of which are loads with a high rate of change of power with respect to the short-circuit capacity at the point of common coupling.

Flicker is the periodic change in the instantaneous light output of a light source caused by fluctuation of the supply voltage. Flicker is a symptom of voltage fluctuation which are typically caused by the use of large fluctuating loads, i.e. loads that have rapidly fluctuating active and reactive power demand. The increased use of high frequency electronic ballasts has greatly decreased the amount of flicker present in the indoor environment.

While power disturbances occur on all electrical systems, the sensitivity of today's electronics makes them more susceptible to them. Improvement of power quality has a positive impact on sustained profitability of the distribution utility on the one hand and customer satisfaction on the other.

2.1 Power quality monitoring

Power quality monitoring is necessary to characterize electromagnetic phenomena at a particular location on an electric power circuit. There are two main reasons for making power quality measurements: plant investigations in order to solve specific problems in an installation; and, utility measurements to obtain the general power quality levels in any part of the network.

Power quality monitors are intended to give all necessary information about significant power quality disturbances over a long period, varying from weeks to months. These instruments have to be able to identify and record the characteristics of many types of disturbances occurring on a timescale of microseconds (transients) to hours (steady state voltage variations). Fast transients require high sample rate analogue-to-digital converters (eg 1-4 MHz) giving a large data throughput. Operating over a long timescale this gives an enormous amount of data to be handled. These instruments must either record very little of the data handled or have large storage facilities (eg hard disk) or have the ability to send the data to a storage facility by means of a modem or similar communication method. To prevent overloading of memory with discrete event type disturbances, monitors have adjustable thresholds which determine the level at which a disturbance is recorded (Gosbell, 2002).

The IEC 61000-4-30 provides common requirements for measurement devices to ensure that analyzers from different manufacturers give the same results for power quality parameters in 50/60 Hz ac power supply systems. This standard ensures that different power quality instruments use the same definitions and measurement techniques, for various power quality parameters, which may take the form of overvoltages and undervoltages, interruptions, sags/dips, swells, frequency, harmonics, phase imbalance, voltage fluctuation, etc. The basic aim of this document is to define measurement methods that will make it possible to obtain reliable, repeatable and comparable results independently of the power quality meter manufacturer. The standard does not provide information about instrumentation design. This kind of information is covered by other standards such as IEC 61000-4-7 and IEC 61000-4-15 (Munõz, 2007).

Power quality instrumentation should not only have good hardware architecture, but also powerful software that permits the measured data to be managed and analyzed.

2.2 Power quality indices

There is an increasing need for performance criteria to assess the quality of the power supplied. Several power quality indices have been proposed to assess the severity of phenomena such as flicker, harmonic content, voltage unbalance, etc in order to analyze the system performance. The quality of the voltage supply, which is a common concern of the power utilities, must be trustworthy and able to supply simple indices in order to provide answers and to settle questions and disputes concerning the quality of supply, and to better address the required investments. This section is concerned with investigating the joint treatment for signal analysis based on both the time and the frequency domains.

To calculate the indices, samples of the measured voltage are used. The voltage signal is recorded with a given sampling rate and resolution. The sampling rate or sampling frequency defines the number of samples per second (or per other unit) taken from a continuous signal to make a discrete signal. From the sampling theorem, it is known that if lower sampling rates are used, the original signal's information may not be completely recoverable from the sampled signal. The resolution is a function of the ability of an instrument to read smaller divisions of a quantity. The voltage resolution can be expressed in bits or volts.

2.2.1 Steady state voltage variation

The rms value is defined as the square root of the arithmetic mean of the squares of the instantaneous values of a quantity taken over a specified time interval and a specified bandwidth. The basic measurement time interval for parameter magnitudes shall be a 10-cycle time interval for a 50 Hz power system or 12-cycle time interval for a 60 Hz power system. The 10/12-cycle values are then aggregated to determine the average rms value every 10 minutes.

In a non-sinusoidal condition, the true rms voltage is evaluated by

$$V_{rms} = \sqrt{\sum_{h=0}^{N} V_h^2} \tag{1}$$

where V_h denotes the rms voltage of order h, and N, the maximum number of harmonics. The current true rms is derived similarly.

2.2.2 Phase displacements between corresponding fundamental voltage and current

To reflect the utilization of the power transferred in a non-sinusoidal and unbalanced condition, Munõz (2007) proposes a new power factor designated as the Orthogonal Current Factor, evaluated by

$$OCF = \frac{I_{a1}sen\varphi_{a1} + I_{b1}sen\varphi_{b1} + I_{c1}sen\varphi_{c1}}{I_{a1} + I_{b1} + I_{c1}} \tag{2}$$

where I_{a1}, I_{b1} and I_{c1} denote the rms values of the phase currents at the fundamental frequency and φ_{a1}, φ_{b1} and φ_{c1} denote the phase differences between the fundamental frequency components of the corresponding phase voltages and currents.

2.2.3 Flicker

The light flicker indices (Pst, Plt), as defined in IEC 61000-4-15, are a measure of perceived light flicker. The short-term flicker index Pst is a statistical quantification of perceived light flicker based on the measurement of source-voltage fluctuations. A Pst value greater than, or equal to 1, causes an irritability for people exposed to light flicker. The long-term flicker index Plt is a function of the Pst index.

2.2.4 System-frequency variation

Frequency deviation is due to an imbalance between generation and consumption in power systems. With the increasing number of distributed generation based on renewable resources, the distribution systems can experience frequency variations due to source intermittency, load variations and equipment malfunctions.

In order to characterize the power system frequency under normal operating conditions, the following procedures are applied. The Relative Frequency Deviation RFD can be defined as a difference between the real frequency value, f, and the rated frequency value, f_r, relative to f_r (50 Hz or 60 Hz).

$$RFD = \frac{\Delta f}{f_r} = \frac{f - f_r}{f_r} \tag{3}$$

The integral deviation during the day is evaluated as (Albert et al, 2008)

$$I_f = \int_0^{24} \Delta f \cdot dt \tag{4}$$

Normally, the frequency is measured every 10s. As power frequency may change within the 10s time interval, the number of cycles may not be an integer number. The fundamental frequency output is the ratio of the number of integral cycles counted during the 10s time interval, divided by the cumulative duration of the integer cycles. The evaluation of system frequency quality relies on the following procedure (Albert et al, 2008):

i. Monitoring the duration t_m over a stated time interval (one day or one week), based on the data obtained on 10s measuring windows;

ii. Determination of the number of 10s intervals N in which the supply voltage had no deviation larger than a specified percentage as given from the contracted voltage;
iii. Determination of the number of 10s intervals N_1 in which the frequency differs by more than 0.5 Hz from the rated value while the voltage is within the percentage of the contracted voltage;
iv. Determination of the number of 10s intervals N_2 in which the frequency is below the lower frequency or over the upper frequency permitted by contract while the voltage is within the percentage of the contracted voltage;
v. Checking conditions $N_1/N \le 0.05$ and $N_2=0$.

2.2.5 Unbalance

The simpler definition of Unbalance Factor (UF) for everyday applications is the IEEE definition, which applies to phase-to-phase voltage measurements only and involves three steps (Gosbell, 2002):
i. Calculate V_{avg}, the average phase-to-phase voltage;
ii Calculate ΔV_{max} maximum deviation of the phase-to-phase voltages from the average calculated in (i) above;
iii Then UF = $\Delta V_{max}/V_{avg}$.
The index can be expressed as,

$$UF = \frac{Maximum\ deviation\ from\ mean\ of\ \{V_{ab}, V_{bc}, V_{ca}\}}{Mean\ of\ \{V_{ab}, V_{bc}, V_{ca}\}} \tag{5}$$

According to IEC 61000-4-30 (2008), the supply voltage unbalance is evaluated using the method of symmetrical components. The negative sequence ratio υ_2 and the zero sequence ratio υ_0, expressed as a percentage, are evaluated by

$$\upsilon_2 = \frac{V_2}{V_1} \times 100 \tag{6}$$

$$\upsilon_0 = \frac{V_0}{V_1} \times 100 \tag{7}$$

where V_1, V_2 and V_0 are positive, negative and zero sequence voltages respectively. The current unbalance is derived similarly.
The negative sequence unbalance and zero sequence unbalance provide more precise and more directly useful values than the algorithms that use only the rms values to calculate unbalance. The zero sequence unbalance by definition is zero when phase-to-phase voltages are measured. However, the phase-to-neutral or phase-to-earth voltages may still contain the zero sequence component in that case IEC 61000-4-30 (2008).
Munõz (2007) proposes another unbalance index in his book using the concept of equivalent negative sequence voltage expressed as

$$V_e^2 = \left(V^+\right)^2 + \left(V^-\right)^2 + \left(V^0\right)^2 \tag{8}$$

where V^+, V^- and V^0 denotes the positive, negative and zero sequence voltage.

The voltage unbalance factor *VUF* is then evaluated by:

$$VUF = \frac{\sqrt{V_e^2 - \left(V^+\right)^2}}{V_e} \tag{9}$$

For balanced three-phase voltages $V_e = V^+$ and *VUF* becomes zero. Identical considerations apply to the three-phase currents where *V* is exchanged by *I*.

2.2.6 Harmonics

Harmonic distortion gives a waveform which is non-sinusoidal and repetitive. The total harmonic distortion of the voltage *VTDH* for single-phase and balanced three-phase systems is defined as

$$VTDH = \frac{\sqrt{\sum_{h \neq 1} V_h^2}}{V_1} \tag{10}$$

where V_h denotes rms values and 1 and h denotes the fundamental and the harmonic order respectively.
To unbalanced three-phase systems, a single equivalent harmonic rms voltage V_{eH} is defined as

$$V_{eH}^2 = \sum_{h \neq 1} \frac{V_{a,h}^2 + V_{b,h}^2 + V_{c,h}^2}{3} \tag{11}$$

Thus, the total harmonic distortion of the voltage *VTDH* for a three-phase unbalanced system is given by

$$VTHD = \frac{V_{eH}}{V_{e1}} \tag{12}$$

The total harmonic distortion of the current is derived similarly.

2.2.7 Voltage sags, swells and interruptions

Voltage variations of a time greater than 0.5 cycles of the power frequency but less than or equal to 1 minute is a major concern when considering the susceptibility of industrial processes based on computers, programmable logic controllers, adjustable speed drives, starters, contactors, etc. Power quality events, referred to as voltage sags, swells and interruptions, or as generally called short-duration voltage variations, are among the most critical events for power utilities. Results of surveys have shown that most of the power quality disturbances are caused by short-duration voltage variation under 1s (Colaço et al, 2008).
To evaluate the grid performance for three-phase voltage sags, a set of indices has been proposed by Thallam and Heydt (2000). The electric power acceptability curves are an empirical set of curves that represent the intensity and duration of bus voltage disturbances. Alternative indices for the assessment of voltage sags, such as voltage sag energy, were proposed.
With the voltage samples captured on each channel of a measuring instrument for each event the rms voltage versus time curve can be plotted. According to the IEC 61000-4-30, the

rms voltage of a short-term voltage variation is found by taking the voltage measured over one cycle of the power-system frequency and refreshed each half-cycle (IEC 61000-4-30, 2008), as evaluated by

$$V_{rms}(n) = \sqrt{\frac{1}{N} \sum_{k=1+n\frac{N}{2}}^{\left(\frac{n}{2}+1\right)N} [\upsilon(k)]^2} \tag{13}$$

with N the number of samples in one cycle, referred to as the window length, $\upsilon(k)$ the voltage sample k, and $V_{rms}(n)$ the rms voltage for multiple windows. The first value of $V_{rms}(0)$ is calculated over the samples k=(1,N), the next one over the samples (½N+1, 1½N) (Bollen, Sabin, Thallan, 2003). This rms voltage value may be a phase-to-phase value or a phase-to-neutral value.

The characteristics of the event are then defined for each channel, i.e., each voltage phase. It is recognized that both voltage magnitude and duration of an event in all three phases should be considered to derive a meaningful voltage variation index (Thallan and Heydt, 2000).

The magnitude of voltage sag, swell and interruption events are defined as the remaining or residual voltage, which is measured in each channel during the event (IEC 61000-4-30, 2008). The residual voltage is the lowest rms voltage for voltage sags and short interruptions and the highest rms voltage for a swell.

In order to estimate the duration of a short-term voltage variation, some voltage threshold are considered as reference values. The duration of a short-term voltage variation is given by the time range that the reference voltage is surpassed. The typical reference values for voltage sags, swells and interruptions are 0.9pu, 0.1pu and 1.1pu in rms voltage, respectively. The voltage threshold is specified with the purpose of detecting the start and the end of a short-term voltage variation. On polyphase systems, a sag begins when the V_{rms} voltage of one or more channels is below the sag threshold and ends when the V_{rms} voltage on all measured channels is equal to, or above, the sag threshold. On polyphase systems, a swell begins when the V_{rms} voltage of one or more channels is above the swell threshold and ends when the V_{rms} voltage on all measured channels is equal to, or below, the swell threshold. On polyphase systems, a voltage interruption begins when the V_{rms} voltages of all channels fall below the voltage interruption threshold and ends when the V_{rms} voltage on any one channel is equal to, or greater than, the voltage interruption threshold. The duration of a short-term voltage variation is the time difference between the beginning and the end of the event (IEC 61000-4-30, 2008).

The magnitude and duration of each event are used to calculate the single-event indices. In the following, two sorts of indices are considered. The first class of index aims to reflect the energy delivered to the load and the second one the severity of an event against a given power acceptability curve.

A. Energy related indices

Voltage swells, sags or short interruptions events at the load terminals will have impact, either because excess of energy is delivered during an occurrence of swell or some energy is not delivered to the load in the event of voltage sag or interruption. The impact depends on how much excess of energy is delivered or how much is not delivered.

The lost energy in a sag event and short interruption event is calculated as (Bollen, Sabin and Thallan, 2003)

$$E_{VS} = \left(1 - V_{pu}^2\right) \times t \tag{14}$$

where t is the duration and V_{pu} the retained voltage of the event.
The index dimension is time, and as such, when calculating the lost energy in a sag event, the index can be interpreted as the duration of an interruption leading to the same loss of energy for an impedance load as the voltage sag (Bollen, Sabin and Thallan, 2003).
The index E_{VS}, when applied to measure voltage swells, becomes

$$E_{VSW} = \left(V_{pu}^2 - 1\right) \times t \tag{15}$$

For three-phase calculation, the related energy for all three phases will be added. Equations (16) and (17) represent the three-phase energy indices for short-duration undervoltages and overvoltages, respectively, where t_i is the phase event duration.

$$\begin{aligned} E_{VS} &= E_{VS,a} + E_{VS,b} + E_{VS,c} \\ &= \left(1 - V_{pu,a}^2\right) \cdot t_a + \left(1 - V_{pu,b}^2\right) \cdot t_b + \left(1 - V_{pu,c}^2\right) \cdot t_c \end{aligned} \tag{16}$$

$$\begin{aligned} E_{VSW} &= E_{VSW,a} + E_{VSW,b} + E_{VSW,c} \\ &= \left(V_{pu,a}^2 - 1\right) \cdot t_a + \left(V_{pu,b}^2 - 1\right) \cdot t_b + \left(V_{pu,c}^2 - 1\right) \cdot t_c \end{aligned} \tag{17}$$

When any of the three-phase voltages do not violate the reference voltage level, the corresponding event duration is nil.
The indices E_{VS} and E_{VSW} measure the characteristic of single events. The site indices are calculated from the single-event indices of all events measured during a certain period of time. The site indices are used for evaluation of the compatibility between the electric power source and the load, and it can be used to assist the mitigation procedures. They are also used to inform the consumers about the quality of the local supply voltage.
The site-energy index (*SEI*) is the sum of the energy indices for all qualified events measured at a given site during a given period. The *SEI* is computed as

$$SEI = \sum_{i=1}^{n} E_{VS}(i) \tag{18}$$

where i is the event number and n the total number of events in a site.
System indices can also be evaluated by taking the average value of the site indices. They are defined as the ratio of the sum of all the site indices within a given region, in a given period. A region can be understood as the utility grid, the network of a given voltage level, a state or province or a country. The system indices can be used to evaluate the system performance and to compare year-to-year variations (Bollen, Salin and Thallan, 2003).

$$SEI_{sistema} = \frac{1}{N} \sum_{i=1}^{N} SEI(i) \tag{19}$$

where i is the site number and N is the total number of sites.

B. Event-severity index based on power acceptability curves

The power acceptability curves so-called CBEMA (Computer Business Equipment Manufacturers Association), ITIC (Information Technology Industry Council), and SEMI (Semiconductor Equipment and Materials International group) are commonly used curves to identify the acceptability of the service supply. When disturbances occur, one may plot the disturbance on the curve, and readily identify whether loss of load is expected. This information may be used to identify whether the distribution system design is adequate, whether power quality enhancement equipment is needed, and whether load vulnerability needs to be lessened (Thallan, Heydt, 2000).

The power acceptability curves may be used as references curves to identify the severity of short-term voltage variations. The event-severity index can be calculated for voltage sags and short interruptions from

$$S_e = \frac{1 - V_{pu}}{1 - V_{curva}(d)} \tag{20}$$

V_{pu} is the least rms voltage in pu in any of the three phases during the event, d is the event duration, and $V_{curva}(d)$ is the retained voltage of the reference curve to the same duration d. The algorithm for calculating the event-severity index S_e by comparing these values with the SEMI curve is given in Table I.

Magnitude	Duration	S_e
0,0 pu	d ≤ 20ms	$S_e = 1 - V$
0,5 pu	20ms < d ≤ 200 ms	$S_e = 2(1 - V)$
0,7 pu	200 ms < d ≤ 500 ms	$S_e = 3{,}3(1 - V)$
0,8 pu	500 ms < d ≤ 10 s	$S_e = 5(1 - V)$
0,9 pu	d > 10 s	$S_e = 10(1 - V)$

Table 1. The Event-Severity Index Definition for the SEMI Curve

As the voltage sag duration increases and the retained voltage decreases, the event-severity index becomes larger. For severity indices equal or smaller than one, it means that the reported voltage sags and short interruptions severities are in the acceptable operation region of the SEMI curve. On the other hand, when the severity indices are greater than one, it denotes that there is no compliance to the SEMI curve.

2.2.8 Composite power quality Indices

Practical composite indices of power quality are desired, which are capable, ultimately, of answering questions related to how much the quality of power is adequate to maintain appropriate performance and productivity of end-user equipment. Poor system quality levels on the other hand, imply either deficiency or excess in the overall system capabilities as designed by its planners, resulting in failure or malfunction of customer equipment and ultimately, dissatisfaction of the customer.

A single measurable index, designed by power quality factor PQF, is proposed in Munõz (2007) to reflect the overall system performance, taking into consideration different power quality aspects formulated previously and appropriate weighting factors ω_i that sum up to one.

$$PQF = \omega_1 RFD + \omega_2 VTHD + \omega_3 ITHD + \omega_4 VUF + \omega_5 IUF + \omega_6 OCF \quad (21)$$

The composite index gives a quick assessment of the power transfer quality at the selected point of the supply. An ideal PQF would be a factor equal to unity, on the other hand, a low value of *PQF* would indicate a high degree of power quality disturbances.

3. Protective relays

The last decades have seen enormous changes in relay technology, evolving from electromechanical, static, digital, to numerical relays. With the last generation of protection relays, these devices have become of great value in the evaluation of the power quality. In the following section the protective relays are described as well as the influence of the power quality on the relay performance. The test types performed on relays to evaluate the relay performance under prescribed voltage supply disturbances and the application of numerical relays for evaluating the power quality are also considered.

The selection of protective devices is based on the safety of personnel and equipment, and the quality and continuity of the electrical supply.

A protection relay is a device which by means of measuring power system quantities (currents and voltages) and processing them through its internal logic, has the capacity to control the operation of a circuit breaker. It is an important device for the safe and reliable operation of power systems.

In recent years, technological progress has led to the use of microprocessors in protective relay manufacture, resulting in the following:

- A significant increase in the amount of information that is processed by the relays.
- Easy calculation of electrical quantities such as for example harmonics.
- Secure and reliable exchange of digital information with remote locations.
- Continuous monitoring of protection relay integrity by self-supervision and auto-diagnostics.

Historically, protection, control, oscillography and metering subsystems were comprised of separate devices, each designed with a specific purpose in mind. Today, protection, control, metering and a wide range of other tasks can be performed by a single item.

The numerical relays have revolutionized protection, control, non-revenue metering, oscillography and communication in power systems. Functional integration, protection multifunction, new methods of communication, reduced physical size, and a vast amount of available information are but a few of the benefits of this revolution.

A protective numerical relay can be explained as a collection of interacting specific purpose blocks (Fig. 1) as Central Processing Unit (CPU) with one or more DSP microprocessors, Memory (RAM, ROM and Flash NVRAM), Analog Inputs, Signal conditioning and Analog to Digital Conversion, Digital Inputs and their Conditioning, Outputs, Communication, Display and Keyboard (HMI), Real Time Clock, the Power Source, and software tools (Carbajal et al, 2009).

By using multiple microprocessors to provide the necessary computational performance, a large number of functions previously implemented in separate items of hardware can be included within a single item. Microprocessor-based protective relays perform, besides multi-protection functions (such as directional/non-directional overcurrent protection, distance protection, undervoltage protection, negative sequence current protection, reclosing, etc.), measurement of power system quantities (current, voltage, etc.), fault/event/disturbance recording, internal fault diagnosis, circuit-breaker monitoring (state, condition, control), voltage transformer supervision, and continuous self-monitoring in order to verify correct operation of most components within the device. Self-monitoring capabilities may include the alternating current (ac) signal inputs, analog measuring circuits, processors and memory for measurement, protection, data communications, trip circuit monitoring, and protection or data communications signals. For those conditions, failure of a self-monitoring routine generates an alarm and may inhibit operation to avoid false trips. Not all functions may be found in a particular relay, nevertheless, the protection relay no longer performs a basic protection function but is becoming an integral and major part within the overall network automation scheme.

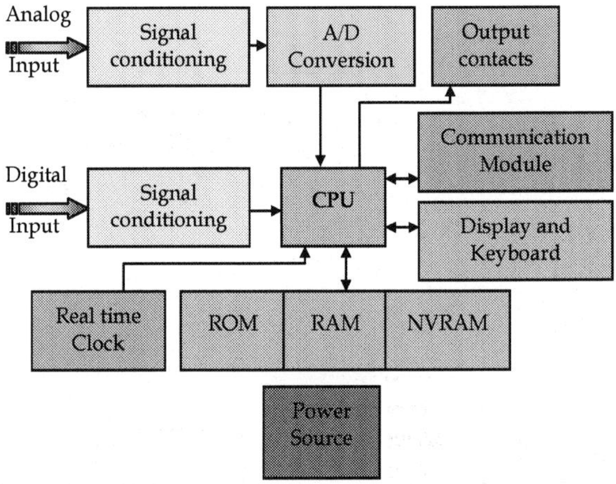

Fig. 1. Numerical relay architecture (Carbajal et alli, 2009)

The main difference between digital and conventional relays (electromechanical and solid-state relays) pertains to the method of input signal processing. In the case of numerical relays, input signals are converted into digital form within the analogue input system before being analysed by the processor. Numerical relays have their logic implemented in software and microprocessor technology.

The numerical relays are equipped with additional functions that have no direct relation with the protective elements but give an additional value to the system. Some of these are metering, monitoring, setting groups, fault record, communication and reports. The communication capability, not only with a center but also with each other, facilitates the deployment of overall-system wide protection and control philosophy. The advanced metering and communication protocol ports allow the relay to become a focal point in a SCADA system.

The numerical relays are hierarchically integrated to the digital system automation (DSA) of the power substations. They are at Level 1, as part of the positioning control unit (PCU), connected straight to the substation power equipment (Level 0). At Level 2, are the substation control unit (SCU) and the substation supervisory system (SCADA), which communicates with Level 3 in the Network Operation Center (NOC).

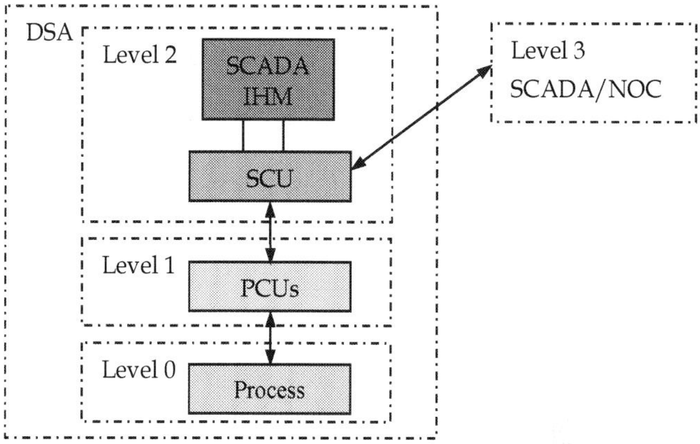

Fig. 2. Functional hierarchy of digital system automation of a power substation.

Although numerical relays are based on signal treatment, they are also affected by the power quality whereby their operation may be affected by disturbances and distortions in the waveform.

3.1 The influence of power quality on protection relays

Relays are exposed to disturbances in currents and voltages in various circumstances. Since the protection relays are required to operate correctly under abnormal system conditions, it is essential that their operation is guaranteed under such conditions (Cease and Kunsman, 2003), (Wang, 2003), (Zamora et al, 2004).

Protective relays are designed to operate when a fault in a power system occurs. However, non-fault voltage and current disturbances may lead to the inadvertent detection of a fault. Relay malfunctions, can result in a failure to trip for a fault condition and subsequently 'cascade' the event to an upstream protective device thus, affecting more customers. Equally a relay may trip when no fault is present which may also have a negative effect upon the customers.

The main steady-state disturbances that can affect the performance of protective relays are harmonics and frequency deviation. The reason why harmonics and frequency deviation are of concern, is that in case of distortion and/or frequency deviation, the reaction to a fault may become slower and the immunity from disturbances may decrease. The question is to what extent the harmonics and frequency deviation can affect the performance of protective relays (Wang, 2003).

The main negative influence that harmonic distortion of either voltages or currents can have on the digital relays, is that they can fail to trip under fault condition, or operate under a no fault condition. When measuring current and voltage, the relays must measure alone, the fundamental component of the signal. For this purpose, it is necessary to filter the harmonic and to calculate the rms value of the fundamental frequency only. The Fourier transform is the more used algorithm to filtrate the harmonic distortion (Cease and Kunsman, 2003).

Frequency variations can influence in the measurement ability of a digital relay and may include some distort in the measurement of some magnitude readings. The magnitude measurement of a relay (voltage, current) is based on making a sampling. This process has a defined number of samples (i.e. from 8 to 64), during one period. From this sampling, the rms value of the signal will be obtained. An error will occur if an increase or decrease of the frequency takes place and the sampling period of time is kept fixed whilst obtaining the rms. In order to achieve correct measurement, the relay needs to estimate power system frequency and must adjust the sampling rate to the network frequency regularly (Cease and Kunsman, 2003), (Zamora et al, 2004).

The main transient disturbances that can affect the performance of protective relays are due to component switching. Although component switching is a short duration disturbance, the relatively high deviations in voltages and currents are a potential threat to the correct operation of protective relays. Tripping, when no faults are present, is prevented in existing equipment by use of low-pass filters and verification intervals. However, this scheme will slow down the generation of a tripping signal in case of a fault (Wang, 2003).

One of the most visible power quality disturbances that have effect in the production sectors is the voltage sags. It almost always results from faults in the transmission and distribution systems, with widespread propagation over the power system. Voltage sags affect the measurements of other important power quality parameters. The power quality measurement standard IEC 61000-4-30 (2008) adopts a concept of flagging the voltage dips to take into consideration their influence on the measurements of power quality parameters. Voltage sags may affect the frequency measurements which may lead to frequency relay malfunction and to unexpected load shedding (Bronzeado, Zimath and Coser, 2008).

3.2 Relaying tests

To study the impact of power system disturbances on protective relays, it is necessary to apply some typical disturbances in protective relay testing. Tests based on IEC61000-4-7, IEC61000-4-15, IEC61000-4-11 and IEC60255-11 will be performed on multifunction numerical relays using a programmable ac/dc, one- and three-phase disturbance source. The relays responses will be presented.

3.2.1 The influence of power quality of the supply voltage on the performance of a numerical relay

To ensure consistent reliability and proper operation, protective relay equipment must be evaluated and tested. The purpose of testing protective relays is to ensure correct operation of the relay for all possible power system conditions and disturbances. The testing of protection relays may be divided into four stages: type tests, routine factory production tests, commissioning tests and periodic maintenance tests. Type tests are required to prove that a relay meets the published specification and complies with all relevant standards. Routine factory production tests are conducted to prove that relays are free from defects during manufacture. Testing will take place at several stages during manufacture. Commissioning tests are designed to prove that a particular protection scheme has been installed correctly, prior to setting to work. Periodic maintenance checks are required to identify equipment failures and degradation in service, so that corrective action can be taken.

The IEC 60255 - Part 11 (2008) specifies type testing on the auxiliary power supply of protective relays. The objective of the tests is to assess whether the protective relay will operate correctly when energized and subjected to voltage sags, voltage short interruptions, alternating components in dc (ripple), gradual shut-down/start-up test and polarity inversion. A programmable source is used to generate the disturbance types applied to the relays input power supplies. The test results are presented according to the acceptance criteria established in the international standard.

The protective relays can be powered by either an a.c. or a d.c. power supply. The auxiliary supply can be supplied from a number of sources or safe supplies such as batteries, uninterruptible power supplies, generators, etc. When the energizing input is a d.c. power supply, the relays generally require a reliable source of d.c. power including measures to prevent damage to internal circuitry. Substation environments are particularly hostile to electronic circuits due to various, commonly found, forms of electrical interference such as switching operations and the effect of faults. The relays must be therefore immune to conducted and radiated interference from the electrically noisy substation environment. While it is possible to arrange for the dc supply to be generated from the measured quantities of the relay, this has the disadvantage of increasing the burden on the current transformers (CT) or voltage transformers (VT), and there will be a minimum primary current or voltage, below which the relay will not operate. This directly affects the possible sensitivity of the relay. So provision of an independent, highly reliable and secure source of relay power supply is an important consideration. To prevent mal-operation or destruction of electronic devices during faults or switching operations, sensitive circuitry is housed in a shielded case to exclude common mode and radiated interference. Since the protection relays are required to operate correctly under abnormal system conditions, it is essential that their operation is guaranteed under such conditions.
A numerical relay commonly found in the power distribution systems is used for testing both a.c. and d.c. auxiliary supplies. The type, levels and duration of tests according to IEC 60255-11 are given in Table 1.

TYPE OF PHENOMENA	TEST SPECIFICATIONS	UNITS
Voltage dips (for dc power supply)	0 10 to 1000	% residual voltage ms
	40 200	% residual voltage ms
	70 500	% residual voltage ms
Voltage dips (for ac power supply)	0 0,5 to 25	% residual voltage cycles
	40 10/12 at 50/60 Hz	% residual voltage cycles
	70 25/30 at 50/60 Hz	% residual voltage cycles
Voltage interruptions (for dc power supply)	0 5	% residual voltage s
Voltage interruptions (for ac power supply)	0 250/300 at 50/60 Hz	% residual voltage cycles
Alternating component in dc (ripple) (for dc power supply)	15 % of rated dc value 100/120 at 50/60 Hz	V Hz, sinusoidal waveform
Gradual shut-down/start-up (for dc power supply)	60 5 60	s, shut-down ramp min, power off s, start-up ramp
Reversal of dc power supply polarity	1	min

Table 1. Type, level and duration of tests

The disturbances should not cause any malfunction in the relay operation. Malfunctions include the operation of output relays and watchdog contacts, the resetting of microprocessors, alarm or trip indication, acceptance of corrupted data over the communication link and the corruption of stored data or settings. The voltage sag and voltage interruption tests are to determine the maximum time-length that a relay can withstand a sag/interruption in the auxiliary supply without de-energising and that when this time is exceeded whereby it does transiently switches off, that no misoperation occurs.
The ac ripple superimposed on dc supply test determines that the relay is able to operate correctly with a superimposed ac voltage on the dc supply. This is caused by the station battery being charged by the battery charger.
The gradual shut-down/start-up test simulates a failed station battery charger, which would result in the auxiliary voltage to the relay slowly ramping down. The ramp up part simulates the battery being recharged after discharging. The relay must power up cleanly when the voltage is applied and it must not malfunction.
The criteria for acceptance are classified in two groups, A and C. The criterion A is related to the relay response to the a.c. and d.c. power supply voltages to 0% at specific time range and alternating component in d.c. for d.c. power supply. The criterion C in turn is related to the other types of phenomena as voltage sags other than to 0%, voltage interruptions for a.c. and d.c. power supply, gradual shut-down/start-up for d.c. power supply and reversal of d.c. power supply polarity (IEC 60255-11, 2008). The effects of the tests should be assessed at the maximum and minimum values of the voltage according to the relay technical documentation.

A. Test of voltage dip applied to the d.c. power supply

Table 2 presents the test results for the applied dc voltage dips to 0%, 40% and 70% to the relay at the lowest and highest voltage operation range of Vmin=87 Vdc and Vmax=300 Vdc, respectively. Different time duration has been tested for voltage sag to 0%.

DC voltage dip to (%)	Test duration (ms)	Test conclusion	
		Vmin	Vmax
0	10	Pass	Pass
	20	Pass	Pass
	30	Pass	Pass
	50	Pass	Pass
	100	Pass	Pass
	200	Fail	Pass
	300	Fail	Pass
	500	Fail	Pass
	1000	Fail	Pass
40	200	Fail	Pass
70	500	Pass	Pass

Table 2. Voltage dip test applied to the d.c. energizing input.

For the lowest voltage, voltage sags to 0% with time duration exceeding 200 ms inclusive, has led the relay to a temporarily loss of function. The relay automatically switched-off resuming the functions at the end of the test, except its communication with the software that had to be manually re-established.

B. Test of voltage dip on the a.c. power supply

Table 3 shows the test conclusion when the a.c. voltage dips to 0%, 40% and 70% has been applied to the relay ac power supply for the minimum and maximum voltage operation range of 80 Vac and 265 Vac, respectively.

DC voltage dip to (%)	Test duration (cycles)	Test conclusion	
		Vmin	Vmax
0	0.5	Pass	Pass
	1	Pass	Pass
	2.5	Pass	Pass
	5	Pass	Pass
	10	Pass	Pass
	25	Fail	Pass
40	12	Pass	Pass
70	30	Pass	Pass

Table 3. Voltage dip test applied to the a.c. energizing input.

The relay has failed, switching off when a voltage dip to 0%, 25 cycles was applied for a reference a.c. voltage of 80 Vac. The relay has operated normally to the other conditions.

C. Test of voltage interruption applied to the d.c. and a.c. power supplies

Short voltage interruption of 5 s has been applied to the relay d.c. and of 300 cycles to the a.c. input, considering the lowest and the highest voltage operation range. The relay has failed during the test, showing no compliance to the standard.

D. Ripple component test on the d.c. power supply

A voltage ripple of 15% of the rated dc voltage, 120 Hz has been applied to the lowest (87 Vdc) and to the highest (300 Vdc) d.c. voltage. The test has been successful.

E. Gradual shut-down and start-up for d.c. power supply

The IEC 60255-11 requires that a 60 s shut-down ramp and a 60 s start-up ramp with 5 min power off in between them, be applied to the d.c. power supply in order to observe the shut-down limit and the lowest start-up voltage. The test is applied to both the dc voltage operating limits, i.e., Vmin=87 Vdc and Vmax=300 Vdc. The results are shown in Table 4.

DC VOLTAGE LIMIT	V_{MIN} (V)	V_{MAX} (V)
Shut-down	63	63
Star-up	81	85

Table 4. Gradual shut-down and start-up test to the dc energizing input.

F. Polarity inversion

Reverse polarity for the power supply input has been applied for 1 min. The relay has shown normal performance.

3.3 Protection relays with power quality analysis function

The importance of power quality assessment is ever increasing due to increased use of equipment and processes more susceptible to power system disturbances. As the power quality measurement process is a demanding one and power quality analyzers have developed into specialized, complex and expensive equipment, the modern numerical relays are able to measure and register values to power quality evaluation.

The protection relays are probably the only devices that are virtually mandatory on circuits of any significant rating, and as they have been equipped with advanced digital processing and statistical software, making them very similar in architecture to the power analyzers, they can to some extent to replace the power quality analyzers.

In order to evaluate the application of protective numerical relays to power quality assessment, two digital relays, named IED 1 and IED 2, with different characteristics and manufactures are investigated. The mass memory of the protective digital relays is, in general, small, with capacity to store only a few numbers of events. The number of events to be likely stored depends on the programmed number of cycles to be captured before and after the disturbance. The larger the estimated event duration, the lesser events will be stored. However, an on-line monitoring system can be applied to accomplish the recurring emptying of the relay mass memory.

The sampling rates of the IED 1 and IED 2 are 24 samples/cycle and 16 samples/cycle, respectively, and of the disturbance analyzer 128 samples/cycle for the programmed mode of operation. The resolution is 16 bits for both IEDs and the disturbance analyzer has 12 bits of resolution and 0.5% of accuracy.

Besides having a greater sampling rate, the IED 1 has larger storage capacity and it can store a larger number of events. Both IEDs have been programmed to capture up to 30 cycles or 500ms in 60Hz of events duration. IED 1 is able to capture voltage samples up to 1s or 60 cycles per event, whilst IED 2 can capture, at most, 30 cycles per event. The number of pre-fault cycles to be captured can also be programmed, and it has been set to 5 cycles for both IEDs.

A programmable disturbance source has been used to apply the voltage disturbances to the relays and to a disturbance analyzer for comparison of the results. A threshold voltage equal to 0.9pu was used to trigger the relays and the disturbance analyzer. From the equipment databases the rms voltages based on (13) and the power quality indices E_{VS} and S_e based on (16) and (20) were calculated for different types and severities of voltage sags and short interruptions. The indices from the relays were compared to those calculated from the disturbance analyzer.

Three-phase and one-phase voltage sags and short-interruptions of 0.7pu and 0.0pu, with duration of 3 and 20 cycles, were applied to the equipment under test following the recommendation of IEC 61000-4-11 (2001). In the following section, some test results are presented and error calculated for each considered event, taking into consideration the disturbance analyzer results as the reference.

3.3.1 Voltage sag – Event 1

Fig. 3 shows the rms voltage in pu versus time to IED 1, IED 2 and the disturbance analyzer to one-phase voltage sag to 0.7pu, 3 cycles of duration. The voltage curves for all three equipment under test (EUT) do not show significant differences.

Table 5 presents the voltage sag magnitudes, durations and the E_{VS} indices for the EUT. It can be seen from Table 5 that the voltage magnitude and the E_{VS} errors of the IED 2 are larger than of the IED 1. Both IEDs have presented the same event duration measurement and therefore the same error. Furthermore, it can be observed that the voltage magnitude errors are larger than the event duration errors.

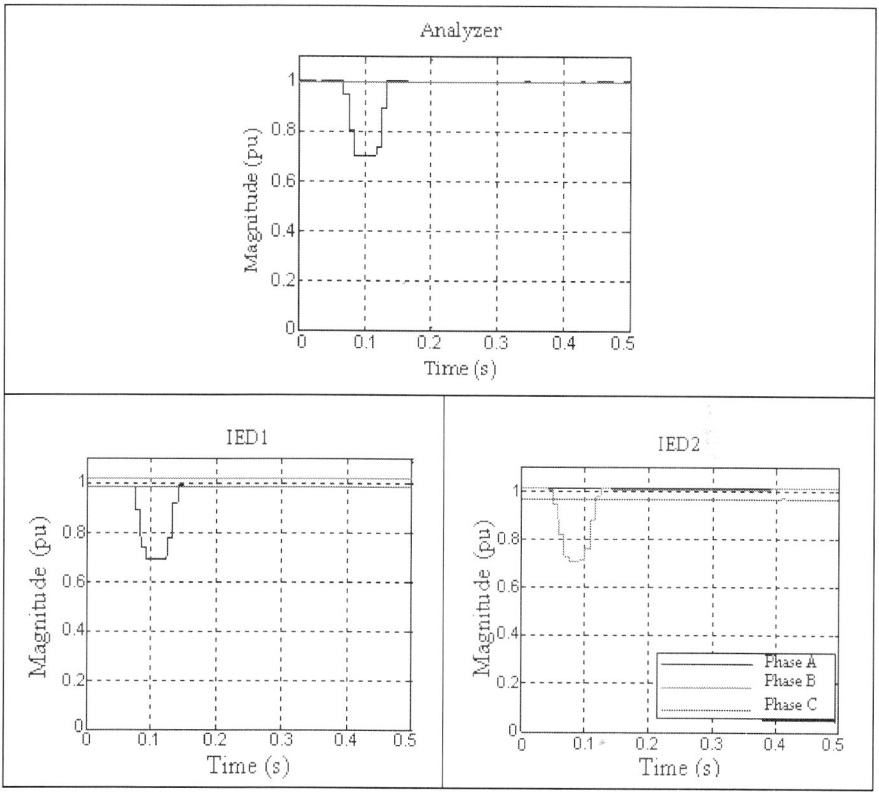

Fig. 3. RMS curve to voltage sag to 0.7pu, 3 cycles.

Event 1					
Event Characteristics	Analyzer Value	IED 1 Value	Error Value	IED 2 Value	Error Value
Magnitude (pu)	0.7006	0.6931	-0.0075	0.7093	0.0087
Duration (s)	0.0583	0.0582	-0.0001	0.0582	-0.0001
E_{VS} (s)	0.0297	0.0302	0.0005	0.0289	-0.0008

Table 5. One-phase Voltage Sag to: 0.7pu / 3 cycles.

3.3.2 Voltage sag – Event 2

The rms voltage curves to one-phase voltage sag to 0.7pu, 20 cycles of duration for the EUT are depicted in Fig. 4. The voltage sag magnitudes and durations and the single-event indices E_{VS} are given in Table 6.

From Table 6 it can be noticed that for a longer event of the same magnitude as the Event 1 the index E_{VS} has increased. Although both IEDs have the same event duration measurement, the error for a longer event has increased.

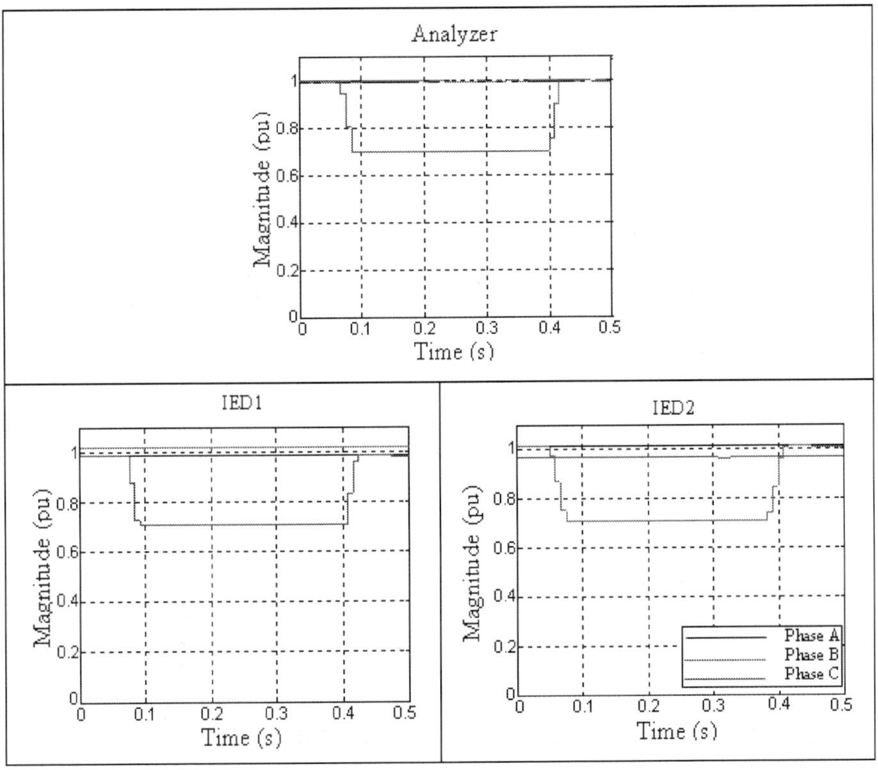

Fig. 4. RMS curves to voltage sag to 0.7pu, 20 cycles.

Event 2					
Event Characteristics	Analyzer	IED 1		IED 2	
	Value	Value	Error	Value	Error
Magnitude (pu)	0.7012	0.7114	0.0102	0.7089	0.0077
Duration (s)	0.3333	0.3410	0.0077	0.3410	0.0077
E_{VS} (s)	0.1694	0.1684	-0.0010	0.1696	0.0002

Table 6. One-phase Voltage Sag to: 0.7pu / 20 cycles.

3.3.3 Short Interruption – Event 3

The test results to one-phase 20 cycles short interruption are presented in Table 7 and the rms voltage curves are shown in Fig. 5.

From Table 7 it can be seen that the E_{VS} indices have increased significantly reflecting, the event severity increasing. In comparison with the analyzer measurement, the errors from the relays measurement have increased when the event magnitude severity increased.

Event 3					
Event Characteristics	Analyzer	IED 1		IED 2	
	Value	Value	Error	Value	Error
Magnitude (pu)	0.0593	0.0036	-0.0557	0.0000	-0.0593
Duration (s)	0.5083	0.4491	-0.0592	0.4408	-0.0675
E_{VS} (s)	0.5065	0.4491	-0.0574	0.4408	-0.0657

Table 7. One-phase short Interruption: 0.0pu / 20 Cycles.

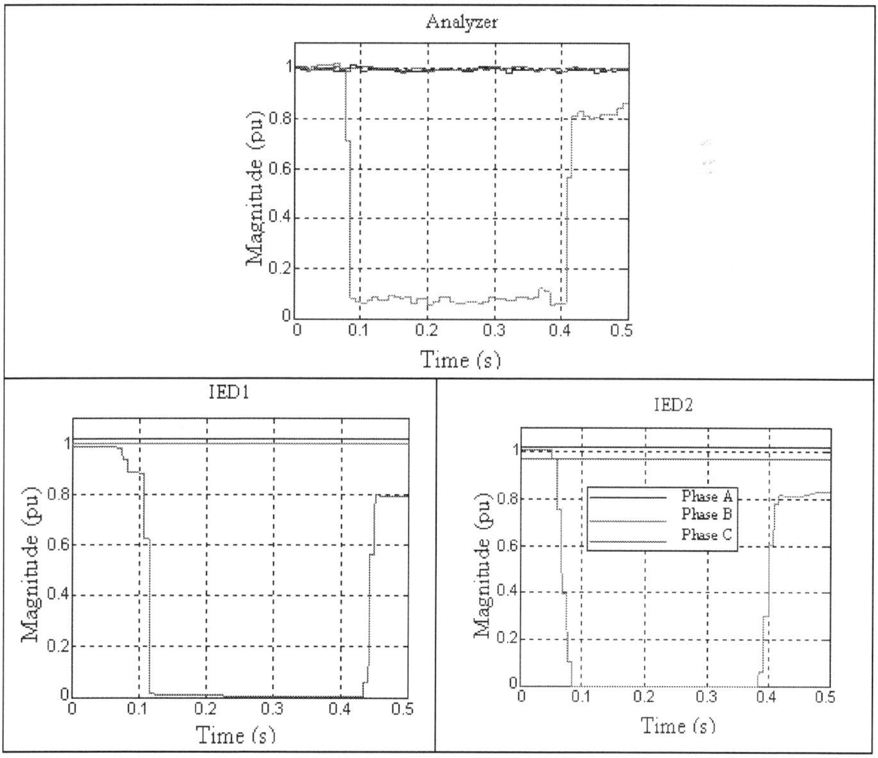

Fig. 5. RMS Voltage to one-phase 20 cycles short interruption.

3.3.4 Event severity index S_e – Event 4

The event-severity indices S_e based on the algorithm presented in Table 1 when calculated to the three events previously described are shown in Table 8.

It can be noted that for event 3, the S_e indices is greater than 1 for all the EUT, which denotes that the event severity is under the SEMI low voltage ride-through curve. In contrast, the indices S_e are smaller than 1 for events 1 and 2, indicating that the SEMI low voltage ride-through curve has not been violated and that the event characteristics are in the normal operating region of the SEMI curve as pointed out in Fig. 6. It is worth observing that each time interval of the SEMI curve corresponds to a different event-severity scale, and as a result the S_e ordinate scale is not shown in Fig. 6.

Event	Analyzer	IED 1		IED 2	
	Value	Value	Error	Value	Error
1	0.5987	0.6137	0.0150	0.5814	-0.0173
2	0.9959	0.9619	-0.0340	0.9705	-0.0254
3	3.1043	3.2881	0.1838	3.3000	0.1957

Table 8. Event-severity Index Se.

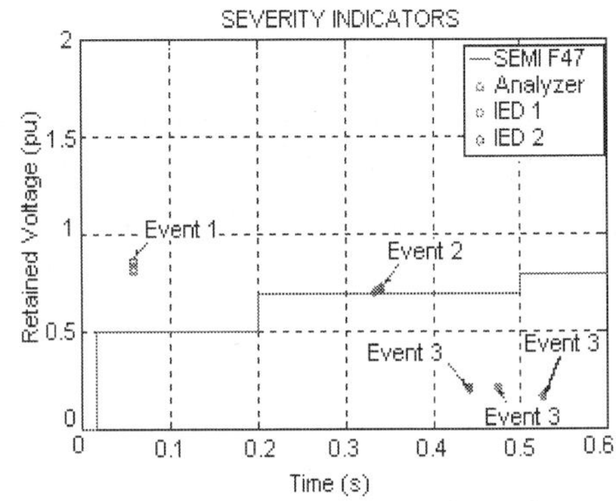

Fig. 6. SEMI curve and the Severity indicators

4. The last development on Wide Area Monitoring, Protection and Control

For improving power system security and performance, new emerging technologies in the area of Wide Area Monitoring, Protection and Control concepts (WAMPC) have been recently proposed. Technological advancements in measuring and communicating time synchronized real-time power system parameters, through highly secure high-speed telecommunication networks, provide opportunities for implementing intelligent protection and control schemes with advance warning mechanisms.

Conventional protection devices are designed to protect individual assets in a power system, for which locally measured values are usually sufficient. In contrast, network control systems with a system-wide view of the processes within the power system require dynamic information about power flows and the phase angles of the electrical quantities at various network nodes in order to safeguard the power system stability. The measurement data can be provided by phasor measurement units (PMU) and open up new fields of application for network control under the concept of Wide Area Monitoring Protection & Control (WAMPC).

Phasor Measurement Units (PMU) have revolutionized the way state estimation is being performed. Their unique ability to measure the voltage and current phasors (magnitude and phase angle) with very high accuracy makes them extremely useful in modern day energy management systems.

5. Conclusions

The quality of the voltage supply is a common concern of the power utilities and customers. The importance of power quality assessment is ever increasing due to increased use of equipment and processes more susceptible to power system disturbances.

Power quality analyzers available in the market are expensive and specialized to be applied on a mass scale. The protection relays are devices that are mandatory on circuits of any significant rating. As they have been equipped with advanced digital processing and statistical software making them very similar in architecture to the power analyzers, they can to some extent, replace the power quality analyzers. Besides the multi functionality of the digital relays, they can additionally contribute to the power quality evaluation of the power system, saving investments in power analyzers.

As a digital technology based device, the protective relays can be sensitive to power quality variations. Performing tests on protection relays is a required practice in order to evaluate the relay performance and characteristics, and to certify the equipment compliance to the requirements of technical standards. The numerical relays perform many functions besides protection which ought to be accurate and promptly operate whenever required. The sensitivity and the immunity of the relays are evaluated through tests performed according to international standards like IEC 60255-11, IEC 61000-4-11, IEC 61000-4-7 and IEC 61000-4-15. Results of tests performed in a well known relay trade name relay when it's a.c. and d.c. power supply are energized and subjected to voltage sags, interruptions, alternating components and gradual shut-down and start-up have shown that the relay failed under d.c. voltage sag and d.c. and a.c. short voltage interruption.

The modern multifunction digital relays are able to measure and register values useful to the power quality assessment. Power providers are building databases to assess utility service quality performance indicators. The widespread use of digital relays can assist in achieving this task. Feasible indices are needed to address the security, quality, reliability and availability of the power system. Tests performed using two numerical relays and a disturbance analyzer have proved that the relays database can be used to calculate indices to identify the severity of disturbances and to evaluate the quality of the site and the system. Indices have been used based on both magnitude and duration of short-duration voltage variation in all three phases. The least rms voltage over the event duration has been used to calculate the index of energy (E_{VS}) and the event severity index (S_e).

Performance indices calculated from the database of two digital relays with sampling rates of 24 and 16 samples per cycle and resolution of 16 bits have been compared to indices calculated from a disturbance analyzer of 128 samples per cycle and resolution of 12 bits. The digital relay IED 1 with 24 sampling rate has presented smaller errors to all events tested. The digital relays have captured correctly all the considered events. As the voltage sag and short interruptions are power frequency events the frequency samplings of the IEDs have proven to be suitable.

6. References

Albert, H.; Golovanov, N.; Kot, A. and Brozek, J. (2008). Frequency Variation. *Handbook of Power Quality*. John Wiley & Sons. Ltd. Edited by Baggini, A. ISBN 978-0-470-06561-7, England.

Bollen, M. H. J.; Sabin, D. D. and Thallam, R. S. (2003). Voltage-sag indices- recent developments in IEEE P1564 Task Force. Invited paper presented at the joint symposium CIGRE/IEEE-PES on "Quality and Security of Electric Power Delivery Systems", Montreal, 7-10 October 2003.

Brozeado, H. S.; Zimath, S. L. and Coser, J. (2008). Effect of Voltage Dips on Frequency Relays: Case Study of Power Quality Improvements in a Petrochemical Plant. *International Conference on Harmonics and Quality of Power* - ICHQP, pp.1-5, Autralia, October 2008, Wollongong.

Carvajal, I. O.; Sedano, E. C. and Jiménez, B. A. R. (2009). An Electric Energy Distribution Systems Protection Microprocessor Based Relay, 52nd *IEEE International Midwest Symposium on Circuits and Systems*, ISBN: 978-1-4244-4479-3, Mexico, August 2009, Cancun.

Cease, T. W. and Kunsman, S. A. (2003). Protective Relaying and Power Quality. IEEE PSRC Working Group Report.

Colaço, A. L. G.; Melo, N. X.; Leão, R. P. S.; Barroso, G. B.; Sampaio, R. F. and Ciarlini, R. C. (2008). Power Quality Assessment and the Application of Protective Relays, *International Conference on Harmonics and Quality of Power* - ICHQP, pp.492-498, Autralia, October 2008, Wollongong.

Duncan, B. K. and Bailey, B.G. (2004). Protection, metering, monitoring, and control of medium-voltage power systems, *IEEE Transactions on Industry Applications*, vol. 40, no. 1, pp. 33 - 40, Jan./Feb. 2004.

Gosbell, G. V. (2002). Power Quality Monitoring - Plant Investigations, Technical Note No.5, Integral Energy Power Quality Centre, University of Wollongong, January 2002.

IEC 61000-4-11 *Testing and Measurement Techniques - Voltage Dips, Short Interruptions and Voltage Variations Immunity Tests*. Edition 1.1 2001-03.

IEC 61000-4-30 *Testing and measurement techniques - Power quality measurement methods*. Edition 2.0 2008-10, ISBN 2-8318-1002-0, 134pp.

IEC 61255-11 *Measuring Relays and Protection Equipment - Voltage dips, short interruptions, variations and ripple on auxiliary power supply port*. Edition 2.0 2008-10. ISBN 2-8318-1003-9, 26pp.

IEEE 1159 *Recommended Practice for Monitoring Electric Power Quality*, 1995, ISBN 1-55937-549-3, 76pp.

Mackrell, A.; Fantana, N.; Steingräber, W.; Kopejtková, D.; Ravetta, C.; Jung, T.; Schumacher, M.; Anguas J.; Ford, G.; Woodcook, D.; Müller, L.; Moore, P.; Mercier, A.; and Skog, J. (2010). Obtaining Value form On-line Substation Condition Monitoring. *Electra*, No.249, April 2010, pp. (21-25), ISSN 1286-1146.

Munõz, A. M-. (2007). *Power quality: mitigation technologies in a distributed environment*, Springer, ISBN 9781846287718.

Thallam, R.S. and Heydt G. T. (2000). Power Acceptability and Voltage Sag Indices in the Three Phase Sense. IEEE Proc. PES Summer Meeting, 16-20 July 2000, Seattle WA, USA, pp. 905-910.

Zamora, I.; Mazón, A. J.; Valverde, V.; Torres, E. and Dyśko, A. (2004). Power Quality and Digital Protection Relays, International Conference on Renewable and Power Quality - ICREPQ 2004, pp.1-8, Barcelona, Spain, April 2004.

Calibration of High Voltage Transducers for Power Quality Measurements in Electric Networks

Hédio Tatizawa[1], Erasmo Silveira Neto[2], Geraldo Francisco Burani[1], Antonio A.C. Arruda[1], Kleiber T. Soletto[1] and Nelson M. Matsuo[1]

[1]*University of São Paulo*

[2]*Cia de Transmissão de Energia Elétrica Paulista - ISA CTEEP*

Brazil

1. Introduction

The term power quality is, in general, closely related to the quality of the voltage. Considering the widespread presence of sensitive loads in the electric grid, and the increasing awareness of the consumers concerning the quality of the power supply, the control and measurement of power quality parameters regarding harmonics, interharmonics, sags, swells, and others are increasingly becoming much more important. Most of the necessary calibration procedures for power quality monitors and power quality analyzers are already defined in the international standards, mainly in the IEC 61000 series (IEC, 1995) and ANSI/IEEE Standards. This chapter presents results of a research which aimed to develop a methodology for the calibration of high voltage transducers for power quality measurements in high voltage networks, considering that such kind of procedures have not been established in the pertinent standards yet (Bradley et al., 1985; Seljeseth et al., 1998). In this research it is also considered that the conventional high voltage laboratory is not suitable for power quality tests. Thus, some improvements are needed regarding such matter. In this development, modelling and computer simulation using ATP - Alternative Transients Program (ATP, 1987) were used to assess both the frequency response of the test setup, and the design of the reactive compensation of the test circuit.

2. Test setup development

Capacitive voltage dividers (CVD) are commonly used for the measurement of power quality parameters in power systems networks (Dugan et al., 2004), thanks to their modularity and easy installation in transmission and distribution substation environment. Fig. 1 shows a typical installation at field, in a 345kV transmission substation. In this Fig. 1 the high voltage branch of the CVD is shown, composed of six 500pF modular capacitances, nominal voltage 50kV. Further to Fig. 2, all the capacitances of the high voltage branch (C1) are identical with nominal value of 500pF. For the measurement, the number of 500pF capacitances can be changed according to the expected voltage to be measured, in order to limit the voltage on the C2 capacitance of the secondary low voltage branch.

This paper shows the development of the test circuit for the calibration of voltage transducers, focused on measurements of power quality disturbances in power systems high voltage networks. This kind of development faces many levels of difficulties, considering that the generation of stabilized and well defined power quality disturbances, in the high voltage range (for example, over 1kV), for calibration purposes, requires the adaptation of the conventional high voltage laboratory equipment. Overall, the conventional high voltage laboratory is only equipped with high voltage sources for generating power frequency (60Hz or 50Hz) and impulse (atmospheric and switching) high voltage waveforms, used in dielectric tests of high voltage equipment insulation (IEC, 1994). For calibration of high voltage transducers used in power quality disturbances measurements, additional waveforms are necessary such as voltage harmonics, sags (or dips), swells, etc. Therefore, in order to achieve the calibration circuit, the test circuit components were defined as follows:

Fig. 1. Capacitive voltage divider (CVD) - installation for measurements in a 345kV substation.

- arbitrary waveform voltage source for generating sinusoidal waveforms, with low harmonic distortion, considering harmonic frequencies up to the 50th order (3000Hz), and generation of composite waveforms (fundamental frequency + harmonics), with enough power capacity for the calibration tests. In this research, a conventional commercial power quality generator was used for such purpose.

- a step-up high voltage test transformer, fed at the low voltage side by the arbitrary waveform voltage source, to produce in the high voltage side the waveforms generated by the source (considering composite waveforms and harmonics), with enough power required by the calibration tests. The option of generating high voltage waveforms in such way was motivated either by absence or by the non availability of high voltage sources for the waveforms required by the research in the calibration tests, mainly considering high voltage levels found in transmission systems, in the hundreds of kV range. The expected load for the test transformer during the calibration tests is supposed to be of capacitive nature, mainly capacitive voltage dividers (CVD) and capacitive voltage transformers (CVT). On the whole, CVTs present very high values of capacitance (few thousands of pF), thus becoming a very heavy load for the high voltage test transformer.
- Capacitive voltage dividers composed of 500pF modules, voltage 50kV.
- Power quality analyzer for the transducers calibrations. In this research, a class A (IEC, 2002) commercial power quality analyzer was used.

At the initial stage tests, a high voltage transformer with rated voltage 220V/100kV and rated power 10kVA was used. The test circuit is shown in Fig. 2.

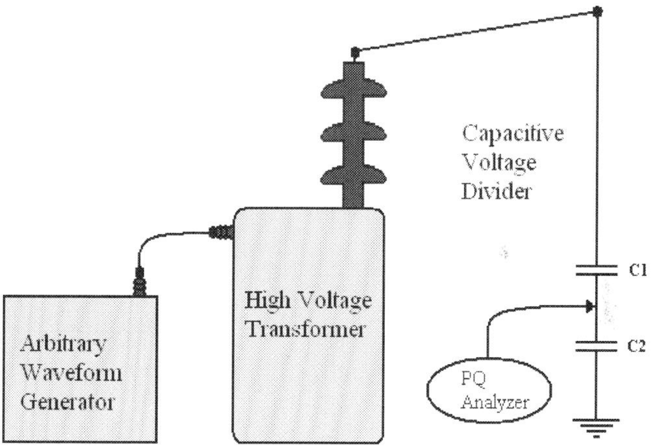

Fig. 2. Tests performed with the arbitrary waveform voltage source - test setup.

2.1 Electric model of the high voltage transformer

The option of using a high voltage transformer to generate high voltage for the calibration tests implies in introducing a series equivalent reactance of the transformer in the test circuit. This option was eventually necessary, considering the non availability of a commercial high voltage source with the capability of generating the required waveforms for the calibration tests. The series association (sum) of this leakage reactance with the load capacitances (CVD - Capacitive Voltage Dividers and CVT - Capacitive Voltage Transformers) results in a resonating (tuned) circuit for certain harmonic frequencies.

For the electrical modelling of the high voltage test transformer, the equivalent circuit was obtained by means of impedance voltage and no-load loss tests. The obtained equivalent circuit, for 60 Hz, is shown in Fig. 3.

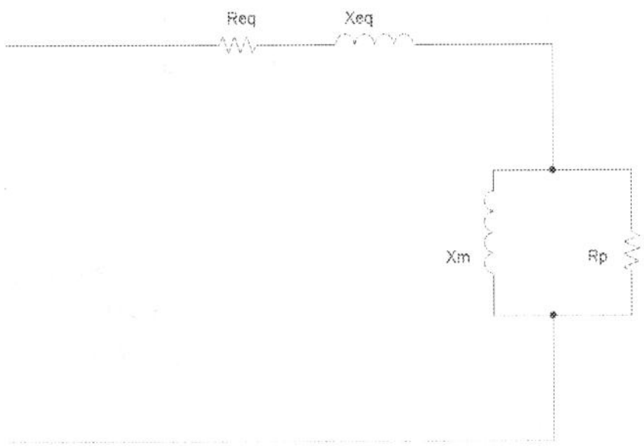

Fig. 3. Equivalent circuit of high voltage transformer where: Req -equivalent resistance, Xeq. - equivalent leakage reactance, Xm - magnetizing reactance, Rp - equivalent no-load loss resistance.

Considering both the impedance voltage and no-load loss tests, the value of the equivalent circuit parameters are shown in Table 1.

Parameter	Values referred to low voltage side (Ω)	Values referred to high voltage side (Ω)
Req (Ω)	0,151	31,2k
Xeq (Ω)	0,673	139,0k
Rp (Ω)	322,7	66,7M
Xm (Ω)	116,3	24,0M

Table 1. Equivalent circuit of the high voltage test transformer

2.2 Electric model of the test setup

Considering the obtained values of the step-up test transformer (referred to the high voltage side) and CVD, the equivalent circuit of the test setup is shown in Fig. 4.

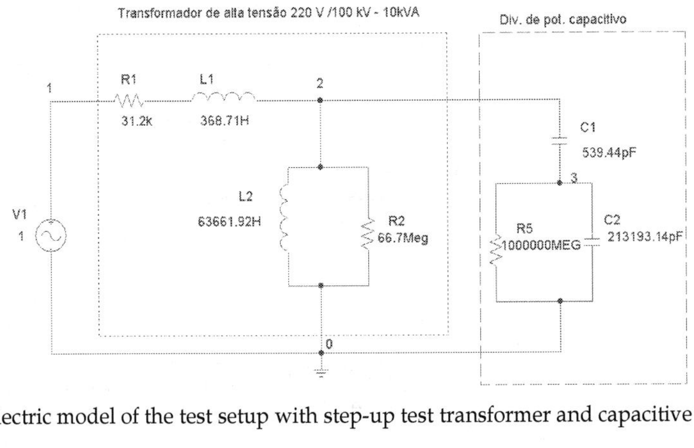

Fig. 4. Electric model of the test setup with step-up test transformer and capacitive voltage divider.

Obs.: Resistance R5 is included for a better numerical stability in computer simulations, without affecting overall results due to its very high value of 1.000.000M Ω.

2.3 Test setup modelling and computer simulation

Aiming to analyze the test circuit behavior under harmonic voltages, modelling and computer simulation were performed using ATP - Alternative Transients Program (ATP, 1987). For the simulations, the model of the test setup shown in Fig. 4 was used. The test setup frequency response, obtained with ATP program, is shown in Fig. 5. In such figure both transformer output voltage and CVD output voltage (multiplied by 100) are shown for each frequency. A resonant frequency can be seen at 350 Hz.

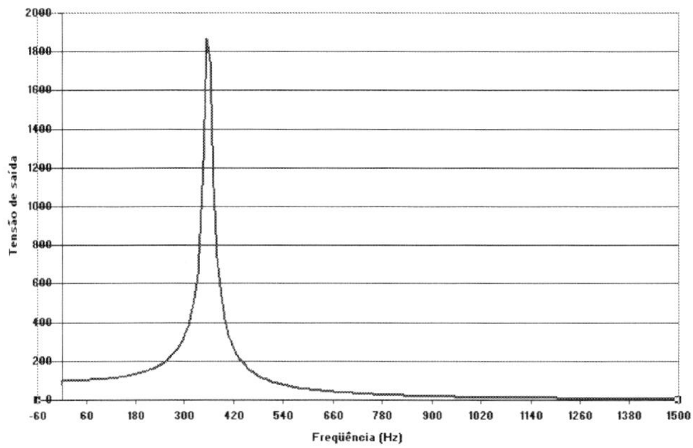

Fig. 5. Test circuit frequency response obtained with ATP program showing transformer high voltage output.

Computer simulation results showed a non flat frequency response of the test circuit, with a resonant frequency at 352Hz (near 5th harmonic). Experimental results of measurements made at the transformers low voltage side, using both a power quality analyzer and spectral analysis demonstrated results which are in accordance with what was previously found, with an amplifying effect at 5th harmonic, caused by the proximity with the resonating frequency.

2.4 Analysis of the circuit

Considering the test setup model, a simplified equivalent circuit is obtained by calculating the series and shunt association of impedances. This simplified circuit is shown in Fig. 6 and 7.

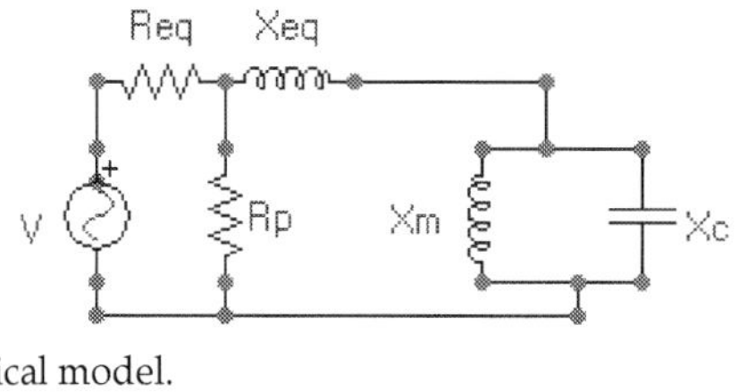

Fig. 6. Test circuit electrical model.

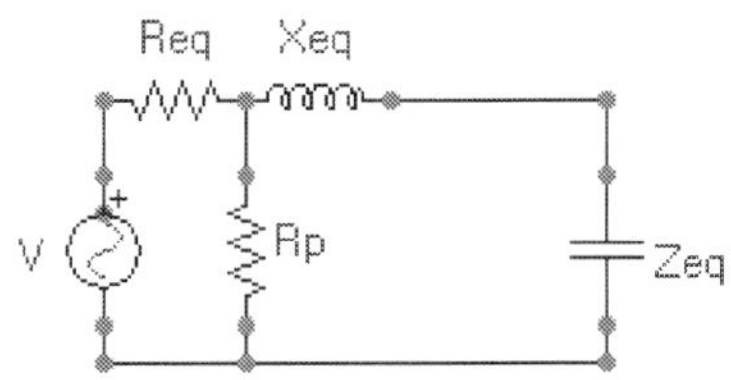

Fig. 7. Test circuit simplified electrical model.

Considering the equivalent circuit, the resonating frequency is caused by the series association of (step-up transformer leakage reactance) and (equivalent impedance of the shunt association of CVD and magnetizing reactance of transformer).
At resonating frequency:

$$
\begin{aligned}
X_{eq} &= Z_{eq} \\
Z_{eq} &= X_m // X_c \\
Z_{eq} &= \frac{X_m . X_c}{X_m - X_c} \\
X_{eq} &= \frac{X_m . X_c}{X_m - X_c}
\end{aligned}
\tag{1}
$$

Where X_m represents the magnetizing reactance of transformer and X_c responds for the capacitive reactance of CVD.
With:

$$
\begin{aligned}
X_{eq} &= j\varpi L_{eq} \\
X_c &= \frac{1}{j\varpi C_{eq}} \\
Xm &= j\varpi L_m
\end{aligned}
\tag{2}
$$

At resonating frequency:

$$\varpi = \sqrt{\frac{1}{L_m C_{eq}}\left(\frac{L_m}{L_{eq}} - 1\right)} \tag{3}$$

Where L_m is the magnetizing inductance of transformer and L_{eq} is the leakage inductance.
By applying the numerical values:

$$\varpi = 2238,58\, rd / s$$

Therefore, the calculated resonating frequency f = 356.28 Hz shows good agreement with the computer simulation results.

2.5 Laboratory tests results – test setup development

Measurement results obtained with a power quality analyzer, and using an arbitrary waveform generator at the high voltage transformer input showed that this test setup can generate high voltage harmonics presenting low harmonic distortion. Fig. 9 shows high voltage transformer output spectrum, when a 60Hz sinusoidal waveform is applied at input. The measurement was performed at the capacitive voltage divider low voltage branch, using a power quality analyzer. In Fig. 8, a low harmonic distortion can be seen, with very small values of higher order voltage harmonics.

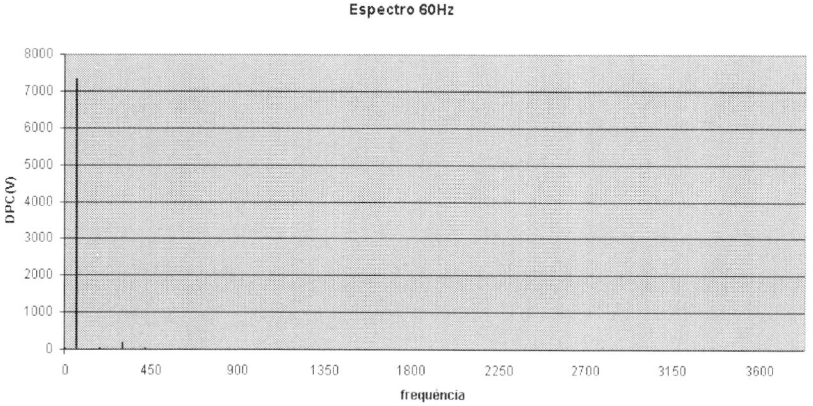

Fig. 8. Test transformer output voltage spectrum, with a 60Hz voltage applied at input.

Fig. 9 shows the high voltage transformer output, when the input arbitrary waveform generator is adjusted for composite waveform generation, with frequencies 60 Hz, 180Hz, 300Hz and 400Hz. The measurement was performed using the power quality analyzer applied to the CVD output.
Fig. 10 presents the output voltage spectrum of the high voltage transformer.

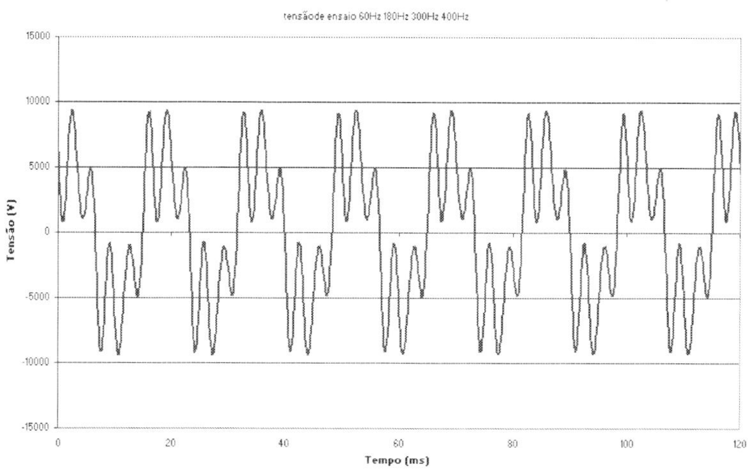

Fig. 9. Test transformer output voltage, for input voltage with 60 Hz, 180Hz, 300Hz and 400Hz harmonic components.

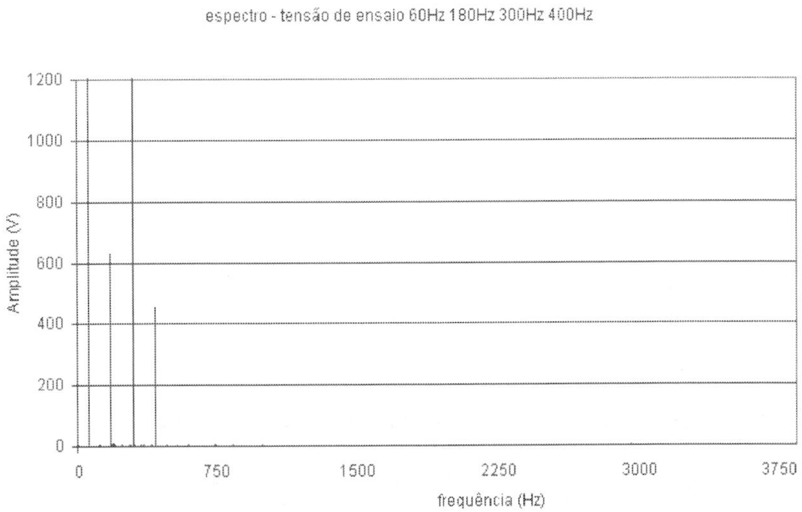

Fig. 10. Test transformer output voltage spectrum, for input voltage with 60 Hz, 180Hz, 300Hz and 400Hz harmonic components.

According to Fig. 10, at the transformer output, only harmonic components actually applied to the input were obtained (60Hz, 180Hz, 300Hz and 400Hz), showing the linear behavior of the transformer. Considering the voltages amplitudes for each harmonic frequency, they depend on the transformer ratio and frequency response of the test setup. In this research, alternatives to improve frequency response where studied, by applying passive components (resistances, capacitances and inductances) at the low voltage side of the test transformer. Such action aimed at generating fundamental and harmonic voltages (high voltages), without causing overflow of the voltage source (arbitrary waveform generator), with rated power 5kVA. Good results were obtained in such studies (see section 4).

2.6 Measurement of CVD capacitances

Capacitance and loss tangent values of the CVD components are of fundamental importance, considering their role in the transformation ratio and phase error during measurements. So, measurements where performed in the high voltage branch capacitances using the Schering Bridge method. It must be considered that in such method measurements are performed applying high voltages, with similar conditions found in actual measurements using CVD. Those capacitance measurements where performed with test voltages of 10kV and 30kV, and similar results where obtained for the capacitance and loss tangent measurements for both test voltages.

In actual conditions, the secondary branch capacitance of the CVD works under a voltage of about 200V (60Hz). For the measurement of that secondary branch capacitance, three methods, which are shown in Table 2, were used.

Measurement results in all three methods were very similar, showing a little influence of test voltage and frequency in the capacitances values.

Measurement method	Applied voltage during measurement (V)	Frequency (Hz)
Schering Bridge (Tettex)	200	60
Resonating Bridge (QuadTech)	1	1k
Volt-ampère Method	200	60

Table 2. Measurement of CVD secondary capacitance

3. Test setup for calibration of voltage transducers

Considering high voltage measurements at field, instrument transformers are commonly used for power frequency voltages. However, for using such transducers in power quality studies, a calibration in a broader frequency range is needed due to the various kinds of power quality disturbances. The capacitive voltage transformer (CVT) is a transducer which is normally found in transmission and distribution substations. Overall, a CVT presents a high capacitance (thousands of pF); therefore a technical difficulty for the test circuit implementation tends to arise.

This high capacitance together with low impedance may become a problem for the voltage source (an arbitrary waveform generator) to feed the test setup, considering its rating of 5kVA. For instance, a 4,000pF CVT, to be used in a 230kV power system, is a 27kVA load at rated voltage. Such load is above the rating of the arbitrary waveform generator with rated power of 5kVA. This difficulty is increased as long as there are higher order harmonics.

Fig. 11 shows the test setup for the calibration of a capacitive voltage transformer (CVT).

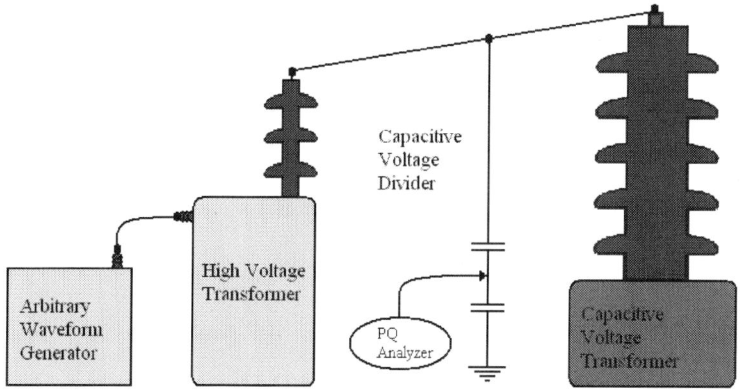

Fig. 11. Test setup for the calibration of a capacitive voltage transformer, with voltage source (arbitrary waveform generator), high voltage transformer, capacitive voltage divider (adopted as Reference Transducer) and test object (CVT).

Fig. 12 shows the electrical equivalent model of the test setup shown in Fig. 11

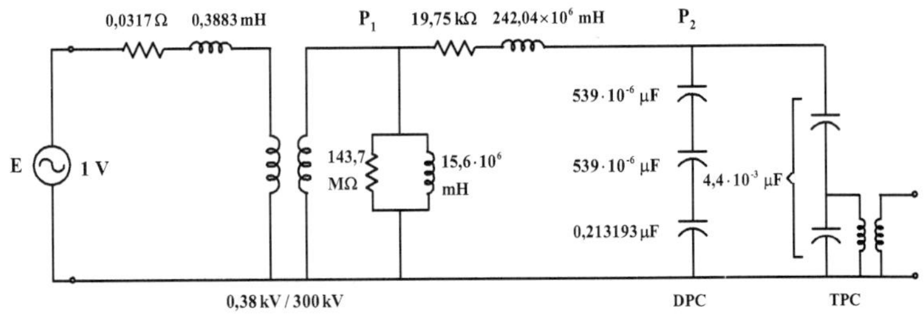

Fig. 12. Electric model of the test setup shown in Fig. 11, with voltage source (arbitrary waveform generator), high voltage transformer, capacitive voltage divider (adopted as Reference Transducer) and test object (CVT).

The test transformer used in this circuit is a 300kV, 70kVA step-up transformer, where the equivalent circuit, again, was obtained by means of no-load and impedance voltage tests. Computer simulation studies were performed, using Alternative Transients Program – ATP, considering a hypothetical voltage source (amplitude 1V) applied at the circuit's input.
Fig. 13 shows the frequency response curve of the circuit of Fig. 12, without reactive compensation.

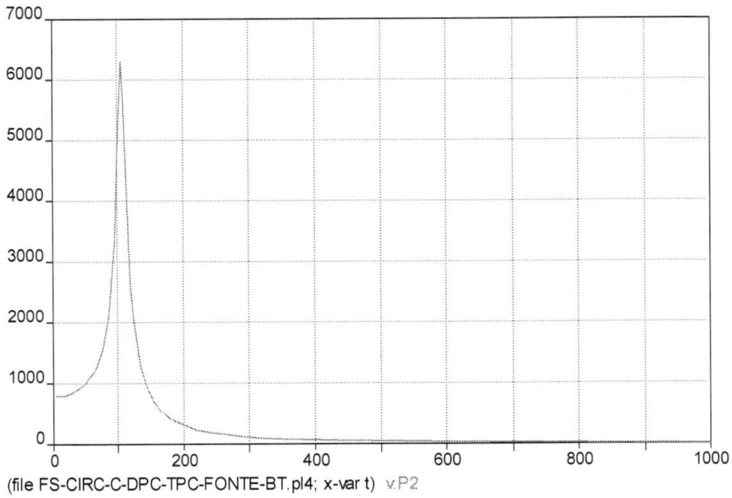

Fig. 13. Frequency response curve of the circuit shown in Fig. 12, without reactive compensation - Output voltage applied to the Capacitive Voltage Transformer (CVT)

The frequency response curve in Fig. 13 is similar to the one in Fig.5, showing a non flat frequency response of the test setup, with a resonant frequency at 125Hz.
Fig. 14 shows this same curve, presented in a log-log scale.

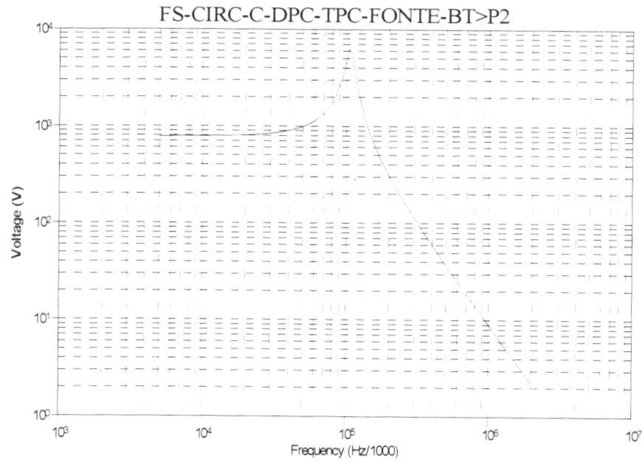

Fig. 14. Frequency response curve of the circuit shown in Fig. 12, without reactive compensation, presented in a log-log scale - Output voltage applied to the Capacitive Voltage Transformer (CVT).

Fig. 15 shows the voltage source electrical current output for each harmonic frequency. In this Fig., a resonant frequency can be seen at about 125Hz, showing a current peak at this frequency.

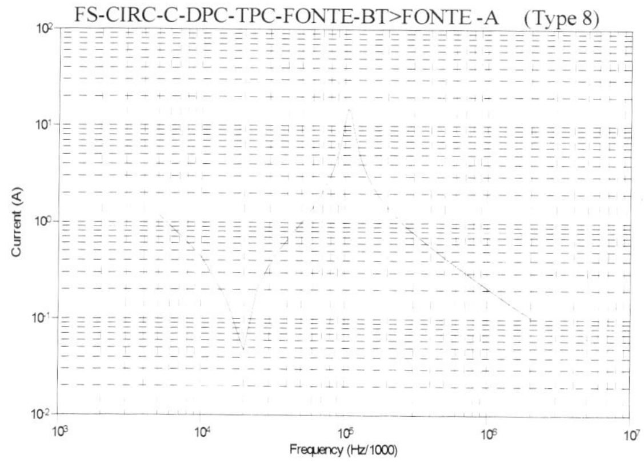

Fig. 15. Frequency response curve of the circuit shown in Fig. 12, without reactive compensation, presented in a log-log scale - Electrical current at voltage source output for each harmonic frequency.

Considering the frequency response curve shown in Fig. 15, high values of electrical current are to be expected at the arbitrary waveform generator output, showing the need for some kind of frequency dependent compensation to be provided. By means of computer simulation using the ATP program, many alternatives of reactive compensation circuits were studied, aiming to generate high values of output voltage applied to the Capacitive Voltage Transformer, and concomitantly, low values electrical currents at the voltage source output.

4. Test setup with shunt reactive compensation

In order to obtain the necessary high voltage waveforms for the calibration tests, additional studies were made, considering the use of passive components (resistances, inductances and capacitances), aiming at reactive compensation considering harmonic frequencies. Those studies were performed using Alternative Transients Program - ATP, considering a hypothetical voltage source (amplitude 1V) applied at the circuit input.

Those computer simulations aimed at evaluating the output voltage applied to the CVT for each harmonic frequency. Also, those studies analyzed the voltage output in many different situations, by applying resistances, capacitances and inductances at the high voltage and/or low voltage sides of the step-up transformer. Such action was done so as to improve the frequency response curve of the test circuit, and therefore obtain higher values of voltage output together with lower values of input electrical current in the test setup.

Fig. 16 shows the electric model of the test setup for the calibration of a 230kV CVT, capacitance 5,300pF, with reactive compensation provided by the shunt capacitance and shunt inductance. This reactive compensation is intended to obtain low intensity of electrical current at 60Hz and, simultaneously, high values of voltage for harmonic voltages applied to the CVT. The dimensioning of the shunt capacitance and inductance was performed with the aid of the ATP program computer simulation.

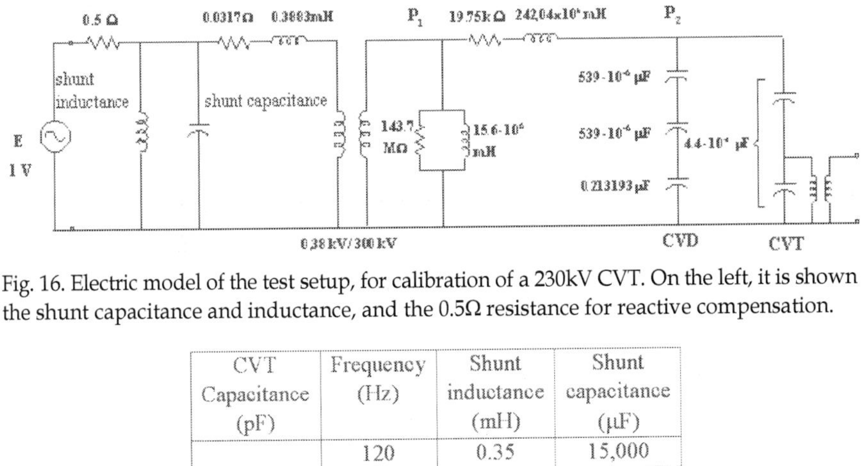

Fig. 16. Electric model of the test setup, for calibration of a 230kV CVT. On the left, it is shown the shunt capacitance and inductance, and the 0.5Ω resistance for reactive compensation.

CVT Capacitance (pF)	Frequency (Hz)	Shunt inductance (mH)	Shunt capacitance (μF)
5,300	120	0.35	15,000
	180	1.2	2,300
	300	1.2	650
	660	1.4	117

Table 3. Reactive compensation for each harmonic frequency, for 5,300pF CVT

In this test setup, the values of the shunt inductance and capacitance are adjusted for each harmonic frequency. For instance, for the calibration of the CVT in harmonic frequency of 300Hz (5th harmonic), a 1.5mH inductance and a 650μF capacitance are used. The 0.5 Ω resistance in series with the voltage source makes the frequency response curve smoother, simplifying the tuning of the shunt capacitance and inductance values for reactive compensation. Table 3 shows values of reactive compensation for other harmonic frequencies, considering a 5,300pF - 230kV capacitive voltage transformer.

Figs. 17 and 18 show the frequency response of the test setup (without the 0.5Ω series resistance), with reactive compensation for 300Hz, for the test transformer voltage (high voltage side) and the voltage source output current, respectively.

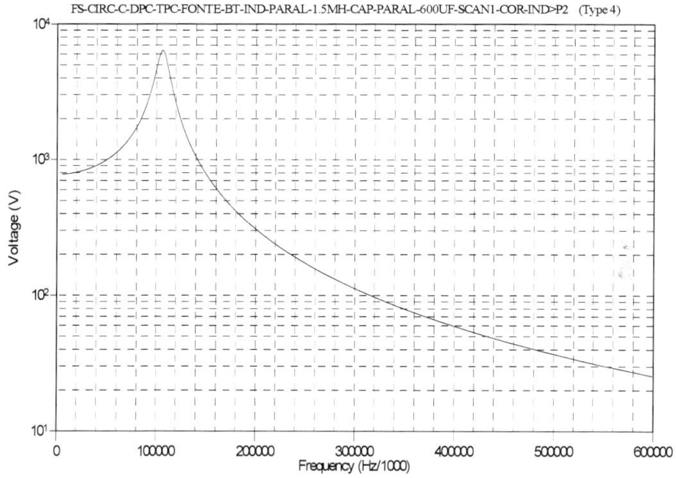

Fig. 17. Test setup frequency response, with reactive compensation for 300Hz, without the 0.5Ω series resistance. Test transformer output voltage (high voltage side).

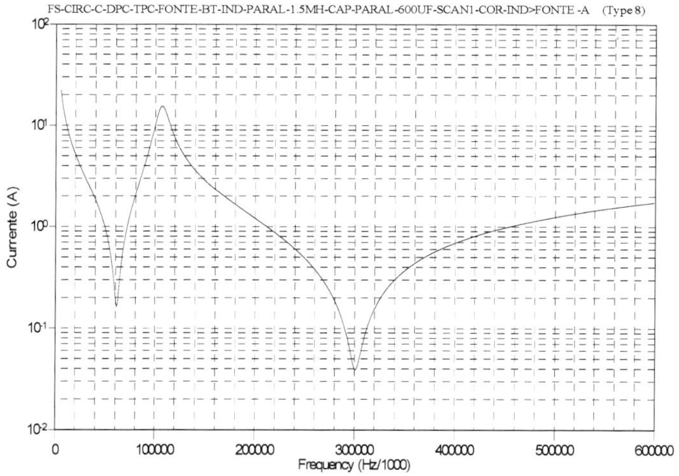

Fig. 18. Test setup frequency response, with reactive compensation for 300Hz, without the 0.5Ω series resistance. Voltage source (arbitrary waveform generator) output current.

Fig. 18 shows that at 300Hz, a minimum value of voltage source output current is obtained. Also, for 60Hz there is another minimum value, making this test setup suitable for applying, during the calibration tests, a composite waveform with a 60Hz and 300Hz voltage harmonic components.

Fig. 19 shows the voltage at the primary side of the step-up transformer, for each harmonic frequency. The primary side voltage remains stable, implying in stability of the arbitrary waveform generator voltage source output.

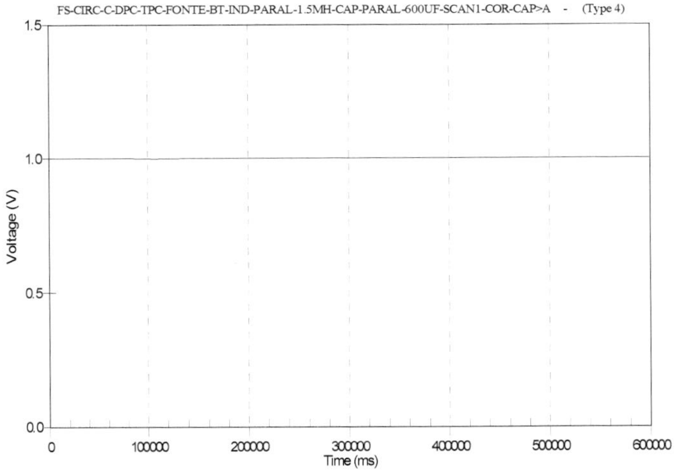

Fig. 19. Voltage at primary (low voltage side) of step-up transformer considering harmonic frequencies.

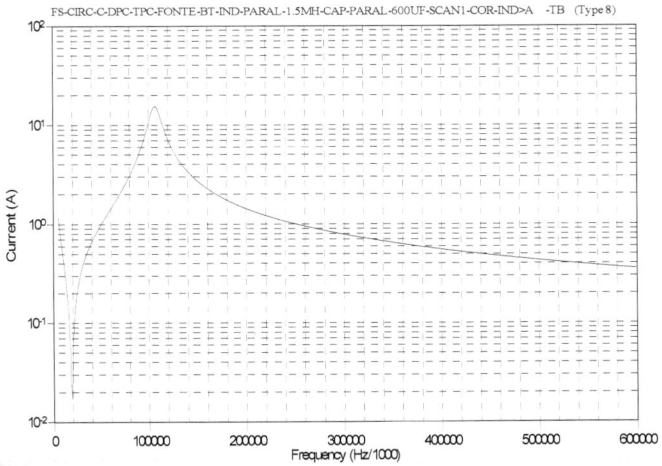

Fig. 20. Electrical current at primary (low voltage side) of step-up transformer.

Fig. 20 shows the electrical current at this same low voltage side of the step-up transformer. High values of electrical current (200A range at test voltage) are expected at 60Hz frequency for the transformer primary side.

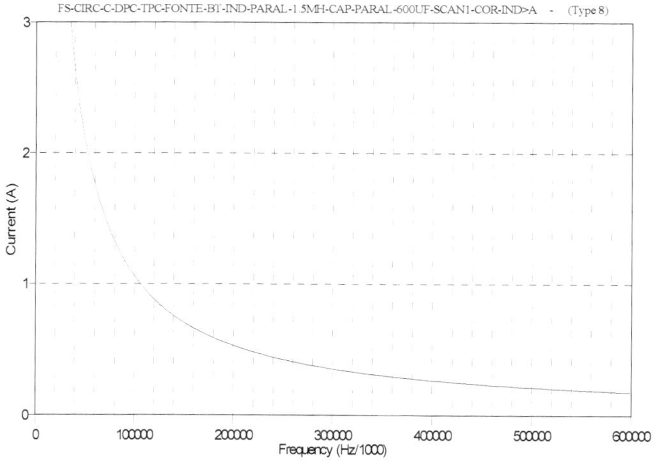

Fig. 21. Electrical current in the shunt inductance, at the primary side of the step-up transformer.

Fig. 21 shows the electrical current in the shunt inductance, at the primary side of the step-up transformer. High values (200A range at test voltage) of electrical current are expected at 60Hz frequency.

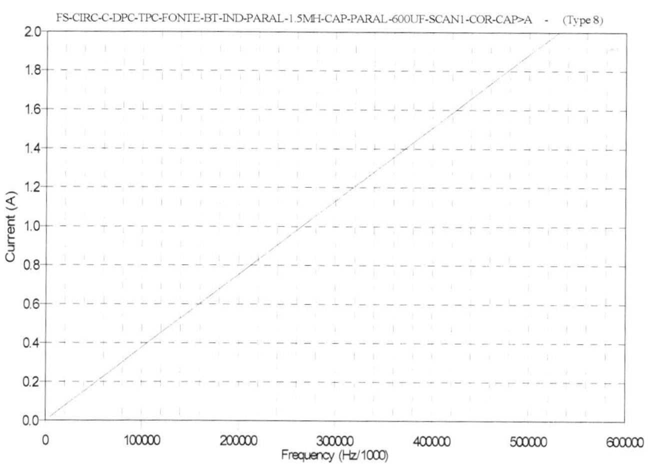

Fig. 22. Electrical current in the shunt capacitance, at the primary side of the step-up transformer.

Fig. 22 shows the electrical current in the shunt capacitance at the low voltage side of transformer.

4.1 Influence of the series resistance

Fig. 23 shows the test setup equivalent circuit, with series 0.5Ω resistance, shunt 1.5mH and 600μF capacitance for passive reactive compensation applied at the low voltage side of the step-up transformer.

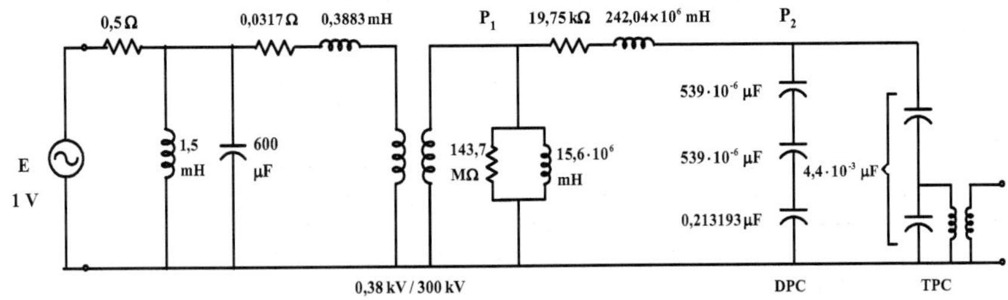

Fig. 23. Electric model of the test setup, with series 0.5Ω resistance, shunt 1.5mH and 600μF capacitance for passive reactive compensation applied at the low voltage side of the step-up transformer.

Fig. 24 shows output voltage of the test setup, applied to CVT.

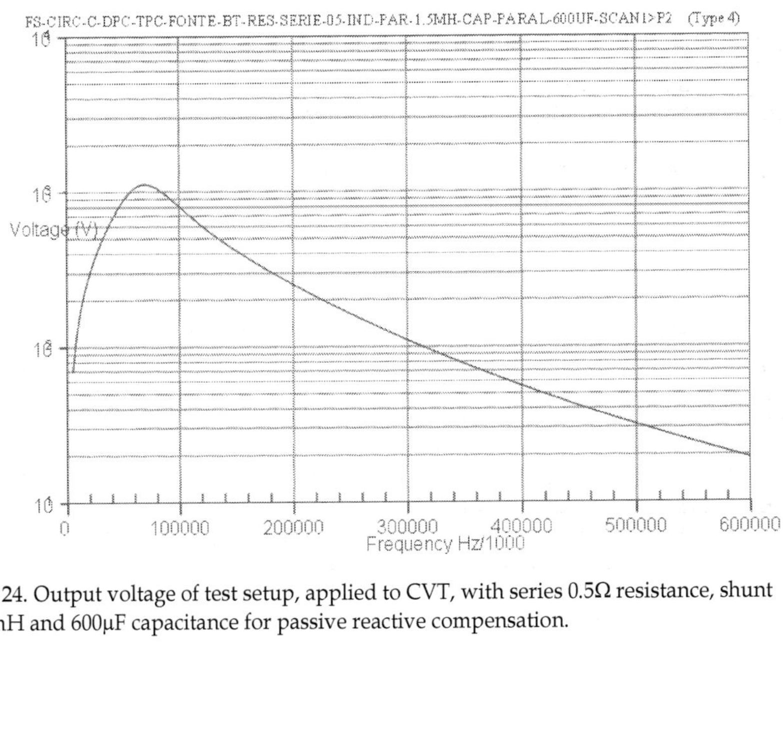

Fig. 24. Output voltage of test setup, applied to CVT, with series 0.5Ω resistance, shunt 1.5mH and 600μF capacitance for passive reactive compensation.

Fig. 25 shows the voltage source output current, with series 0.5Ω resistance, shunt 1.5mH and 600μF capacitance for passive reactive compensation.

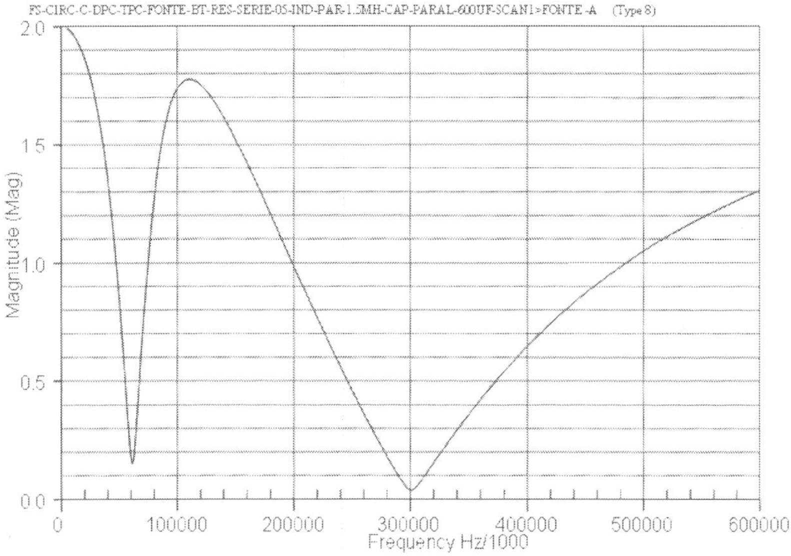

Fig. 25. Voltage source output current, with series 0.5Ω resistance, shunt 1.5mH and 600μF capacitance for passive reactive compensation.

Figs. 24 and 25 show that the presence of the series 0.5Ω resistance provides an smoothing effect on the peaks of the frequency response curves (compare with Figs. 17 and 18) at the resonant frequencies. This feature is advantageous, making the tuning of the test circuit easier, in the sense that the specification and adjustment of the values of the capacitances and inductances used for reactive compensation are less strict, without the necessity of being too precise concerning the design specification of such values.

5. Conclusions

Considering the obtained experimental results, it was possible to assure the effectiveness of this test setup for generating sinusoidal high voltage waveforms, keeping under control, at low values, the total harmonic distortion. Also, with this test setup, composite waveforms were produced, at high voltage level, keeping the harmonic distortion under control, with the aid of reactive compensation. Additionally, with reactive compensation, it was possible to keep under acceptable low values the output current of voltage source, considering 60Hz and other higher order harmonic current components. This test setup also showed the feasibility of testing, at high voltage levels, test objects with high capacitance (capacitive voltage transformers with thousands of pF capacitance), for 60Hz and higher order harmonics, using reactive compensation. For testing CVT's with higher capacitance, or for to generate harmonics of higher frequency, the use of a higher output power arbitrary waveforms generator with may become necessary. This solution is feasible, but may imply in higher costs.

6. References

ATP - Alternative transients program - Rule Book, Bonneville Power Administration, 1987.

Bradley DA, Bodger PS, Hyland PR, Harmonic response tests of voltage transducers for the New Zealand power systems, IEEE T-PAS, vol PAS-104, nº 7, 1985.

Dugan, RC, McGranaghan, MF,Santoso, S, Beaty, HW, Electrical power systems quality, 2nd Edition, McGraw-Hill, 2004.

IEC - International Eletrotechnical Commission, IEC 60060-1 Standard, High voltage test techniques 1994.

IEC - International Eletrotechnical Commission, IEC 61000 Standard series, Electromagnetic compatibility (EMC) - (various parts), 1995.

IEC - International Eletrotechnical Commission, IEC 61000-4-7 Standard, Electromagnetic compatibility (EMC) - Test and measurement techniques - General guide on harmonics and interharmonics measurements and interpretation, for power supply systems and equipment connected thereto 2° edition 2002.

Seljeseth H, Saethre EA, Ohnstad T, Voltage transformer frequency response. Measuring harmonics in Norwegian 300kV and 132kV power systems. Proceedings IEEE 8° International Conference on Harmonics and Quality of Power ICQHP '98, Athens, 1998.

13

Methods for Estimation of Voltage Harmonic Components

Tomáš Radil[1] and Pedro M. Ramos[2]
[1]Instituto de Telecomunições, Lisbon
[2]Instituto de Telecomunições and Department of Electrical and Computer Engineering, IST, Technical University of Lisbon, Lisbon
Portugal

1. Introduction

The estimation of harmonic components of power voltage signals is one of the measurements prescribed by the power quality (PQ) standards (IEC 61000-4-30, 2008). In general, the harmonic components originate from non-linear elements connected to the power system such as non-linear loads (*e.g.*, switched-mode power supplies) or transformers. An example of a distorted voltage measured in an office building is shown in Fig. 1. The negative effects associated with the presence of voltage harmonics include, for example, overheating and increased losses of transformers, malfunction of electronic instruments, additional losses in rotating machines or overheating of capacitor banks.

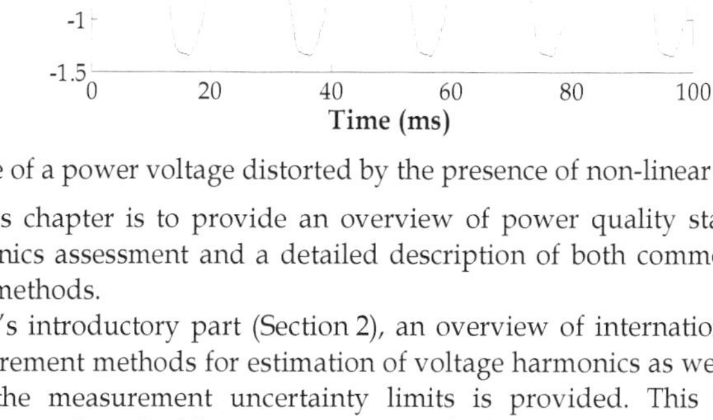

Fig. 1. Example of a power voltage distorted by the presence of non-linear loads

The aim of this chapter is to provide an overview of power quality standards related to voltage harmonics assessment and a detailed description of both common and alternative measurement methods.

In this chapter's introductory part (Section 2), an overview of international standards that concern measurement methods for estimation of voltage harmonics as well as the standards that address the measurement uncertainty limits is provided. This is followed by a description of several applicable measurement methods. The description is divided in two parts: Section 3 deals with methods working in the frequency domain (including the standard method described in the standard (IEC 61000-4-7, 2009)) while Section 4 contains the description of several time-domain based methods. Each section presents a detailed description of the respective method including its strong and weak points and potential limitations. Section 5 focuses on the calculation of the total harmonic distortion. In Section 6, the performance of the methods described in Section 3 and Section 4 is discussed. The chapter is concluded by Section 7, in which the characteristics of the described algorithms will be summarized.

2. Power quality standards and harmonic estimation

Several international standards deal with the issue of harmonic estimation. In this section, a short overview of these standards and of the requirements imposed in them is provided.
In the standard (IEEE 1159-2009, 2009), the harmonics are classified as one of the waveform distortions that typically have a steady state nature, their frequency band is from 0 Hz up to 9 kHz and their magnitude can reach up to 20% of the fundamental.
The standard (IEC 61000–4–7, 2009) describes a general instrument for harmonic estimation. The instrument is based on the Discrete Fourier Transform (DFT); however, the application of other algorithms is also allowed. The DFT algorithm and its application according to the standard are described in Section 3.1. In the standard (IEC 61000–4–30, 2008), it is required that at least 50 harmonics are estimated.
Standard (IEC 61000–4–7, 2009) also includes the accuracy requirements for harmonic estimation. The requirements are divided into two classes: Class I of the IEC 61000–4–7 corresponds to Class A of IEC 61000–4–30, while Class II of the IEC 61000–4–7 corresponds to Class S of IEC 61000–4–30. The requirements are based on the relation between the magnitudes of the measured harmonics (U_h) and the nominal voltage range (U_{nom}) as shown in Table 1.

Class	Condition	Maximum error
I	$U_h \geq 1\% \; U_{\mathrm{nom}}$	$\pm 5\% \; U_h$
	$U_h < 1\% \; U_{\mathrm{nom}}$	$\pm 0.05\% \; U_{\mathrm{nom}}$
II	$U_h \geq 3\% \; U_{\mathrm{nom}}$	$\pm 5\% \; U_h$
	$U_h < 3\% \; U_{\mathrm{nom}}$	$\pm 0.15\% \; U_{\mathrm{nom}}$

Table 1. Accuracy requirements for voltage harmonics measurement

The measuring range is specified in (IEC 61000–4–30, 2008) using the compatibility levels (maximum disturbance levels to which a device is likely to be subjected) for low-frequency disturbances in industrial plants, which are standardized in (IEC 61000–2–4, 2002). The measuring range should be from 10% to 200% of the class 3 compatibility levels specified in (IEC 61000–2–4, 2002) for Class A instruments and as 10% to 100% of these compatibility levels for Class S instruments.
The class 3 compatibility levels according to (IEC 61000–2–4, 2002) are shown in Table 2. Note that the compatibility levels of odd harmonics are higher than the compatibility levels of even harmonics. This reflects the fact that in power systems, the odd harmonics are usually dominant.
The compatibility level for total harmonic distortion (*THD*) is 10% in the class 3.

Harmonic order h	Class 3 compatibility level % of fundamental
2	3
3	6
4	1.5
5	8
6	1
7	7
8	1
9	2.5
10	1
11	5
13	4.5
15	2
17	4
21	1.75
$10 < h \leq 50$ (h even)	1
$21 < h \leq 45$ (h odd multiples of three)	1
$17 < h \leq 49$ (h odd)	$4.5 \cdot (17/h) - 0.5$

Table 2. Voltage harmonics compatibility levels

3. Frequency domain methods

One approach to harmonic estimation is to use some kind of a transform to decompose the time series of measured voltage signal samples into frequency components. Most commonly, methods based on the Discrete Fourier Transform (DFT) are used but, for example, the Discrete Wavelet Transform (DWT) is also sometimes applied as well (Pham & Wong, 1999), (Gaouda et al., 2002).

In Section 3.1, the application of the DFT for harmonic estimation according to the standard (IEC 61000-4-7, 2009) is described. In Section 3.2, an alternative method based on the Goertzel algorithm (Goertzel, 1958) and its properties are described.

3.1 Discrete Fourier Transform

The Discrete Fourier Transform (DFT) and its optimized implementation called the Fast Fourier Transform (FFT) is arguably the most used method for harmonic estimation. The harmonic measuring instrument described in (IEC 61000–4–7, 2009) is based on this method. The DFT of a voltage signal u whose length is N samples is described as (Oppenheim et al., 1999)

$$X[k] = \sum_{n=0}^{N-1} u[n] e^{-2\pi i k n / N} \qquad k = 0, \ldots, N-1 \,. \tag{1}$$

The result of (1) is a complex frequency spectrum $X[k]$ with frequency resolution of

$$\Delta f = f_S / N \tag{2}$$

where f_s is the sampling frequency.
The amplitudes of individual frequency components are then calculated as

$$U[k]=\frac{2}{N}\sqrt{Re(X[k])^2+Im(X[k])^2} \tag{3}$$

In (3), the factor $1/N$ is a normalization factor and the multiplication by two is used to take into account the symmetry of a real-input DFT ($|X[k]|=|X[N-k]|$).
In (IEC 61000-4-7, 2009), the DFT is applied to 10 cycles (in case of 50 Hz power systems) or 12 cycles (in case of 60 Hz power systems) of the power system's fundamental frequency. Since the power system's frequency is varying, the length of the window to which the DFT is applied has to be adjusted accordingly. Standard (IEC 61000-4-7, 2009) allows a maximum error of this adjustment of ±0.03%. The window adjustment can be done *e.g.*, by using a phase-locked loop (PLL) to generate the sampling frequency based on the actual power system's frequency. Alternatively, when the sampling frequency is high enough, the window can be adjusted by selecting the number of samples that correspond to 10 (or 12) cycles at the measured fundamental frequency. In a 50-Hz system, at least $f_s \cong 10$ kS/s are required to ensure the 0.03% maximum error specification.
An example of an amplitude DFT spectrum calculated from 10 cycles of the signal shown in Fig. 1 is depicted in Fig. 2.

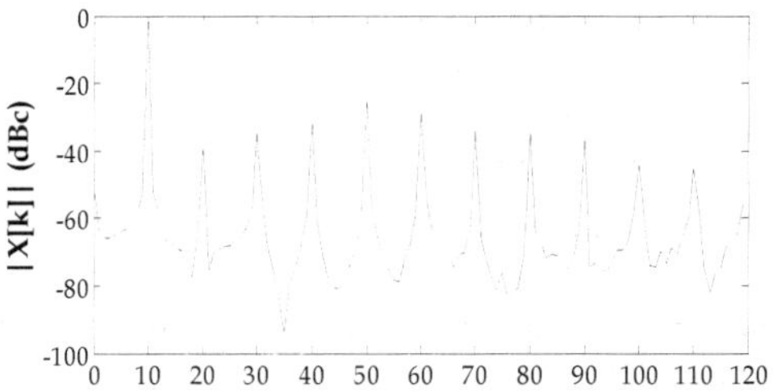

Fig. 2. Detail of an amplitude DFT spectrum of a measured power voltage signal

Note that when processing signals according to the (IEC 61000-4-7, 2009), the frequency resolution of the spectrum is 5 Hz for both 50 Hz and 60 Hz power systems. This means that the harmonic components (the fundamental, the 2nd and the higher harmonics) can be found on indices $k=10,20,30,\ldots$ for 50 Hz power systems (*i.e.*, $U_1=U[10]$, $U_2=U[20]$, etc.) and on $k=12,24,36,\ldots$ for 60 Hz power systems.

3.2 The Goertzel algorithm

The Goertzel algorithm (Goertzel, 1958) is an efficient algorithm for calculation of individual lines of the DFT spectrum. The algorithm applies a second-order infinite impulse response (IIR) filter to the samples of the voltage signal in order to calculate one spectrum line.
The Goertzel algorithm calculates the line k of the spectrum using

$$X[k] = e^{2\pi k/N} s[N-1] - s[N-2] \tag{4}$$

where

$$s[n] = u[n] + 2\cos(2\pi k / N)\, s[n-1] - s[n-2], \tag{5}$$

$s[-1] = s[-2] = 0$, $u[n]$ is the measured voltage signal, N is the number of samples being processed and $n \in [0; N-1]$.

The Goertzel algorithm is more efficient than the Fast Fourier Transform (FFT) when the number of spectrum lines to be calculated (H) meets the condition

$$H \le \log_2 (N). \tag{6}$$

However, even when the number of spectrum lines to be calculated does not fulfill the condition (6), the application of the Goertzel algorithm can be advantageous in some cases. It is faster than implementing DFT according to its definition (1) and unlike most of the implementations of the FFT, the Goertzel algorithm does not require N to be an integer power of two. Although algorithms for fast calculation of DFT when the number of samples is not equal to power of two exist, *e.g.* (Rader, 1968), many libraries for digital signal processing include only the radix-2 FFT. For the Goertzel algorithm, it is sufficient to ensure that N contains an integer number of fundamental periods (to avoid problems with spectrum leakage).

4. Time domain methods

Section 3 described methods for estimation of voltage harmonics in the frequency domain. However, it is possible to estimate the harmonics in the time domain as well. The time-domain approaches are based on a least-square fitting procedure that attempts to estimate the parameters of the voltage signal's model so the root-mean-square error between the model and the measured signal is minimized.

The time domain methods that try to fit one or more single-tone harmonic signals on the measured signal are in general called sine fitting algorithms.

The general model of a signal that contains multiple harmonic components can be written as

$$u = \sum_{h=1}^{H} \left[U_h \cos\left(2\pi f_h t + \varphi_h\right) \right] + C \tag{7}$$

where U_h are the amplitudes of individual harmonics, f_h their frequencies (expressed as an integer multiple of the fundamental's frequency: $f_h = h \cdot f_1$), φ_h their phases, C is the dc component and H is the number of harmonics included in the model.

For the purpose of the sine fitting algorithms, it is convenient to re-write (7) as

$$u = \sum_{h=1}^{H} \left[A_h \cos\left(2\pi f_h t\right) + B_h \sin\left(2\pi f_h t\right) \right] + C \tag{8}$$

where A_h are the in-phase and B_h are the quadrature components.

The amplitudes U_h and phases φ_h can be then calculated as

$$U_h = \sqrt{A_h^2 + B_h^2}, \tag{9}$$

$$\varphi_h = \operatorname{atan2}\left(-B_h; A_h\right). \tag{10}$$

4.1 The 3- and 4-parameter sine fitting algorithm

The 3- and 4-parameter sine fitting algorithms are described in (IEEE Std. 1057-2007, 2008) where they are used for testing of analog to digital converters in waveform recorders.
The 3-parameter algorithm estimates the amplitude and the phase of a signal whose frequency is known. With the frequency known, model (8) is a linear function of the remaining unknown parameters. Thus, the calculation using the 3-parameter sine fitting algorithm is non-iterative and is based on solving

$$\begin{bmatrix} A & B & C \end{bmatrix}^{\mathrm{T}} = \left(\mathbf{D}^{\mathrm{T}}\mathbf{D}\right)^{-1}\mathbf{D}^{\mathrm{T}}\mathbf{u} \tag{11}$$

where $\mathbf{u}$ is the column vector of measured voltage samples, $\mathbf{D}$ is a matrix

$$\mathbf{D} = \begin{bmatrix} \cos(2\pi f t_0) & \sin(2\pi f t_0) & 1 \\ \cos(2\pi f t_1) & \sin(2\pi f t_1) & 1 \\ \vdots & \vdots & \vdots \\ \cos(2\pi f t_{N-1}) & \sin(2\pi f t_{N-1}) & 1 \end{bmatrix} \tag{12}$$

and t_n are the timestamps of voltage samples.
The accuracy of the 3-parameter sine fitting algorithm depends on the frequency estimate. The frequency can be estimated using algorithms such as the Interpolated DFT (IpDFT) algorithm (Renders et al., 1984); however, many algorithms are applicable for this task (Slepička et al., 2010).
In case the frequency estimate is not sufficiently accurate (Andersson & Händle, 2006), the 4-parameter algorithm can be used. Including the frequency in the algorithm makes the least-square procedure non-linear, which means that the algorithm has to use an iterative optimization process to find the optimum value of the estimated parameters.
The 4-parameter sine fitting algorithm solves the equation

$$\begin{bmatrix} A^{(i)} & B^{(i)} & C^{(i)} & \Delta\omega^{(i)} \end{bmatrix}^{\mathrm{T}} = \left[\left(\mathbf{D}^{(i)}\right)^{\mathrm{T}}\mathbf{D}^{(i)}\right]^{-1}\left[\mathbf{D}^{(i)}\right]^{\mathrm{T}}\mathbf{u} \tag{13}$$

where i is the iteration number, ω is the angular frequency $\omega = 2\pi f$; $\Delta\omega^{(i)}$ is the change of the angular frequency from the previous iteration, matrix $\mathbf{D}^{(i)}$ is

$$\mathbf{D}^{(i)} = \begin{bmatrix} \cos\left(\omega^{(i-1)}t_0\right) & \sin\left(\omega^{(i-1)}t_0\right) & 1 & \alpha^{(i-1)}(t_0) \\ \cos\left(\omega^{(i-1)}t_1\right) & \sin\left(\omega^{(i-1)}t_1\right) & 1 & \alpha^{(i-1)}(t_1) \\ \vdots & \vdots & \vdots & \vdots \\ \cos\left(\omega^{(i-1)}t_{N-1}\right) & \sin\left(\omega^{(i-1)}t_{N-1}\right) & 1 & \alpha^{(i-1)}(t_{N-1}) \end{bmatrix}, \tag{14}$$

and $\alpha^{(i-1)}(t) = -A^{(i-1)}t \times \sin\left(\omega^{(i-1)}t\right) + B^{(i-1)}t \times \cos\left(\omega^{(i-1)}t\right)$.

The iterative computation continues until the absolute relative change of the estimated frequency drops below a predefined threshold or until the maximum number of allowed iterations is exceeded.

In order to estimate the voltage harmonics, first, the 4-parameter sine fitting algorithm is applied to the voltage signal and the signal's fundamental frequency, amplitude and phase are estimated. In the second step, the 3-parameter algorithm is repeatedly applied to the residuals after estimation of the fundamental to estimate the individual higher harmonics. This means that the frequency supplied to the 3-paramater algorithm is an integer multiple of the fundamental's estimated frequency.

4.2 The multiharmonic fitting algorithm

The previously described combination of the 4- and 3-parameter sine fitting algorithms estimates the harmonic amplitudes and phases one by one. The advantage of this approach is that it keeps the computational requirements low because in each step only operations with small matrices are required. Its weak point is that it relies on the accuracy of the estimation of the fundamental frequency using the 4-paramater algorithm. Since the 3- and 4-parameter algorithms take into account only one frequency at a time, the other frequencies contained in the signal act as disturbances that affect the final estimate of the frequency and of the harmonic amplitudes.

The multiharmonic fitting algorithm (Ramos et al., 2006) provides more accurate but also computationally heavier approach. It uses an optimization procedure in which all the parameters (*i.e.*, the fundamental's frequency and the amplitudes and phases of all harmonics) are estimated at the same time.

There are two versions of the multiharmonic fitting algorithm: non-iterative and iterative version.

The non-iterative version is similar to the 3-parameter sine fitting algorithm. It assumes that the signal's fundamental frequency is known and estimates the remaining parameters (the components A_h and B_h of the harmonic amplitudes and the dc component C)

$$\begin{bmatrix} A_1 & B_1 & A_2 & B_2 & \cdots & A_H & B_H & C \end{bmatrix}^{\mathrm{T}} = \left(\mathbf{D}^{\mathrm{T}}\mathbf{D}\right)^{-1}\mathbf{D}^{\mathrm{T}}\mathbf{u} \tag{15}$$

where $\mathbf{D}$ is a matrix

$$\mathbf{D} = \begin{bmatrix} \cos(\omega t_0) & \sin(\omega t_0) & \cos(2\omega t_0) & \sin(2\omega t_0) & \cdots & \cos(H\omega t_0) & \sin(H\omega t_0) & 1 \\ \cos(\omega t_1) & \sin(\omega t_1) & \cos(2\omega t_1) & \sin(2\omega t_1) & \cdots & \cos(H\omega t_1) & \sin(H\omega t_1) & 1 \\ \vdots & \vdots & \vdots & \vdots & \ddots & \vdots & \vdots & \vdots \\ \cos(\omega t_{N-1}) & \sin(\omega t_{N-1}) & \cos(2\omega t_{N-1}) & \sin(2\omega t_{N-1}) & \cdots & \cos(H\omega t_{N-1}) & \sin(H\omega t_{N-1}) & 1 \end{bmatrix} \tag{16}$$

and ω is the angular frequency of the fundamental.

As in the case of the 3-parameter sine fitting algorithm, the non-iterative multiharmonic algorithm relies on the initial frequency estimate. However, the initial estimate can be improved using an iterative optimization procedure.

The iterative multiharmonic fitting algorithm adds the frequency into the calculations

$$\begin{bmatrix} A_1^{(i)} & B_1^{(i)} & A_2^{(i)} & B_2^{(i)} & \cdots & A_H^{(i)} & B_H^{(i)} & C^{(i)} & \Delta\omega^{(i-1)} \end{bmatrix}^{\mathrm{T}} = \left[\left(\mathbf{D}^{(i)}\right)^{\mathrm{T}}\mathbf{D}^{(i)}\right]^{-1}\left[\mathbf{D}^{(i)}\right]^{\mathrm{T}}\mathbf{u} \tag{17}$$

where $\mathbf{D}^{(i)}$ is a matrix

$$\mathbf{D}^{(i)}=\begin{bmatrix}\cos(\omega t_0) & \sin(\omega t_0) & \cdots & \cos(H\omega t_0) & \sin(H\omega t_0) & 1 & \alpha^{(i-1)}(t_0)\\ \cos(\omega t_1) & \sin(\omega t_1) & \cdots & \cos(H\omega t_1) & \sin(H\omega t_1) & 1 & \alpha^{(i-1)}(t_1)\\ \vdots & \vdots & \ddots & \vdots & \vdots & \vdots & \vdots\\ \cos(\omega t_{N-1}) & \sin(\omega t_{N-1}) & \cdots & \cos(H\omega t_{N-1}) & \sin(H\omega t_{N-1}) & 1 & \alpha^{(i-1)}(t_{N-1})\end{bmatrix} \quad (18)$$

and $\alpha^{(i-1)}(t)=\sum_{h=1}^{H}\left[-A_h^{(i-1)}ht\sin\left(h\omega^{(i-1)}t\right)+B_h^{(i-1)}ht\cos\left(h\omega^{(i-1)}t\right)\right]$.

5. Calculation of the Total Harmonic Distortion

The Total Harmonic Distortion (*THD*) is an important indicator used to express the total amount of harmonic components. It is defined as

$$THD=\frac{\sqrt{\sum_{h=2}^{H}U_h^2}}{U_1} \quad (19)$$

and expressed in relative units or in percents.

For stationary signals whose length is exactly 10 cycles (or 12 in case of 60 Hz power systems) the whole energy of a harmonic component is concentrated in one frequency bin. However, if the signal's parameters such as its fundamental frequency vary, the energy will leak into neighbouring frequency bins (spectral leakage). To take into account this effect, the standard (IEC 61000-4-7, 2009) defines, besides the *THD*, two more indicators: the group total harmonic distortion (*THDG*) and the subgroup total harmonic distortion (*THDS*).

The group total harmonic distortion is defined as

$$THDG=\frac{\sqrt{\sum_{h=2}^{H}U_{g,h}^2}}{U_{g,1}} \quad (20)$$

where

$$U_{g,h}^2=\frac{1}{2}U^2\left[P\cdot h-P/2\right]+\sum_{k=-P/2+1}^{P/2-1}U^2\left[P\cdot h-k\right]+\frac{1}{2}U^2\left[P\cdot h+P/2\right] \quad (21)$$

and P is the number of fundamental periods within the signal ($P=10$ for 50 Hz power systems and $P=12$ for 60 Hz power systems).

The subgroup total harmonic distortion is defined as

$$THDS=\frac{\sqrt{\sum_{h=2}^{H}U_{sg,h}^2}}{U_{sg,1}} \quad (22)$$

where

$$U_{sg,h}^2=\sum_{k=-1}^{1}U^2\left[P\cdot h+k\right]. \quad (23)$$

Note that only the *THD* can be calculated when using the time-domain fitting algorithms. The Goertzel algorithm is not suitable when *THDG* and *THDS* have to be estimated, because as the number of spectrum lines that have to be calculated increases, the algorithm's computational requirements also increase significantly.

6. Comparison of the methods

In this section, the above described methods for harmonics estimation are compared. Their accuracy is discussed in Section 6.1, while Section 6.2 focuses on the computational requirements of individual algorithms.

6.1 Accuracy

The standard (IEC 61000-4-30, 2008) specifies the conditions under which power quality measuring instruments should be tested. The standard recognizes three classes of instruments: Class A, Class S and Class B instruments. While for classes A and S the tests and required performance is described in the standard, the performance of Class B instruments is specified by manufacturer.

As it was described in Section 2 of this chapter, the levels of harmonics in signals used for testing should be up to 200% of the values in Table 1 for Class A instruments and 100% of these values for Class S instruments. Furthermore, the test signals should contain other disturbances and variations of parameters as described by the three testing states. The most important parameters of these testing states are summarized in Table 3.

Influence quantity	Testing state 1	Testing state 2	Testing state 3
Frequency	$f_{nom} \pm 0.5$ Hz	$f_{nom} - 1$ Hz ± 0.5 Hz	$f_{nom} + 1$ Hz ± 0.5 Hz
Flicker	$P_{st} < 0.1$	$P_{st} = 1 \pm 0.1$	$P_{st} = 4 \pm 0.1$
Voltage	$U_{din} \pm 1\%$	determined by flicker and interharmonics	determined by flicker and interharmonics
Interharmonics	0% to 0.5% U_{din}	1% $\pm$ 0.5% U_{din} at 7.5 f_{nom}	1% $\pm$ 0.5% U_{din} at 3.5 f_{nom}

Table 3. Testing conditions for Class A and S instruments according to IEC 61000-4-30

In Table 3, f_{nom} designates the nominal power frequency, P_{st} is the short-term flicker severity and U_{din} is the nominal input voltage.

Test signals according to the three testing states were simulated in order to test the above described algorithms for harmonic estimation. Harmonics were added to the signals to achieve signals with $THD = 20\%$. The phases of harmonics were random and the distribution of harmonic amplitudes was as follows from Table 4.

The testing signals also contained white Gaussian noise corresponding to a signal to noise ratio of 75 dB (the noise was added to simulate the equivalent noise of an ideal 12-bit analog to digital converter). The Goertzel algorithm, the combined 4- and 3- parameter sine fitting algorithm, the non-iterative multiharmonic fitting and the iterative multiharmonic fitting algorithm were applied to 10 000 of such test signals and the maximum error of estimation of individual harmonic amplitudes was calculated. The maximum allowed error for individual harmonics calculated using the Table 1 are shown in Fig. 3a and Fig. 3b for Class A and Class S instruments, respectively. The simulation results for the three testing states are shown in Fig. 4, Fig. 5 and Fig. 6.

Harmonic order h	Amplitude (in % of the fundamental)
2nd	4%
3rd	12%
4th	2%
5th	10%
7th	4%
h odd from 9th to 17th	2%
h even from 6th to 18th	1.6%
h from 19th to 50th	1.6%

Table 4. Distribution of harmonic amplitudes in the test signals

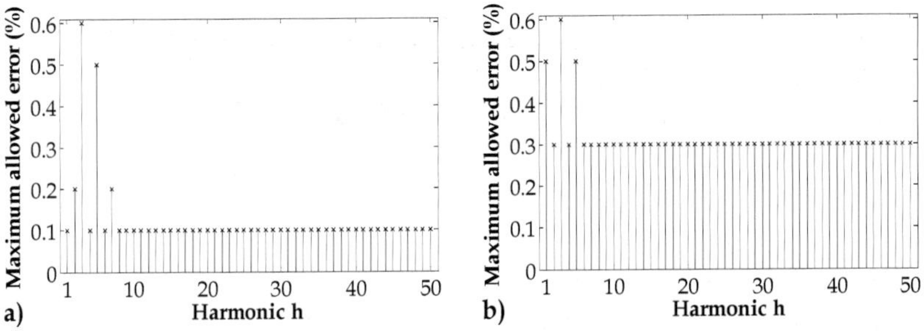

Fig. 3. Maximum allowed error of harmonic estimation for a) Class A and b) Class S instruments

From the comparison of Fig. 4a, Fig. 5a and Fig. 6a with Fig. 3a it can be seen that the DFT (*i.e.*, also the Goertzel algorithm), the non-iterative multiharmonic fitting and the iterative harmonic fitting are all suitable for a Class A instrument according to the IEC 61000-4-30 specification. The combined 4- and 3-paramter sine fitting algorithm produces worse results. However, this algorithm can still be used in Class S or Class B instruments (compare Fig. 4b, Fig. 5b and Fig. 6b with Fig. 3b). Note that the harmonic levels employed in the test corresponded to Class A testing; for Class S lower levels should be applied (*e.g.*, the *THD* should be up to 10%).

From Fig. 4a, Fig. 5a and Fig. 6a it can be seen that the multiharmonic algorithms are more accurate than the DFT calculation and that the difference between the results provided by the non-iterative and the iterative multiharmonic algorithm is negligible. The main difference between these two algorithms is in the estimates of the phases and in the frequency estimate, which are not employed when estimating only harmonic amplitudes.

In the following test, the accuracy of estimation of the *THD* was investigated. Signals with *THD* ranging from 0.5% up to 20% were simulated. The signals contained influencing quantities according to the testing state 3 (see Table 3) and normally distributed additive noise corresponding to the equivalent noise of an ideal 12-bit analog to digital converter. In total, 10 000 of such signals were simulated and the considered algorithms were applied to them. The DFT was used to calculate the values of *THD* (19), *THDG* (20) and *THDS* (22). The maximum absolute errors of estimation of the *THD* using all the algorithms are shown in Fig. 7.

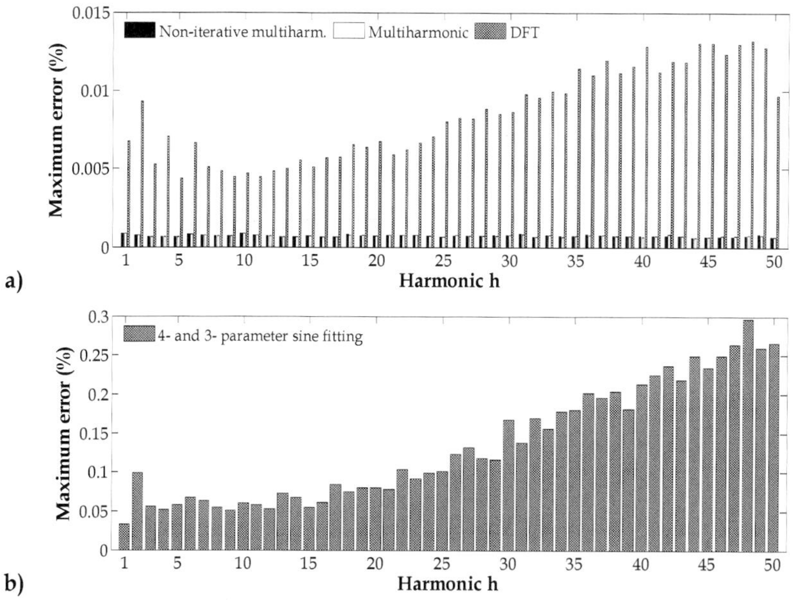

Fig. 4. Accuracy of estimation of harmonic amplitudes - testing state 1

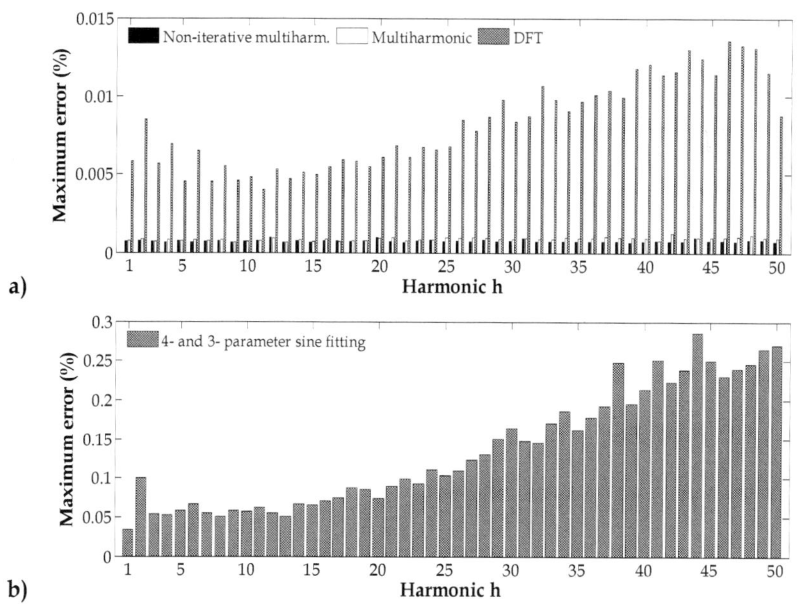

Fig. 5. Accuracy of estimation of harmonic amplitudes - testing state 2

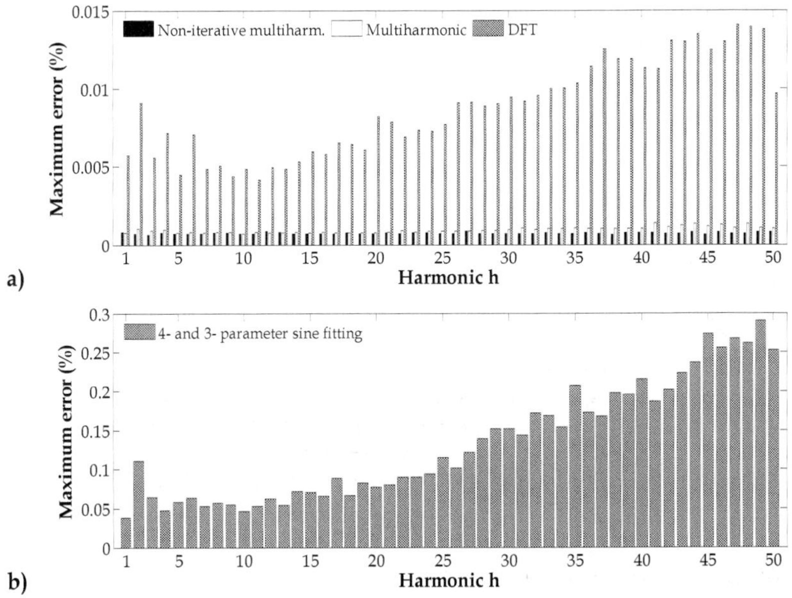

Fig. 6. Accuracy of estimation of harmonic amplitudes - testing state 3

The calculation of the *THDG* values includes all the spectrum lines of the DFT spectrum. This means, that it includes also the frequencies that contain spurious components (*e.g.*, the interharmonic component) and the noise. This explains the poor results of the *THDG* shown in Fig. 7. The rest of the algorithms produced almost identical results. Only the combined 4- and 3-parameter sine fitting algorithm performs slightly worse for higher values of the *THD*.

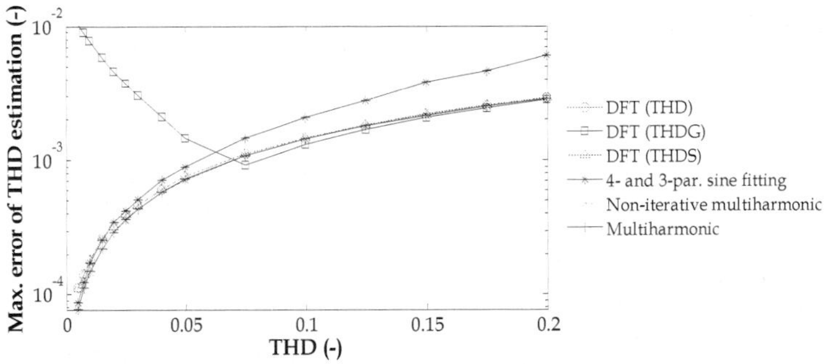

Fig. 7. Accuracy of estimation of the THD - testing state 3

6.2 Computational requirements

In this section, the memory and processing time requirements of the previously described algorithms are discussed.

The memory requirements of the algorithms for fast calculation of the DFT depend on the particular implementation and can differ substantially. For example, the real radix-2 FFT available as a library function in the VisualDSP++ development environment for digital signal processors (Analog Devices, 2010) requires $2 \times N$ memory spaces to store the real and imaginary part of the result and $3 \times N/4$ memory space to store the twiddle factors.

The Goertzel algorithm has very low memory requirements which do not depend on the number of processed samples N. A typical implementation requires only 4 memory spaces: 2 for the variable s; one for the multiplication constant in (5) and one auxiliary variable.

When implementing the 3- and 4-parameter sine fitting algorithms by the definition (see (11) and (13), respectively), it is necessary to construct the matrix $\mathbf{D}$ whose size is $N \times 3$ and $N \times 4$, respectively. However, both algorithms can be optimized by constructing directly the resulting matrices $\mathbf{D}^{\mathrm{T}}\mathbf{D}$ and $\mathbf{D}^{\mathrm{T}}\mathbf{u}$. This way, only matrices 3×3 and 3×1 (in the case 3-parameter sine fitting algorithm) or 4×4 and 4×1 (in the case 4-parameter sine fitting algorithm) have to be stored in the memory. Besides saving memory space, this approach is also faster because some of elements of these matrices are identical (Radil, 2009). This way, the 3-parameter algorithm requires 24 memory spaces and the 4-paramater algorithm requires 40 memory spaces independent of the length of the processed signal. The values include the space for intermediate results and exclude memory space required to calculate the initial estimate.

The multiharmonic fitting algorithms are more complex and attempts to construct the matrices $\mathbf{D}^{\mathrm{T}}\mathbf{D}$ and $\mathbf{D}^{\mathrm{T}}\mathbf{u}$ directly lead to higher computational burden. The non-iterative multiharmonic algorithm requires $(N+2)\times(2H+1)+2\times(2H+1)^2$ memory space and the iterative algorithm requires $(N+2)\times(2H+2)+2\times(2H+2)^2$ (including the space for intermediate results and excluding memory space required to calculate the initial estimate).

Another important parameter of methods for estimation of harmonics is their time consumption because in power quality monitoring it is usually required that all the processing, including the harmonic estimation, is performed in real-time.

To test the processing time required by individual algorithm, 10 000 signals with frequency $f = 49.95\,\mathrm{Hz}$, $THD = 2\%$ and random normally distributed noise were simulated. The signals were 10 fundamental cycles long (10 010 samples at $f_s = 50\,\mathrm{kS/s}$). The average processing time of individual algorithms implemented in Matlab was then evaluated and is shown in Table 5.

From Table 5 it can be seen that, as expected, the FFT is the fastest of the considered algorithms. However, the Goertzel algorithm, the combined 4- and 3-parameter sine fitting algorithm and the non-iterative multiharmonic fitting are also able to work in real-time (the length of the processed signal was approximately 200 ms).

Furthermore, some of the algorithms were implemented in a digital signal processor (DSP) Analog Devices ADSP-21369 running at the clock frequency of 264 MHz. A DSP like this one can be used for real-time processing in a power quality analyzer (Radil, 2009). Only the algorithms, whose implementation fits into the DSP's internal memory, were selected. The algorithms are: 2048-point FFT, the Goertzel algorithm and the combined 4- and 3-parameter sine fitting algorithm. The DSP was acquiring a voltage signal from a 230 V/50 Hz power system at a sampling rate f_s=10kS/s.. The average processing times are shown in Table 6.

The results in Table 6 are similar to the results shown in Table 5 and all three considered algorithms are suitable for real-time operation.
Note that the processing time of the 4-parameter algorithm shown in Table 6 is composed of two parts: initial calculations whose length depends only on the number of samples and the iterative parts which depends on the number of iterations required by the algorithm to converge. In the DSP implementation, the algorithm required on average 3 iterations. Significantly higher number of iterations indicates that the processed signal is disrupted by *e.g.*, sag or interruption.

Method	Average processing time (ms)
FFT	1.0
Goertzel algorihm	10.2
Combined 4- and 3-parameter sine fitting algorithm	143
Non-iterative multiharmonic fitting algorithm	85
Multiharmonic fitting algorithm	683

Table 5. Average processing times of the algorithms implemented in Matlab

Method	Average processing time (ms)
FFT	0.5
Goertzel algorihm	1.53
Combined 4- and 3-parameter sine fitting algorithm	100.2

Table 6. Average processing times of the algorithms implemented in a DSP

7. Summary and conclusions

In this chapter, an overview of several methods applicable for estimation of voltage harmonics was provided. The methods include: DFT, the Goertzel algorithm, method based on the 4- and 3-parameter sine fitting algorithms, the non-iterative multiharmonic fitting and the iterative multiharmonic fitting.
The DFT algorithm is a standard algorithm recommended by the power quality standards. However, as it was shown in Section 6 of this chapter, other methods can provided higher accuracy and/or lower computational and implementation requirements. The summary of the properties of the discussed methods is shown in Table 7. The selection of the most suitable method then depends on the requirements of each application and on the available resources.

Method	Pros	Cons
DFT (FFT)	• very fast • moderate memory requirements	• many implementations work only with number of samples that is an integer power of two
Goertzel algorihm	• fast • very low memory requirements	• signal has to include an integer number of periods
Combined 4- and 3-parameter sine fitting algorithm	• moderate memory requirements • signal not limited to an integer number of periods	• low accuracy (but can be used for Class S or Class B instruments) • slow (but still suitable for real-time operation)
Non-iterative multiharmonic fitting algorithm	• accurate • reasonably fast • signal not limited to an integer number of periods	• high memory requirements
Multiharmonic fitting algorithm	• accurate • signal not limited to an integer number of periods	• very slow • high memory requirements

Table 7. Summary of the properties of the described algorithms for harmonic estimation

8. References

Analog Devices (2010). *VisualDSP++ 5.0 Run-Time Library Manual for SHARC Processors (Edition 1.4)*, Analog Devices, Inc.

Andersson, T.; Händel, P. (2006). IEEE Standard 1057, Cramér-Rao bound and the parsimony principle, *IEEE Transactions on Instrumentation and Measurement*, vol. 55, no. 1, February 2006, pp. 44 - 53, ISSN 0018-9456

Gaouda, A. M.; Kanoun, S. H.; Salama, M. M. A.; Chikhani, A. Y. (2002). Wavelet-based signal processing for disturbance classification and measurement, *IEE Proceedings - Generation, Transmission and Distribution*, vol. 149, no. 3, May 2002, pp. 310 - 318, ISSN 1350-2360

Goertzel, G. (1958). An algorithm for the evaluation of finite trigonometric series, *The American Mathematical Monthly*, vol. 65, no. 1, January 1958, pp. 34 - 35, ISSN 0002-9890

IEC 61000-2-4 (2002). *IEC 61000-2-4 Electromagnetic compatibility (EMC) - Part 2-4: Environment - Compatibility levels in industrial plants for low-frequency conducted disturbances, Edition 2.0*, IEC, ISBN 2-8318-6413-5

IEC 61000-4-7 (2009). *IEC 61000-4-7 Electromagnetic compatibility (EMC) - Part 4-7: Testing and measurement techniques - General guide on harmonics and interharmonics measurements and instrumentation, for power supply systems and equipment connected thereto, Edition 2.1*, IEC, ISBN 2-8318-1062-6

IEC 61000-4-30 (2008). *IEC 61000-4-30 Electromagnetic compatibility (EMC) - Part 4-30: Testing and measurement techniques - Power quality measurement methods, Edition 2.0*, IEC, ISBN 2-8318-1002-0

IEEE Std. 1057-2007 (2008). *IEEE Standard for Digitizing Waveform Recorders*, IEEE Instrumentation and Measurement Society, ISBN 978-0-7381-5350-6

IEEE Std. 1159-2009 (2009). *IEEE Recommended Practice for Monitoring Electric Power Quality*, IEEE Power & Energy Society, ISBN 978-0-7381-5939-3

Oppenheim, A. V.; Schafer, R. W.; Buck, J. R. (1999). *Discrete-time signal processing*. Upper Saddle River, N.J.: Prentice Hall. ISBN 0-13-754920-2

Pham, V.L.; Wong, K.P. (1999). Wavelet-transform-based algorithm for harmonic analysis of power system waveforms, *IEE Proceedings on Generation, Transmission and Distribution*, vol. 146, no. 3, May 1999, pp. 249-254, ISSN 1350-2360

Rader, C. M. (1968). Discrete Fourier transforms when the number of data samples is prime, *Proceedings of the IEEE*, vol. 56, no. 6, June 1968, pp. 1107- 1108, ISSN 0018-9219

Radil, T.; Ramos, P. M.; Serra, A. C. (2009). Single-phase power quality analyzer based on a new detection and classification algorithm, *Procedings of IMEKO World Congress*, pp. 917 - 922, ISBN 978-963-88410-0-1, Lisbon, Portugal, September 2009

Ramos, P. M.; da Silva, M. F.; Martins, R. C.; Serra, A. C. (2006). Simulation and experimental results of multiharmonic least-squares fitting algorithms applied to periodic signals, *IEEE Transactions on Instrumentation and Measurement*, vol. 55, no. 2, April 2006, pp. 646 - 651, ISSN 0018-9456

Renders, H.; Schoukens, J.; Vilain, G. (1984). High-Accuracy Spectrum Analysis of Sampled Discrete Frequency Signals by Analytical Leakage Compensation", *IEEE Transactions on Instrumentation and Measurement*, vol. 33, no. 4, December 1984, pp. 287 - 292, ISSN 0018-9456

Slepička, D.; Petri, D.; Agrež, D.; Radil, T.; Lapuh, R.; Schoukens, J.; Nunzi, E.; Sedláček, M. (2010). Comparison of Nonparametric Frequency Estimators, *Proceedings of the IEEE International Instrumentation and Technology Conf. - I2MTC*, pp. 73 - 77, ISBN 978-1-4244-2833-5, Austin, USA, May 2010

14

Improved Power Quality AC/DC Converters

Abdul Hamid Bhat and Pramod Agarwal
National Institute of Technology Srinagar, Kashmir
India

1. Introduction

In this chapter, various power quality problems created by the widespread use of conventional diode bridge rectifiers and line-commutated AC/DC converters have been discussed. The impact of these problems on the health of power systems and working of sensitive equipments has also been discussed. Need for addressing these burning power quality issues has been emphasized. Recent trends of addressing these issues have been briefly discussed. A new breed of improved power quality AC/DC converters has been discussed in details. Development of various topologies and state-of-art of this breed of converters has been discussed briefly. The state-of-the-art multilevel converters used as improved power quality converters have been covered in this chapter with emphasis on three-level neutral-point clamped converters. This converter has been investigated for improved power quality and various simulation results have been presented to prove their effectiveness in terms of excellent power quality like nearly unity input power factor and negligible harmonic distortion of source current. The simulation results have been obtained with sinusoidal PWM and sapce-vector PWM modualtion algorithms. Simulation restls have been validated through experimental results which are obtained on a three-level converter by real-time implementation of space-vector PWM technique using a real-time DSP board. The performance investigation of the converter proves the effectiveness of three-level converter in elegantly addressing the burning power quality issues. We know that power systems are designed to operate at frequencies of 50 or 60 Hz. However, certain types of loads produce harmonic currents in the power system. The power system harmonics are not a new phenomenon. Concern over harmonic distortions has ebbed and flowed during the history of electrical power systems. Traditionally the saturated iron in transformers and induction machines, electric arc furnaces, welding equipment, fluorescent lamps (with magnetic ballasts), etc. have been responsible for the generation of harmonics in electric power systems. Most of these equipments also cause the flow of reactive component of current in the system. In recent years, many power electronic converters utilizing switching devices are being widely used in domestic, commercial and industrial applications, ranging from few watts to MWs. However these converters suffer from the drawbacks of harmonic generation and reactive power flow from the source and offer highly non-linear characteristics. The generation of harmonics and reactive power flow in the power systems has given rise to the 'Electric Power Quality' problems. Any significant deviation in the magnitude of the voltage, current and frequency, or their waveform purity may result in a potential power quality problem. Power quality problems arise when these deviations exceed beyond the tolerable limit and can occur in three different ways as frequency events, voltage events and waveform events. Distortion of the voltage/current waveforms from the normal sinusoidal waveshape is considered as waveform event. *One of the most harmful waveform events are the harmonic distortions. Harmonics are basically the additional frequency components present in the mains voltage or current which are integer multiples of the mains (fundamental) frequency. Harmonic distortion originates due to the nonlinear characteristics of devices and loads on the power system.* The inter-harmonics are due to the presence of additional frequencies which are non-integral multiples of the mains frequency. Moreover, notching is a periodic voltage disturbance caused by the normal operation of power electronic devices when current is commutated from one phase to another.

All these disturbances may originate problems to both utility and customers. Among them, harmonic distortions are considered one of the most significant reasons for power quality problems. Harmonic problems counter many of the conventional rules of the power system design and operation that consider only the fundamental frequency. Harmonic distortions are mainly caused by the nonlinear devices in which the current is not proportional to the applied voltage as shown in Fig. 1, where a nonlinear resistor is supplied by a sinusoidal voltage source. The resulting current is distorted while the applied voltage is perfectly sinusoidal. Increasing the voltage by a few percent may cause the current to double and take a different waveshape. This, in essence, is the source of harmonic distortion in the power system.

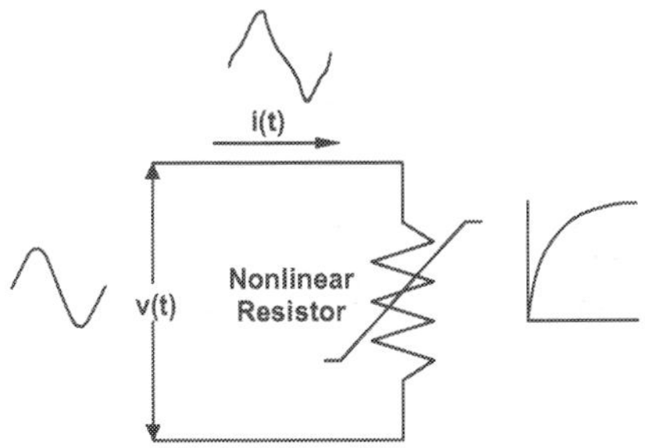

Fig. 1. Current distortion caused by nonlinear resistor

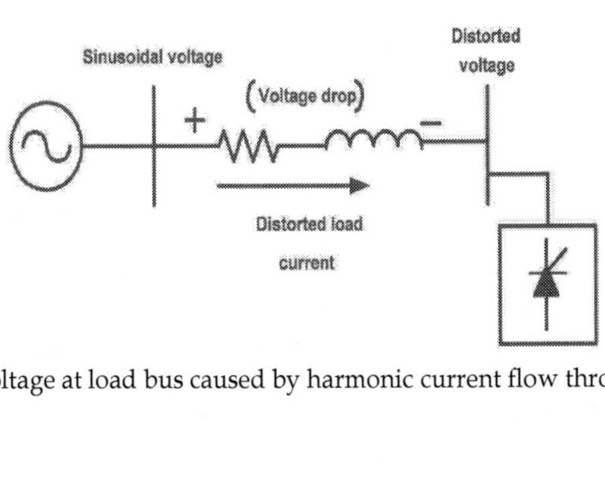

Fig. 2. Distorted voltage at load bus caused by harmonic current flow through the system impedance

2. Causes and effects of harmonics and reactive power

Any device with nonlinear characteristics that derives input power from a sinusoidal electrical system may be responsible for injecting harmonic currents and voltages into the electrical power system. Developments in digital electronics and power semiconducting devices have led to a rapid increase in the use of nonlinear devices. Power converters, most widely used in industrial, commercial and domestic applications are considered the primary source of undesired harmonics. In converter theory, the DC current is considered to be constant and the line currents at the AC side will consist of abrupt pulses instead of a smooth sinusoidal wave. Power, the product of voltage and current, at the DC side contains harmonics in the current/voltage. Since no energy storage can take place in the elements of a converter, the power balance of the input and output requires harmonics in the input power, and thus harmonic currents will flow in the supply lines. Energy balance considerations show and Fourier analysis of the square waves confirm that for a 6-pulse converter, each 6n harmonics in the DC voltage requires harmonic currents of frequencies 6n±1 in the AC lines. Harmonics are the integral multiples of fundamental frequency superimposed on the fundamental frequency. These harmonics combine with the fundamental to form distorted waveshapes.

Harmonics are caused by the loads in which the current waveform does not conform to the fundamental waveform of the supply voltage. These loads are called the nonlinear loads and are the source of harmonic current and voltage distortion. Current harmonics generated by these nonlinear loads are propagated throughout the power network. Voltage distortion is the result of these distorted currents passing through the series impedance of the system, as shown in Fig. 2. Harmonic current passing through the system impedance causes a voltage drop for each harmonic and results in voltage harmonics appearing at the load bus and leads to power quality problems. The most common measure of distortion is the total harmonic distortion, THD. THD applies to both current and voltage and is defined as the rms value of harmonics divided by the rms value of the fundamental, multiplied by 100. THD of current varies from a few percent to more than 100%. THD of voltage is usually less than 5% but values above 10% are definitely unacceptable and cause problems for the sensitive equipment and loads. Most of the power electronic loads include AC/DC converters which are the primary source of harmonics and reactive power flow in a power system.

Some of the commonly used AC/DC converters have been simulated using MATLAB/Simulink software. The waveforms of source voltage and source current in all the cases have been plotted and the harmonic spectrum of source current in all the cases has been obtained to prove the power quality problems (like injection of harmonics in source current and deterioration of input power factor) created by these converters. Fig. 3 depicts the load voltage and load current waveforms of a single-phase diode bridge rectifier driving an R-L load. Fig. 4 shows the corresponding source voltage and source current waveforms. It can be clearly seen from the harmonic spectrum of source current in Fig. 5 that the source current in highly distorted with a THD of about 43%. Fig. 6 shows the waveforms of source voltage and source current drawn by a single-phase practical bridge rectifier with a filter capacitor connected across the load and Fig. 7 shows the corrrsponding harmonic spectrum of the source current which shows the THD of source current as high as 54.24%. Fig. 8 shows the waveforms for source voltage and source current for a single-phase fully-controlled converter driving an R-L load at a firing angle of 90° and Fig. 9 shows source current harmonic spectrum with a THD of 54.25%. Fig. 10 shows the phase *A* source voltage and source current waveforms for a three-phase fully-controlled converter driving an R-L load at a firing angle of 90°. Fig. 11 shows the source current harmonic spectrum with a THD of about 42%. Thus the above simulated results show that the diode bridge rectifiers and phase-controlled converters create serious power quality problems in terms of distortion of the source current and deterioration of input power factor.

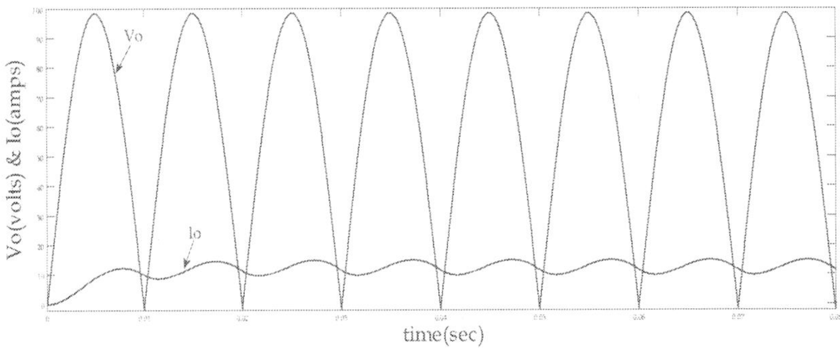

Fig. 3. Load voltage and load current waveforms of a single-phase bridge rectifier with R-L load

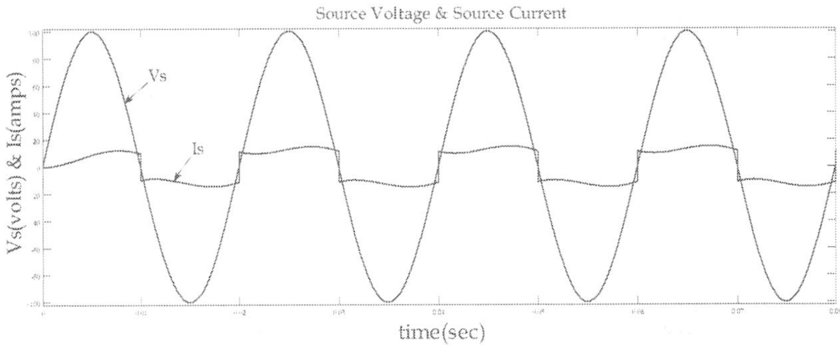

Fig. 4. Source voltage and source current waveforms of a single-phase bridge rectifier with R-L load

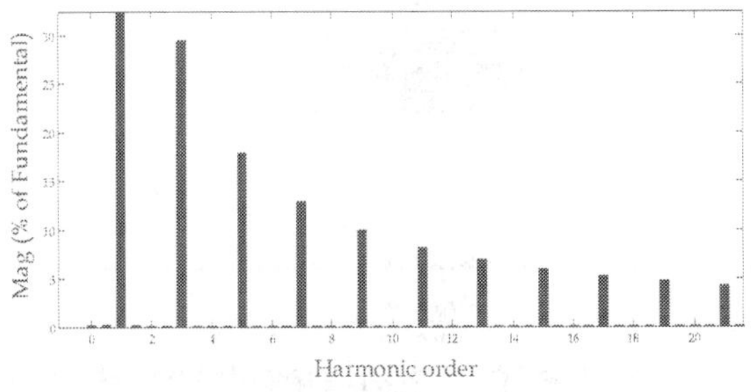

Fig. 5. Harmonic spectrum of source current

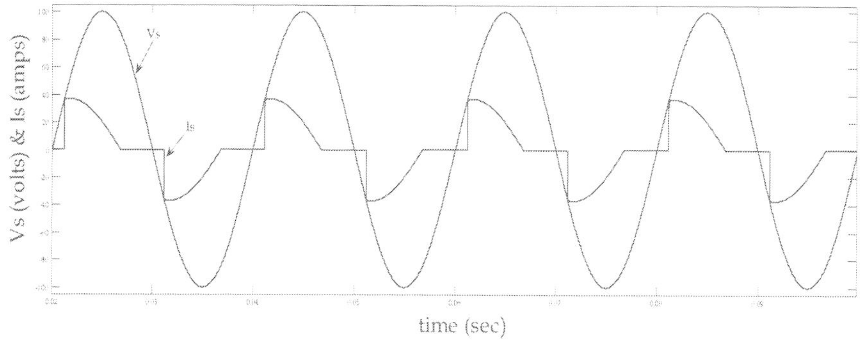

Fig. 6. Source voltage and source current waveforms of a single-phase practical bridge rectifier

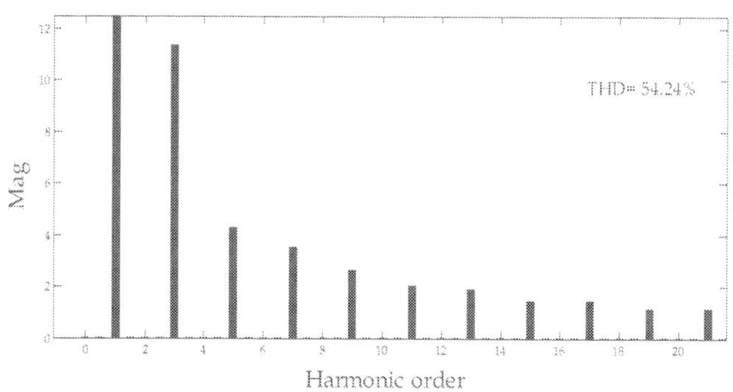

Fig. 7. Harmonic spectrum of source current

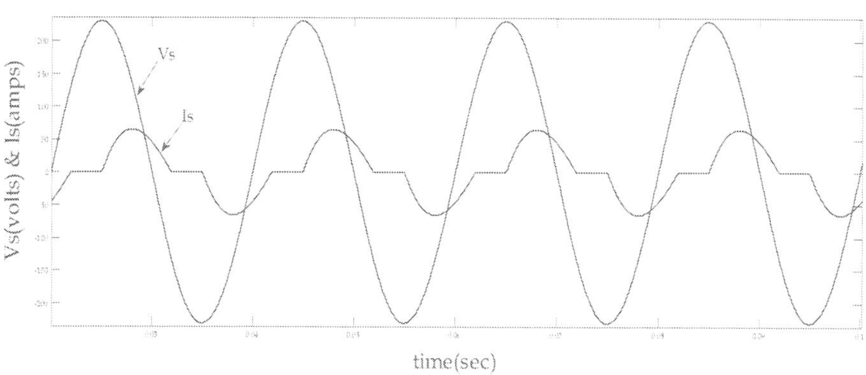

Fig. 8. Source voltage and source current waveforms of a single-phase fully-controlled converter at α=90°

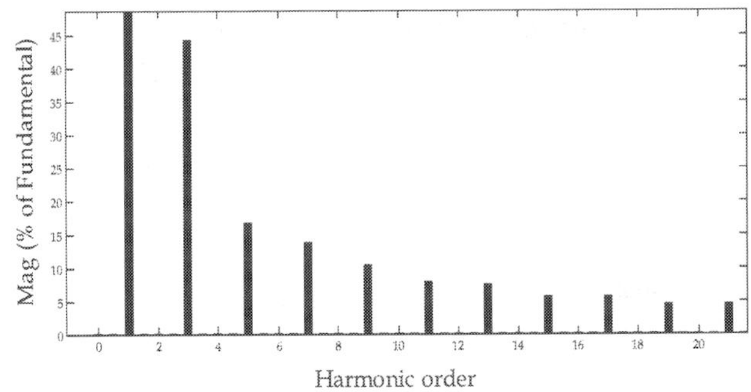

Fig. 9. Harmonic spectrum of source current

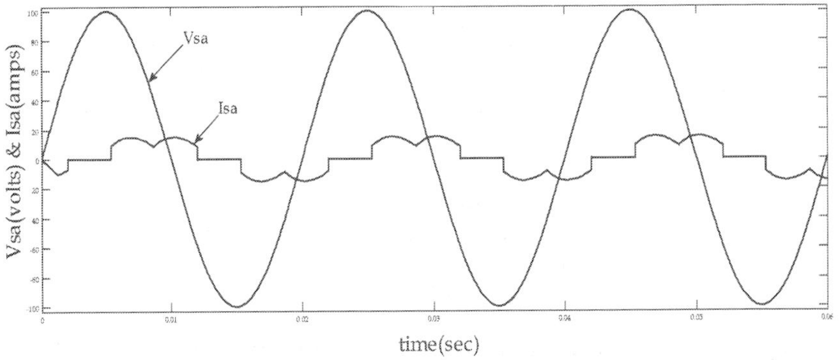

Fig. 10. Source voltage and source current waveforms of a three-phase line- commutated converter at α=90°

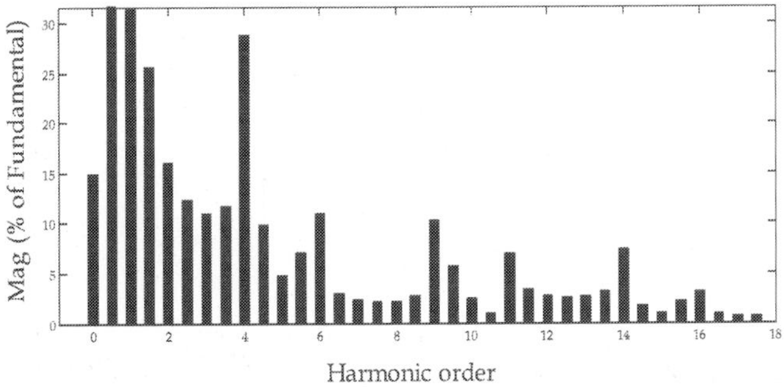

Fig. 11. Harmonic spectrum of source current

AC/DC power converters are extensively used in various applications like power supplies, DC motor drives, front-end converters in adjustable-speed AC drives, HVDC transmission, SMPS, fluorescent lights (with electronic ballasts), utility interface with non-conventional energy sources, in process technology like welding, power supplies for telecommunications systems, aerospace, military environment and so on. Traditionally, AC-DC power conversion has been dominated by diode or phase-controlled rectifiers which act as non-linear loads on the power systems and draw input currents which are rich in harmonics and have poor supply power factor, thus creating the power quality problem for the power distribution network and for other electrical systems in the vicinity of rectifier. The other associated problems with these converters include:

- Large reactive power drawn by rectifiers from the power system which requires that the distribution equipment handle large power, thus increasing its volt-ampere ratings;
- Voltage drops at the buses;
- Higher input current harmonics resulting in the distorted line current which tends to distort the line voltage waveform. This often creates problems in the reliable operation of sensitive equipment operating on the same bus;
- Increased losses in the equipments (due to harmonics) such as transformers and motors connected to the utility;
- Electromagnetic interference with the nearby communications circuits;
- Excessive neutral current, resulting in overheated neutrals. The currents of triplen harmonics, especially odd harmonics (3rd, 9th, 15th,...) are actually additive in the neutral of three-phase Y-connected circuits;
- Incorrect reading meters, including induction disc-type W-hr meters and averaging type current meters;
- Blown-fuses on power factor correction capacitors due to high voltages and currents from resonance with line impedance and capacitor bank failures;
- Mal-operation of equipments such as computers, telephone systems, and electronic controllers;
- Nuisance operation of protective devices including false tripping of relays and failure of a UPS to transfer properly, especially if the controls incorporate zero-crossing sensing circuits;
- Damaging dielectric heating in cables and so on.

SCR=I/I_1	<II	II<h<17	17<h<23	23<h<35	35<h	TDD
<20	4.0	2.0	1.5	0.6	0.3	5.0
20-50	7.0	3.5	2.5	1.0	0.5	8.0
50-100	10.0	4.5	4.0	1.5	0.7	12.0
100-1000	12.0	5.5	5.0	2.0	1.0	15.0
>1000	15.0	7.0	6.0	2.5	1.4	20.0

Table 1. IEEE 519 Current Distortion Limits

Various standards are set to limit the harmonics by nonlinear loads. IEEE standard 519 was first issued in 1991. It gave the first guidelines for system harmonics limitations and was revised in 1992 [10]. IEEE 519-1992 Recommended Practices and Requirements for Harmonic Control in Electrical Power Systems provide the guidelines for determining what are the acceptable limits. The harmonic limits for current depend on the ratio of Short Circuit Current (SCC) at the Point of Common Coupling (PCC) to the average Load Current of maximum demand over one year, as illustrated in Table 1. Thus the basic philosophy of IEEE 519-1992 is to limit the harmonic current injected into the power system and to make utility responsible for maintaining the voltage distortion. The IEC 61000 series is an internationally accepted set of standards and comprises of IEC 61000-3-2, 61000-3-3, 61000-3-4, 61000-3-5, and 61000-3-6. However, IEC 61000-3-4 is the most relevant one for industrial installations. The IEEE 519 standard limits the harmonics primarily at the service entrance, while IEC-3-2 is applied at the terminals of end-user equipment.

3. Classical solutions and recent trends

Classically, shunt passive filters consisting of tuned LC filters and/or high pass filters are used to suppress the harmonics and power capacitors are employed to improve the power factor of the utility/mains. The shunt passive filters are tuned most of the time to a particular harmonic frequency to be eliminated so that a low impedance is offered at the tuned frequency than the source impedance in order to reduce the harmonic current flowing into the source. Thus, the filtering characteristics are determined by the impedance ratio of the source and passive filter. Therefore the shunt passive filters suffer from the following drawbacks:

1. The source impedance, which is not accurately known and varies with the system configuration, strongly influences the filtering characteristics of the shunt passive filter.
2. The shunt passive filter acts as a sink to the harmonic current flowing from the source. In the worst case, the filter may fall in series resonance with the source impedance.
3. At a specific frequency, an anti-resonance or parallel resonance may occur between the source impedance and the shunt passive filter, which is also called the harmonic amplification.
4. As both the harmonics and the fundamental current component flow into the filter, the capacity of filter must be rated by taking into account both the currents.
5. Increase in harmonic current component can overload the filter.
6. If a good level of compensation is required, one needs as many filters as the number of harmonics to be eliminated

Conventional methods of var compensation are based on the vars generated or absorbed by the passive elements having energy storage capability. Dynamic compensation is achieved either by Switched Capacitor Var Compensators or Switched Capacitor and Thyristor Controlled Reactors. As the vars generated or absorbed are directly proportional to the energy storage capability of the passive elements used, their size increases with the increment in the vars to be compensated. Introduction of the sizeable inductors and capacitors into the system may lead to resonance created by the peripheral low frequency current sources. Moreover, the capability of this class of compensators to manipulate vars depends on the voltage level prevailing at the point where they are connected. Since the bus voltage reduces as the var demand increases, the compensators fail to perform when their participation is most needed. Also, they pollute the utility with low order harmonics which are difficult to filter. The sensitivity of these problems has attracted the attention of researchers to develop the techniques with adjustable and dynamic components. Extensive research is being carried out in the field of harmonics and reactive power compensation to overcome these limitations. With the continuous proliferation of nonlinear type loads, the requirements of power compensation involved avoidance of harmonic current generation also in addition to compensating for the reactive power generation. The equipments used for the harmonic compensation in addition to var compensation are known as Active Power Filters (APFs) [1,5,11,22]. These filters have provided the required harmonic filtering and control performance in comparison to conventional shunt passive filters and static var compensators. The objectives of active filtering are to solve these problems by combining the advantages of regulated systems with reduced rating of the necessary passive components. The APFs are generally built around a PWM converter with capacitor/inductor on its DC side. The PWM converter switches are controlled to draw/supply a compensating current from/to the utility so that it cancels the current harmonics on the AC side by generating the nonlinearities opposite to the load nonlinearities and makes the source current almost sinusoidal which may be in phase or phase displaced with mains voltage, based on both harmonic and reactive power compensation requirements or only harmonics compensation requirement. In addition to the harmonics and reactive power compensation, APFs are also used to eliminate voltage harmonics, for load balancing, to regulate the terminal voltage, to suppress the voltage flickers, etc. These wide range of objectives are achieved either individually or in combination depending upon the requirements, control strategy and configuration, which is to be selected appropriately.

Based on the objectives, the APFs are broadly classified as Shunt Active Power Filter, Series Active Power Filter, and Hybrid Power Filter. However, the APFs suffer from the drawback of large size and rating (in some cases, the filter rating may be comparable with that of the load), complexity in the control and cost.

4. Improved power quality AC/DC converters

A new breed of AC/DC Power Converters has been developed to overcome all the drawbacks of passive filters, var compensators and active power filters used for harmonics and reactive power compensation. This new breed of converters is specifically known as Power Factor Correction Converters (PFCs), Switched Mode Rectifiers (SMRs), PWM Converters, Improved Power Quality Converters (IPQCs), and High Power Factor Converters (HPFCs). They are included as an inherent part of the AC-DC conversion system which produces excellent power quality at the line-side and load-side, higher efficiency, and reduced size. The power quality issues created by the use of conventional AC/DC converters are elegantly addressed by IPQCs. The output voltage is regulated even under the fluctuations of source voltage and sudden load changes. The PWM switching pattern controls the switchings of the power devices for input current waveshaping so that it becomes almost harmonic-pollution free and in phase with the source voltage, thus producing a nearly sinusoidal supply current at unity power factor without the need of any passive or active filter for harmonics and reactive power compensation. The reduced size of magnetics used in the converter system and the single-stage power conversion techniques have resulted in the development of reduced size, high power density, efficient, and reduced cost power converters. They have been made possible mainly because of the use of modern solid state, self-commutating power semiconducting devices such as Power MOSFETs, IGBTs, IGCTs, GTOs, etc. Remarkable progress in the capacity and switching speed of these devices has made it possible to develop the IPQCs for medium and large power applications. The parallel progress in the processors and high-speed DSPs has made it possible to implement the complex and computation-intensive control algorithms at very high speeds for the control of IPQCs. In fact, the development and progress in the fields of power semiconducting devices and DSPs has revolutionized the field of Power Electronics in recent past.

Improved Power Quality Converters are being developed with unidirectional and bidirectional power flow capabilities. Three-phase unidirectional IPQCs are realized using a three-phase diode bridge followed by step-down chopper, step-up chopper, step-down/up chopper, isolated, forward, flyback, push-pull, half-bridge, full-bridge, SEPIC, Cuk, Zeta, and multilevel converters. A high-frequency isolation transformer offers reduced size, weight, cost, appropriate voltage matching and isolation. On the other hand, three-phase bidirectional IPQCs consist of basic converters such as push-pull, half-bridge, voltage source converter (VSC) topology, or current source converter (CSC) topology. Four-quadrant three-phase AC/DC power converters are normally implemented using matrix converters.

Due to all these advantages, IPQCs have generated tremendous interest among the researchers and application engineers to solve the increasing power quality problems. In fact, when an application engineer is at a decision stage, the active solution is advantageous over the passive filtering.

5. Converter topologies

Broadly, Three-phase Improved Power Quality Converters have been classified on the basis of the converter topology as Boost, Buck, Buck-Boost and Multilevel converters with unidirectional and bi-directional power flow and the type of converter used as unidirectional and bi-directional converters. B. Singh, et. al. [23] presented the broad classification of three-phase IPQCs. In case of three-phase boost converters, the output voltage is greater than the peak input voltage. Unlike a single-phase boost converter, the voltage across the output capacitor does not have low-frequency ripple in balanced conditions. Thus a wide bandwidth voltage feedback loop can be used resulting in fast voltage control without distorting the input current references.

Fig. 12 shows one of the topologies of three-phase unidirectional boost converters [19]. High power-factor can be easily obtained when three-phase unidirectional boost converters are operated in discontinuous conduction mode (DCM) with constant duty cycles [28]. This is because the basic types of DC-DC converters, when operating in DCM, have self-power factor correction (PFC) property, that is, if these converters are connected to the rectified AC line, they have the capability to give higher power factor by the nature of their topologies. The peak of the supply-side inductor current is sampling the line-voltage automatically, giving boost converter the self-PFC property because no control loop is required from its input side. This is an advantage over continuous conduction mode (CCM) PFC circuit in which multi-loop control strategy is essential. However the input inductor operating in DCM cannot hold the excessive input energy because it must release all its stored energy before the end of each switching cycle. As a result, a bulky capacitor is used to balance the instantaneous power between the input and output. Also since the input current is normally a train of triangular pulses with nearly constant duty ratio, an input filter is necessary for smoothing the pulsating input current into a continuous one. Three-phase, unidirectional boost converters are widely used nowadays as a replacement of conventional diode rectifiers to provide unity input pf, reduced THD at AC mains and constant, regulated DC output voltage even under fluctuations of AC voltage and DC load.

Fig. 13 depicts one of the topologies of bidirectional boost converters. In case of bidirectional boost converters operating in CCM [9], since the input current is the inductor current, it can be easily programmed by current-mode control. Various current control techniques are available for controlling the input current so as to make the input current THD negligible associated with a unity input pf. We know that VSIs can reverse the power flow from load to DC link as a rectifier. However a standalone voltage source rectifier requires a special DC bus able to keep voltage constant without the requirement of a voltage supply. This is accomplished with a DC capacitor and a feedback control loop. Boost converters operating in CCM give low dv/dt stress and hence produce low EMI emissions as compared to those operating in DCM. The softswitching techniques reduce the di/dt and dv/dt and hence improve the performance of bi-directional boost converters by causing low EMI emissions. The three-phase bi-directional boost PFCs are suitable for high power applications with improved performance as front-end converters with regeneration capability for variable-speed AC motor drives and also for hoists, cranes, lifts, BESS, line-interactive UPS, etc. [23]. Three-phase, buck converters produce output voltages less than the converter input voltage [8]. They have some attractive features compared to boost rectifiers such as meeting the requirement of varying controllable output DC voltage, inherent short-circuit protection, and easy inrush current. One of the topologies of three-phase unidirectional buck converters is shown in Fig. 14. Their input currents can be controlled in the open loop and much wider voltage loop bandwidth can be achieved. A unidirectional buck converter is a replacement of the thyristor semi-converter with improved power quality at AC mains and output DC bus.

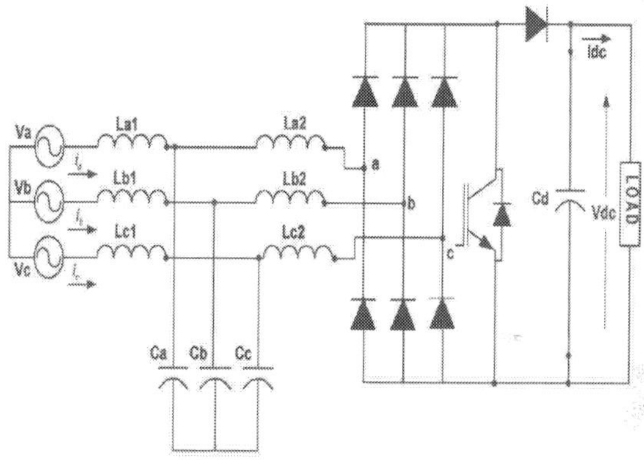

Fig. 12. Three-phase unidirectional boost converter

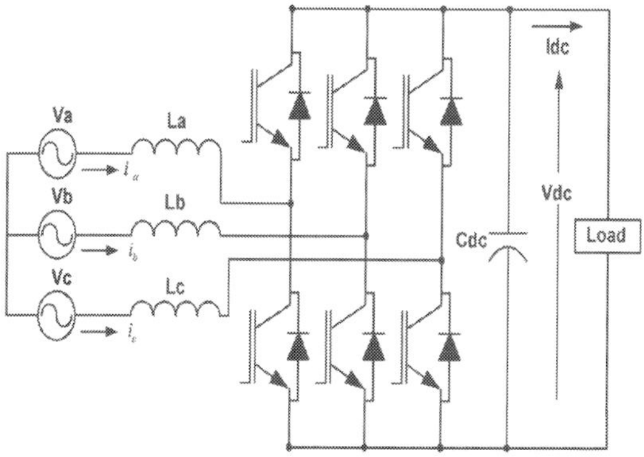

Fig. 13. VSI-bridge-based bidirectional boost converter

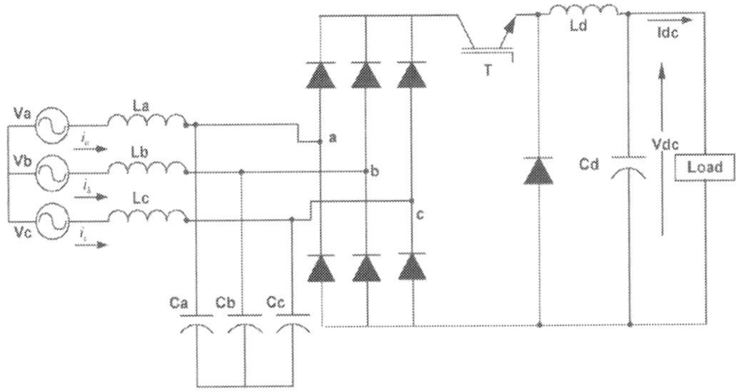

Fig. 14. Single-switch unidirectional buck converter

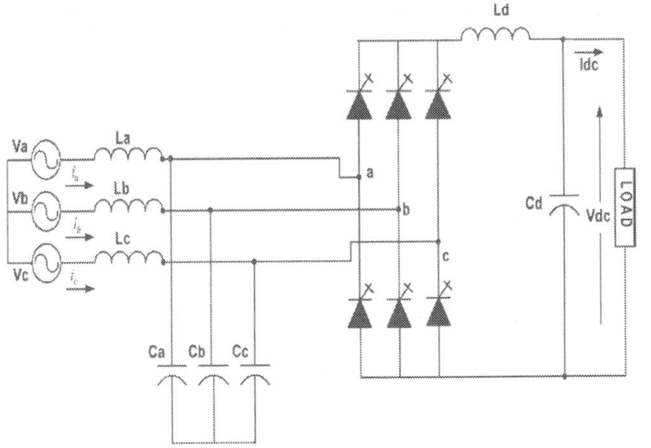

Fig. 15. GTO-based bidirectional buck converter

Fig. 15 shows one of the topologies of the bidirectional buck converter [23]. A three-phase bidirectional buck converter provides a similar function as a conventional thyristor bridge converter but with improved power quality such as high power factor and reduced harmonic currents at AC mains and fast regulated output voltage with reversible power flow [6,14,27]. Two buck converters connected in anti-parallel provide the behaviour similar to a dual converter for four-quadrant operation with improved power quality and fast response.

In three-phase boost converters, the output voltages lower than the supply voltage cannot be achieved. Also in three-phase buck converters, the output voltages higher than the supply voltage cannot be achieved. However, it has the inherent DC short-circuit current and inrush current limitation capability. The three-phase buck-boost type AC/DC converters have step-up or step-down output voltage characteristics and also the capability of limiting the inrush and DC short-circuit currents. Therefore this type of converter is convenient for several power supplies and is highly suitable for input pf correction [7,12,13]. Fig. 16 depicts one of the topologies of three-phase unidirectional buck-boost converters.

There are some applications which require output DC voltage widely varying from low voltage to high voltage with bidirectional DC current as four-quadrant operation and bidirectional power flow. As discussed by B. Singh, et. al. [23], the simplest way of realizing a three-phase bidirectional buck-boost converter is by using a matrix converter as shown in Fig. 17. The three-phase bidirectional buck-boost converters can be used for medium power applications in telecommunications and also for motor drive control.

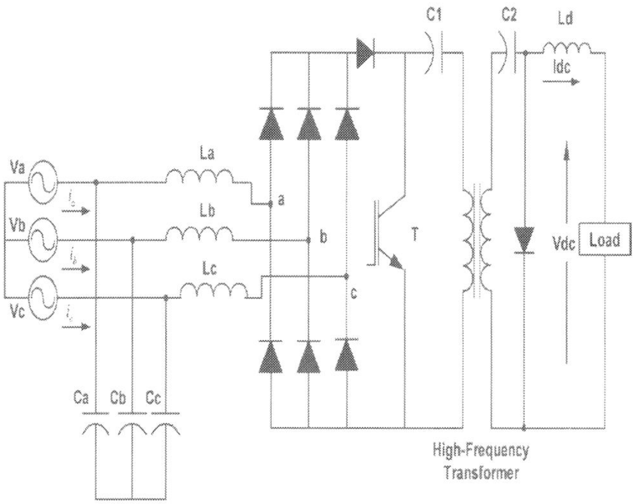

Fig. 16. Isolated cuk-derived unidirectional buck-boost converter

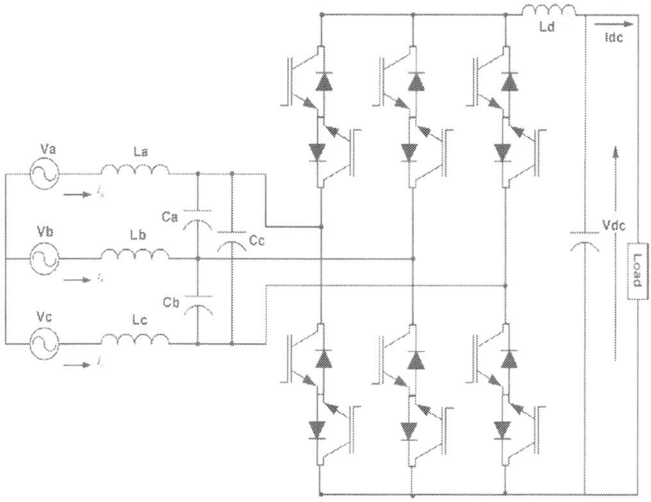

Fig. 17. Matrix-converter-based bidirectional buck-boost converter

Multilevel Converters (MLCs) are gaining widespread popularity because of their excellent performance with reduced THD of input current, high supply power factor, ripple-free regulated DC output voltage, reduced voltage stress of devices, reduced dv/dt stresses, and hence lower EMI emissions [3,15,16,17,18,20,21,24,25]. They also avoid the use of transformers in some applications which further enhances the efficiency of these converters. The sinusoidal source currents at unity power factor are produced at reduced switching frequencies in comparison with their two-level counterparts. Moreover since an MLC itself consists of series connection of switching power devices and each device is clamped to the DC-link capacitor voltage through the clamping diodes, it does not require special consideration to balance the voltages of power devices.

On the other hand, the series connection of power devices is a big issue in two-level converters. Moreover, in case of a multilevel converter, each device is stressed to a voltage $V_{dc}/(n-1)$, where V_{dc} is the DC-bus voltage and n is the number of levels. Hence the device stress is considerably reduced as the number of levels increases [15]. This makes multilevel converters the best choice for the high-voltage and high-power applications and they have invited a lot of attention for high-power industrial applications. Nevertheless, the neutral point of the neutral point clamped converter is prone to fluctuations due to irregular charging and discharging of the output capacitors [20]. Thus the terminal voltage applied at the switches on DC side can exceed that imposed by the manufacturer. Moreover the device count is large in multilevel converters and complex control is involved [15].

Fig. 18 depicts one of the topologies of three-phase, unidirectional multi-level converters [23]. These converters also offer boost operation for the output voltage with unidirectional power flow.

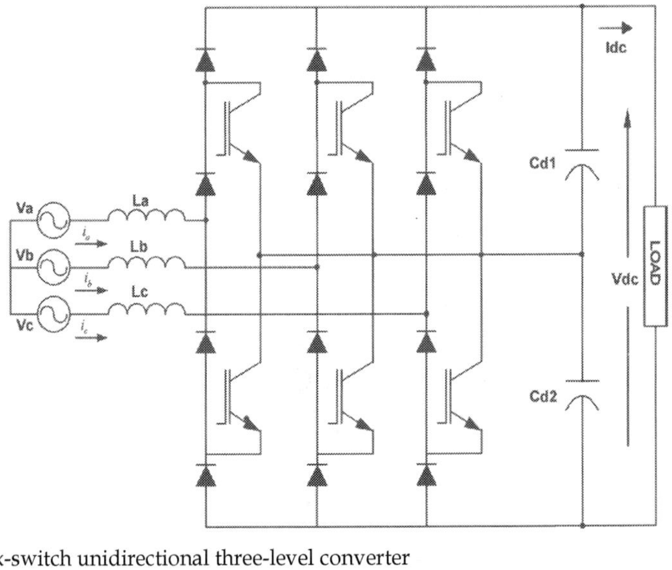

Fig. 18. Six-switch unidirectional three-level converter

J. S. Lai and F. Z. Peng [15] classified the bidirectional MLCs into three main categories as diode-clamped MLC, flying capacitor MLC, and cascaded MLC as shown in Fig. 19. In fact, all the three types of multilevel bidirectional converters shown in this Figure can be used in reactive power compensation without having voltage unbalance problem.

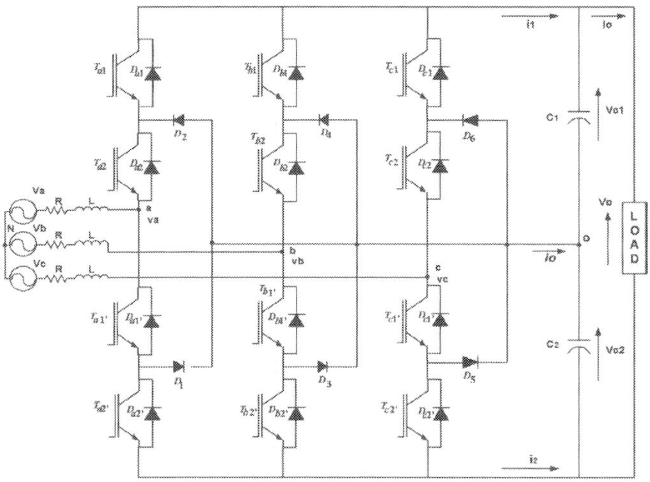

(a)

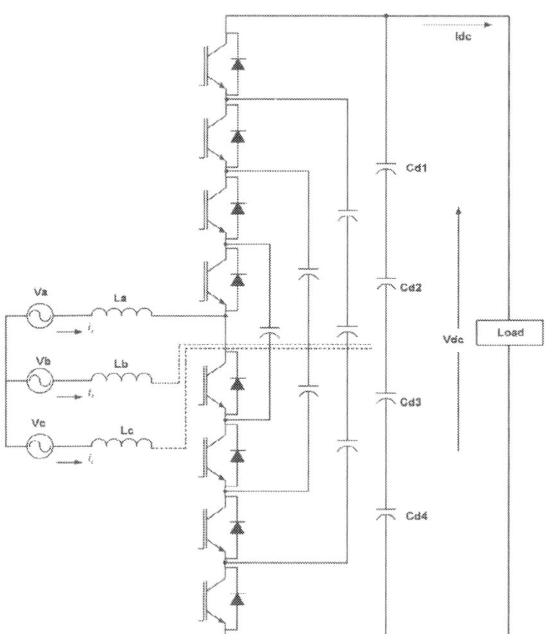

(b)

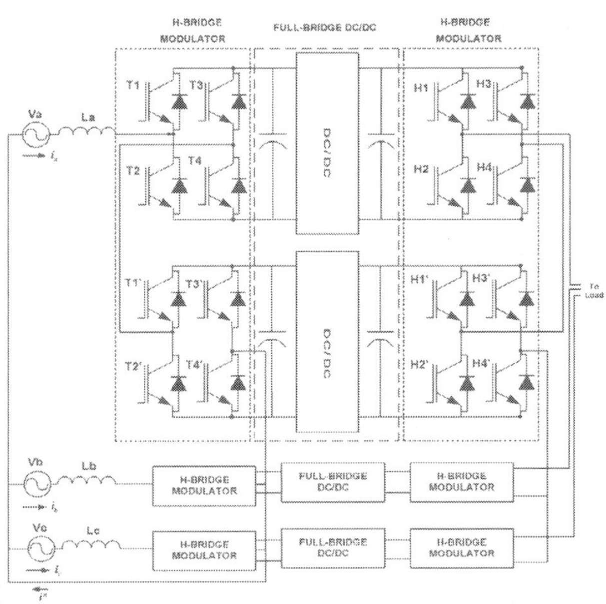

(c)

Fig. 19. (a) Three-level diode-clamped bidirectional converter (b) Five-level flying capacitor bidirectional converter (c) Three-level converter using H-bridge (cascade) modules

In case of diode-clamped multilevel converters, the reactive power flow control is easier. But the main drawback of this converter is that excessive clamping diodes are required when the number of levels is high. In case of flying capacitor multilevel converters, large amount of storage capacitors provides extra ride through capabilities during power outage. But the main drawback is that a large number of capacitors is required when the number of levels is high which makes the system less reliable and bulky and thus more difficult to package. In case of cascaded multilevel converters, least number of components is required and modularized circuit layout and packaging is possible because each level has the same structure and there are no extra clamping diodes or voltage balancing capacitors. But the main drawback is that it needs separate DC sources, thus making its applications somewhat limited. A comparison of different types of three-phase MLCs, in terms of power components required in each type of converter has been given in a tabular form in Table 2.

	Device Count		
Converter Type	Diode-Clamped MLC	Flying-Capacitor MLC	Cascaded MLC
Main Power Switches	$(n-1)\times 2$	$(n-1)\times 2$	$(n-1)\times 2$
Clamping Diodes	$(n-1)\times(n-2)$	0	0
DC-Bus Capacitors	$(n-1)$	$(n-1)$	$(n-1)/2$
Balancing Capacitors	0	$(n-1)\times(n-2)/2$	0

Table 2. Comparison of different types of Multilevel Converters

In this table, *n* specifies the number of levels. Multilevel bidirectional converters are used at high power ratings at high voltages with boost voltage for bidirectional power flow. Since the multilevel rectifiers offer a number of advantages over their two-level counterparts and are a promising alternative to medium and high voltage and high power industrial applications, they are a subject of intense research these days.

6. Three-phase neutral-point clamped bidirectional rectifier

Improved Power Quality Converters are now seen as a viable alternative over the conventional methods of improving the power quality. This new breed of AC/DC converters gives excellent power quality indices like nearly unity input power factor, negligible THD in source current, reduced ripple factor of load voltage and fast-regulated load voltage. Among the various topologies of improved power quality converters developed so far, multi-level converters provide the viable solution for medium to high power industrial applications at high voltages and have recently developed a great interest among the researchers. Diode-clamped multilevel converter is the most popularly used topology among the multilevel power converters [15,17,21,24,26].

Various modulation strategies have been researched for the control of multilevel bidirectional converters. Among the various modulation strategies, space vector PWM (SVPWM) has been found the best for the control of these converters as this modulation technique results in lower switching frequency, better utilization of DC-bus voltage, negligible input current THD and addresses the issue of DC-bus capacitor voltage unbalance in the neutral-point clamped converters. In fact, the large number of redundant switching states in SVPWM is being exploited for the balance of DC-bus capacitor voltages. The development of high speed DSPs has made possible the implementation of computation-intensive algorithm like SVPWM in multilevel converters where the complexity and computational burden increases excessively, especially for higher number of levels. The development of DSPs for real-time simulation has added a new dimension in the easy implementation of very complex control algorithms for the control of multilevel converters.

The performance of a three-phase, three-level bidirectional rectifier using sinusoidal pulse-width modulation (SPWM) technique and space vector pulse-width modulation (SVPWM) technique is evaluated in this chapter. A comparative evaluation of the three-level converter using above modulation techniques is performed to emphasize the advantages offered by SVPWM technique over SPWM technique for the control of these converters.

Fig. 20 shows the power circuit of a three-phase three-level (neutral-point clamped) bidirectional rectifier. In this circuit topology, each power switch is stressed to half the DC bus voltage instead of full DC bus voltage as is the case with two-level converters.

The independent power switches (T_{x1} and T_{x2}, $x = a, b, c$) are controlled in each leg of the converter. The constraints for four power switches in an arm of the converter are defined so as to avoid the power switches conducting at the same time.

$$T_{xi} + T_{xi'} = 1 \tag{1}$$

where $T_{xi} = 1$(or 0) if the power switch T_{xi} is turned on (or off) and $x = a,b,c$ and $i = 1,2$.

Four switching states are possible for each rectifier leg. However, only three valid switching states can be generated to achieve three voltage levels on the AC terminals of rectifier leg. The equivalent switching functions of the rectifier are defined below:

$$S_a = \begin{cases} 1 & if\ T_{a1}\ and\ T_{a2}\ are\ turned\ on \\ 0 & if\ T_{a1'}\ and\ T_{a2}\ are\ turned\ on \\ -1 & if\ T_{a1'}\ and\ T_{a2'}\ are\ turned\ on \end{cases} \tag{2}$$

$$S_b = \begin{cases} 1 & if\ T_{b1}\ and\ T_{b2}\ are\ turned\ on \\ 0 & if\ T_{b1'}\ and\ T_{b2}\ are\ turned\ on \\ -1 & if\ T_{b1'}\ and\ T_{b2'}\ are\ turned\ on \end{cases} \tag{3}$$

$$S_c = \begin{cases} 1 & if\ T_{c1}\ and\ T_{c2}\ are\ turned\ on \\ 0 & if\ T_{c1'}\ and\ T_{c2}\ are\ turned\ on \\ -1 & if\ T_{c1'}\ and\ T_{c2'}\ are\ turned\ on \end{cases} \tag{4}$$

Three valid operation modes in leg-*A* of the converter are as:

Operation mode 1 (S_a = 1): In this mode, the power switches T_{a1} and T_{a2} are turned on. The AC terminal voltage v_a' (or v_{ao}) is equal to $V_o/2$ (assuming $V_{c1} = V_{c2}$). In this case, the boost inductor voltage $v_L = v_a - V_o/2 < 0$ if the voltage drop across the equivalent series resistance is neglected. Therefore, the line-current i_a decreases and the current slope is $(v_a - V_o/2)/L$. The line-current i_a will charge or discharge the DC-bus capacitor C_1 if the AC system voltage V_a is positive or negative, respectively.

Operation mode 2 (S_a = 0): In this mode, the power switches T_{a1}' and T_{a2} are turned on and the AC terminal voltage v_{ao} is equal to zero. The boost inductor voltage $v_L = v_a$. The line-current increases or decreases during the positive or negative half cycle of mains voltage v_a, respectively. The converter input current i_a will not charge or discharge any one of the DC bus capacitors in this mode of operation. In fact, the input power is stored in the boost inductor during this mode.

Operation mode 3 (S_a = -1): In this mode, the power switches T_{a1}' and T_{a2}' are turned on and a voltage level of $-v_o/2$ is generated on the AC terminal voltage v_{ao}. The boost inductor voltage $v_L = v_a + V_o/2 > 0$ and the line current increases. The current slope is $(v_a + V_o/2)/L$. The line current i_a will charge or discharge the DC bus capacitor C_2 during the positive or negative half of mains voltage v_a, respectively. The operation modes are similar in converter legs *B* and *C* to control the line currents i_b and i_c.

Table 3 shows the valid switching states of the power switches of three legs and the corresponding voltages on the ac side of the rectifier.

S	T_{x1}	T_{x2}	$T_{x1'}$	$T_{x2'}$	v_{xn}
1	1	1	0	0	$V_1 = V_o/2$
0	0	1	1	0	0
-1	0	0	1	1	$V_2 = -V_o/2$

$x = a, b, c.$

Table 3. Valid switching states and corresponding voltages

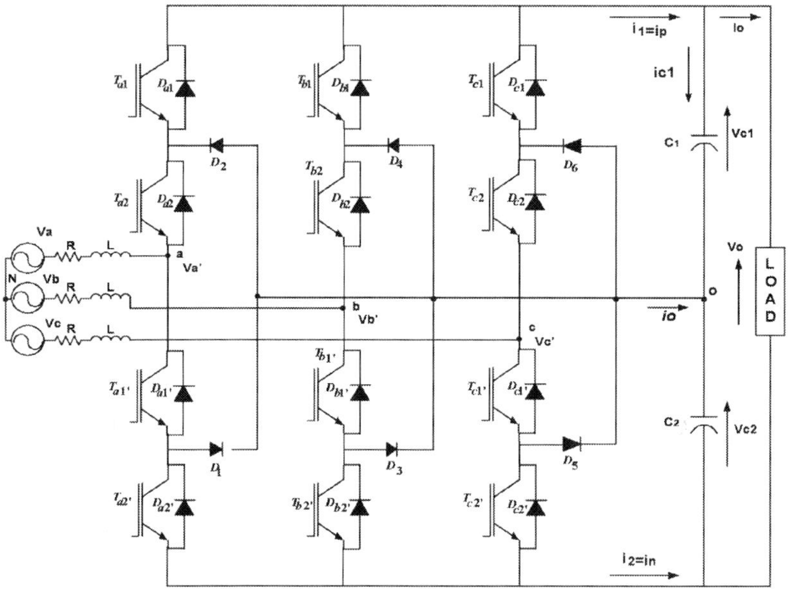

Fig. 20. Three-level diode-clamped bidirectional rectifier

7. Sinusoidal pulse width modulation (SPWM)

In this technique, the neutral point of DC bus is connected to the neutral of three-phase AC source as shown in Fig. 21. Three unipolar PWM waveforms are generated on the three phases on input side of the converter based on carrier-based PWM scheme.
The equivalent switching function of the rectifier is defined as:

$$s = \begin{cases} 1 & if \;\; T_{x1} = T_{x2} = 1 \\ 0 & if \;\; T_{x1'} = T_{x2} = 1 \\ -1 & if \;\; T_{x1'} = T_{x2'} = 1 \end{cases} \tag{5}$$

The supply side line-to-neutral voltage of the rectifier can be expressed as,

$$\begin{aligned} v_{xo} &= \frac{s(s+1)}{2} V_{c1} - \frac{s(s-1)}{2} V_{c2} = \\ &= \frac{s^2}{2} \Delta V + \frac{s}{2} V_0 \end{aligned} \tag{6}$$

If the two capacitor voltages V_{c1} and V_{c2} are equal, i.e., $\Delta V=0$, then there are three voltage levels, $V_o/2$, 0 *and* $- V_o/2$, on the AC side (line-to-neutral voltages) of the rectifier. By proper combinations of the power switches of any arm, three different voltage levels are generated in the line-to-neutral voltage by the rectifier. The three valid modes for phase-leg *A* of the rectifier are described schematically in Fig. 22.

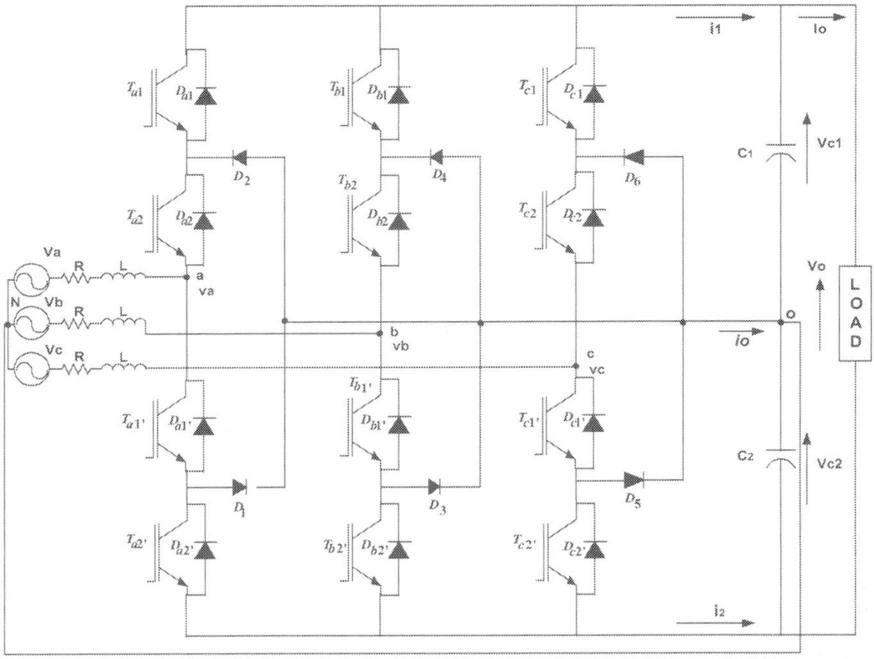

Fig. 21. Adopted Three-Phase Neutral-Point Clamped Converter for SPWM Technique

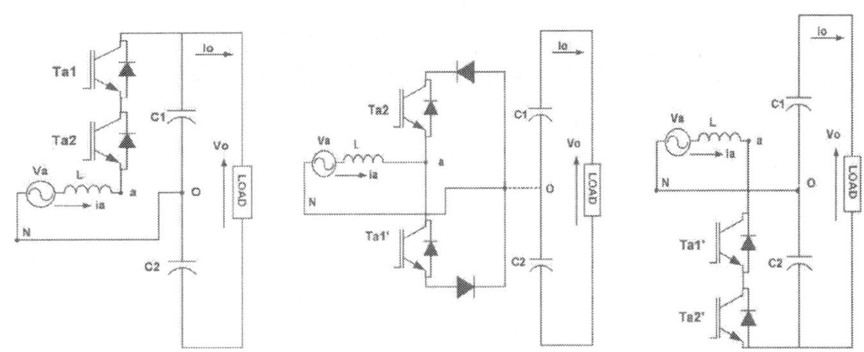

(a) (b) (c)

Fig. 22. Three valid operating modes for phase *A* of NPC rectifier (a) Mode 1 ($v_{ao} = V_o/2$) (b) Mode 2 ($v_{ao} = 0$) (c) Mode 3 ($v_{ao} = -V_o/2$)

In the same manner, during the negative half cycle of mains voltage of phase *A*, two voltage levels, *0* and $-V_o/2$, are produced in v_{ao}. In this case, the switch T_{a1}' is turned on and the switches T_{a2} and T_{a2}' are turned on to achieve $v_{ao} = 0$ and $-V_o/2$ ($-V_{c2}$) respectively.
The sinusoidal PWM controller block and the generation of gating pulses is shown in Fig. 23.

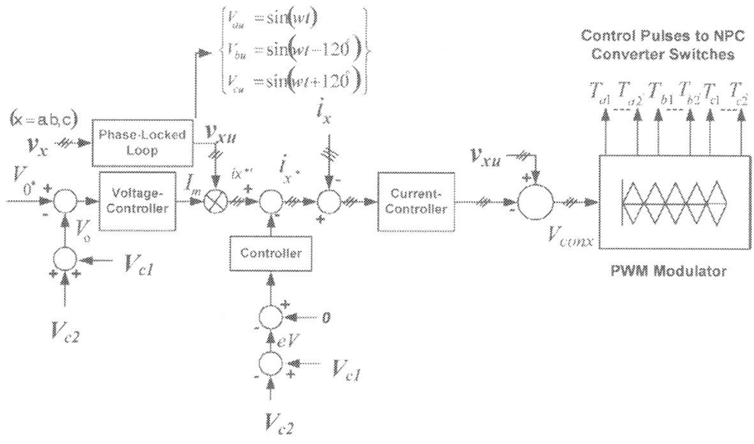

(a)

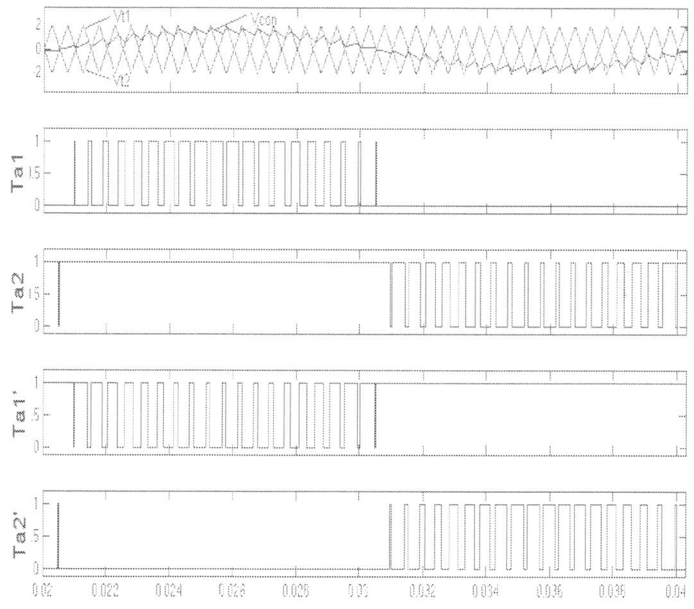

(b)

Fig. 23. (a) Block diagram of the PWM controller (b) Carrier-based PWM scheme (Control Pulses)

To regulate the DC bus voltage, a voltage controller is used to maintain the voltage at the desired reference value. The output of the voltage controller is multiplied by unit three-phase sinusoidal waveforms in phase with the mains three-phase voltages (obtained from a phase locked loop) to form three line-current commands for three phases of the rectifier. To

compensate the voltage unbalance between two capacitors in the DC link, a voltage compensator is added to the line-current commands. The line-current error between the current commands i_{sx}^* and actual line current i_{sx} of each phase is fed to the current controller of that phase based on sine-triangle PWM scheme to track the line-current commands. The neutral point voltage on the DC link is controlled by power switches by adjusting the neutral point current i_0. By appropriate control, unipolar PWM voltage waveforms are generated on the three line-to-neutral voltages v_{ao}, v_{bo} and v_{co}.
Apply Kirchhoff's voltage law (KVL) on phase a (AC side) of the rectifier, we have

$$v_a = Ri_a + L\frac{di_a}{dt} + v_{ao} \tag{7}$$

From (6), for phase a (x = a),

$$v_{ao} = \frac{s^2}{2}\Delta V + \frac{s}{2}V_0 \tag{8}$$

According to (7), we have

$$v_a = Ri_a + L\frac{di_a}{dt} + \frac{s}{2}V_0 + \frac{s^2}{2}\Delta V \tag{9}$$

If the dc-link capacitor voltages are equal, equation (9) can be written as,

$$v_a = Ri_a + L\frac{di_a}{dt} + \frac{s}{2}V_0 \tag{10}$$

Assuming the ideal power switches, no power loss occurs in the converter and the instantaneous power at the input and output of converter are equal.
To obtain a general control law for the neutral point current, the DC side quantities can be given as,

$$i_1 = C\frac{dV_{c1}}{dt} + \frac{V_{c1} + V_{c2}}{R} \tag{11}$$

$$i_2 = -C\frac{dV_{c2}}{dt} - \frac{V_{c1} + V_{c2}}{R} \tag{12}$$

Neutral point current i_o is given by,

$$i_0 = -i_1 - i_2$$

$$i_o = -C\frac{d(V_{c1} - V_{c2})}{dt} \tag{13}$$

This equation yields,

$$\Delta V = (V_{c1} - V_{c2}) = -\frac{1}{C}\int i_o dt + constant \tag{14}$$

This signifies that a DC component in the neutral current i_o can be used to compensate the voltage unbalance on the dc side of converter (neutral point voltage). The adopted controller for the rectifier as shown in Fig. 23(a) is supposed to fulfill all the control objectives. A proportional-integral (PI) voltage controller is employed in the outer loop control to maintain the DC-link voltage at the desired reference value. The line-current command is derived from the output of PI controller and the phase-locked loop (PLL) circuit and it is given by,

$$i_{x*'} = \left(k_p \Delta V_0 + k_i \int \Delta V_0 dt\right) \sin(\omega t + \theta) \tag{15}$$

where

$$\begin{cases} \theta = 0^\circ & for\ \ phase\ \ a \\ \theta = -120^\circ & for\ \ phase\ \ b \\ \theta = 120^\circ & for\ \ phase\ \ c \end{cases}$$

and $x = a,b,c$.

A three-phase PLL is used to generate three unit sinusoidal waves in phase with their corresponding supply voltages. To balance the neutral point voltage, the output of capacitor voltage balance controller is added to the line-current commands. The sensed line currents are compared with the respective reference line currents and the current errors thus generated are fed to the current controllers to track the source current commands. The carrier-based sinusoidal PWM scheme is employed for generating proper switching signals. Neglecting the high-frequency switching terms, we can write

$$v_{sx} = L.\frac{di_x}{dt} + V_{conx} \tag{16}$$

where V_{conx} $(x=a,b,c)$ is the modulating control signal of PWM converter derived from the closed-loop control scheme of the system.

From the control and gating signals as shown in Fig. 23(b), the switching signals of the power devices can be defined as,

$$s = \begin{cases} 1 & if\ \ V_{conx} \rangle V_{t1} \\ 0 & if\ \ V_{t1} \rangle V_{conx} \rangle V_{t2} \\ -1 & if\ \ V_{t2} \rangle V_{conx} \end{cases} \tag{17}$$

Hence,

$$T_{x1} = \frac{s(s+1)}{2} \tag{18}$$

$$T_{x1'} = 1 - T_{x1} \tag{19}$$

$$T_{x2'} = \frac{s(s-1)}{2} \tag{20}$$

$$T_{x2} = 1 - T_{x2'} \tag{21}$$

For a particular phase, in the positive half of the control signal V_{conx}, the power switch T_{x2} is turned on and the line current is controlled by turning on or off the power switch T_{x1}. In the negative half of V_{conx}, T_{x1} is turned off and turning on or off T_{x2} can control the line current to follow the current command. For equal capacitor voltages (V_{c1} = V_{c2} = $V_0/2$), three voltage levels ($V_0/2$, 0, and - $V_0/2$) are generated on the AC side of the rectifier phase voltage $V_{xo'}$.

8. Performance evaluation

The performance of the NPC converter using SPWM technique is evaluated under ideal mains conditions and various performance indices are obtained after exhaustive simulations. MATLAB/Simulink and SimPowerSystems software has been used for simulation purposes. The system parameters chosen for simulations are given in the following Table 4.

System Parameters	
Supply-side parameters	**Load-side parameters**
Supply voltage: 110 volts (rms), 50 Hz	Load: R_o = 15 Ω, L_o = 5 mH
Boost Inductor: L = 4.5 mH, R = 0.4 Ω	DC-Bus capacitors: C_1 = C_2 = 4700 μF
Reference DC voltage: V_o^* = 400 volts	
Sampling frequency, f_s = 5 kHz	

Table 4. System parameters for Simulation

Fig. 24 shows the phase *A* source voltage and line-current waveforms for the rectification mode of operation. It is observed from this figure that the rectifier draws a sinusoidal current from the source at nearly unity power factor. Fig. 25 shows the frequency spectrum of line-current drawn by the rectifier. The line-current THD is 3.2%. Fig. 26 shows the source voltage and line-current waveforms when the converter is operated in inversion mode.

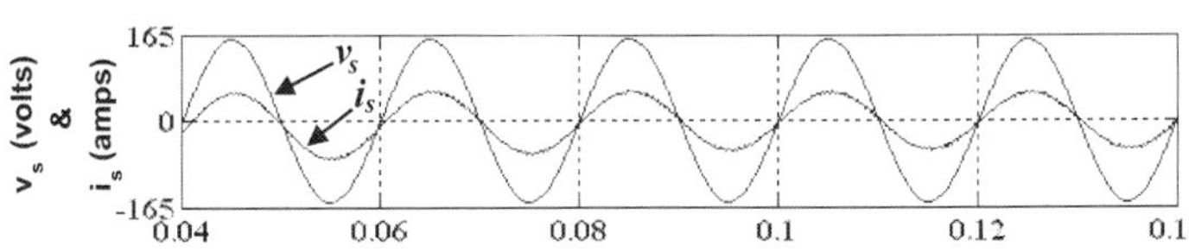

Fig. 24. Phase *A* voltage and line-current waveforms for rectification mode of operation

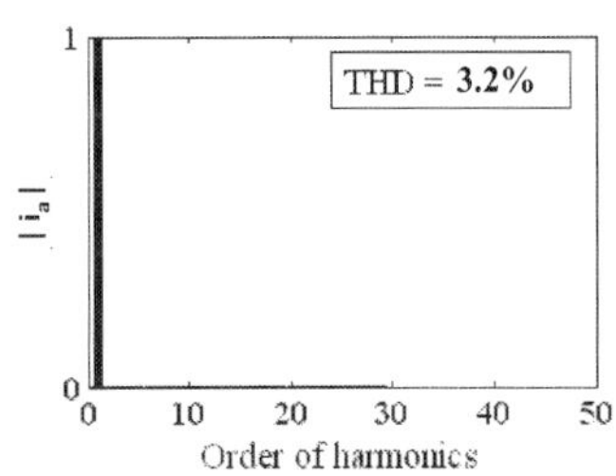

Fig. 25. Frequency spectrum of phase *A* current

The line current is still sinusoidal in nature with very low THD. Fig. 27 shows the load voltage waveform. It is observed that the load voltage is regulated at the desired reference value of 400 volts with a smaller ripple factor of only 0.7 volt. Fig. 28 shows the DC-bus capacitor voltages, V_{c1} and V_{c2}. The use of capacitor voltage unbalance controller in the control scheme results in a smaller DC-bus capacitor voltage unbalancing problem as shown in Fig. 28, when the rectifier feeds balanced loads.

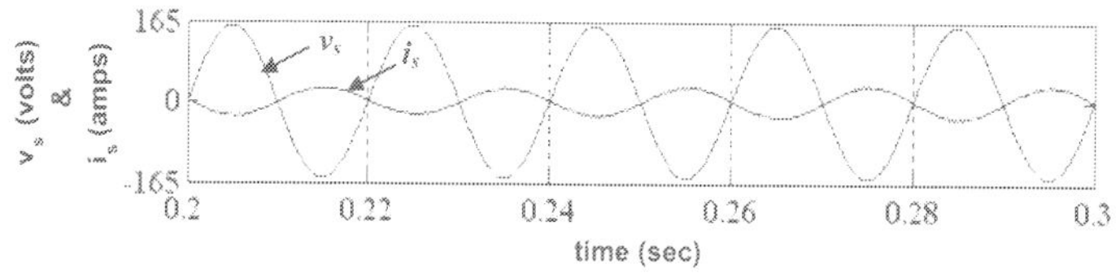

Fig. 26. Phase *A* voltage and line-current waveforms for inversion mode of operation

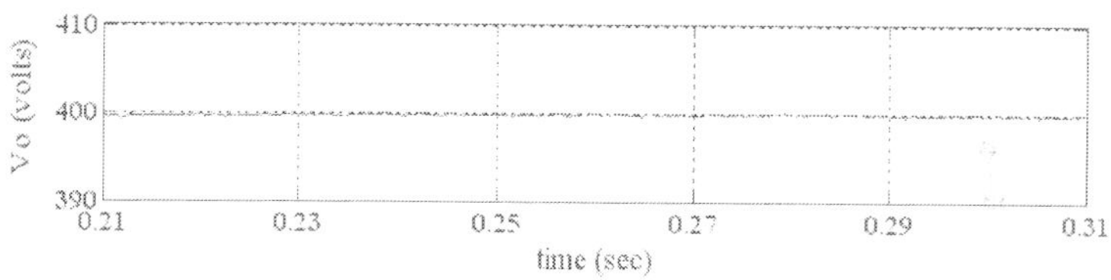

Fig. 27. DC-bus voltage of rectifier

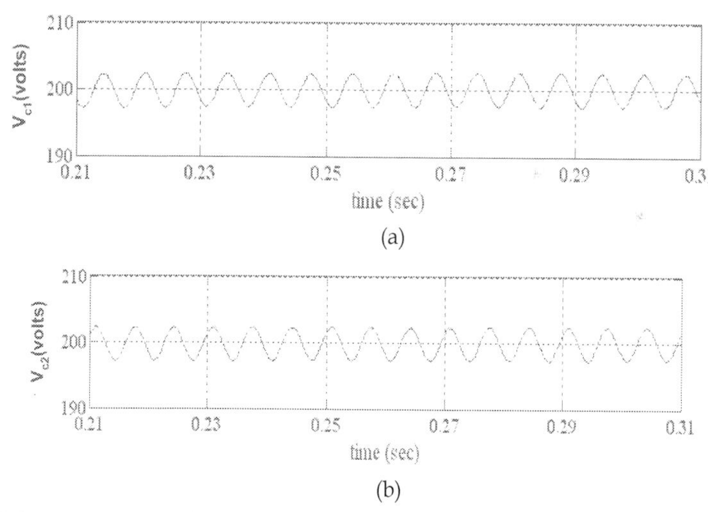

Fig. 28. DC-Bus capacitor voltages (a) Voltage across C_1 (b) Voltage across C_2

9. Space vector pulse width modulation (SVPWM)

A standard space vector modulation technique [2] has been used for the three-phase neutral-point clamped bidirectional rectifier of Fig. 20. The basic operating principle of the rectifier employing SVPWM is discussed below.

As shown in Fig. 20, the input terminals of the rectifier *a, b* and *c* are connected to the terminals of three-phase source *A, B* and *C* through the filter inductances *L*. Each power switch has a voltage stress of half the DC bus voltage instead of full DC bus voltage in the two-level PFCs. The power switches are commutated with a high switching frequency to generate the PWM voltages v_a', v_b' and v_c'.

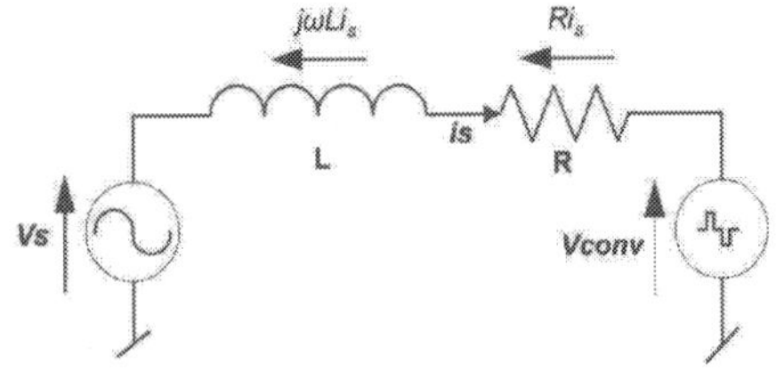

Fig. 29. Single-phase representation of NPC rectifier circuit

Fig. 29 shows a single-phase representation of the three-phase NPC rectifier of the Fig. 20. Again, L and R represent the line inductor. v_s is the source voltage and v_{conv} is the bridge converter voltage controllable from the DC-side. Magnitude of v_{conv} depends on the modulation index and DC voltage level of the converter. The inductors connected between input of the rectifier and supply lines are integral part of the circuit. They provide the boost feature of converter. The line-current, i_s is controlled by voltage drop across the inductance L interconnecting two voltage sources (line and converter). It means that the inductance voltage v_L equals the difference between the line voltage v_s and the converter voltage v_{conv}. Upon controlling the phase angle δ and the amplitude of the converter voltage v_{conv}, the phase and amplitude of the line-current are indirectly controlled. In this way, the average value and sign of DC current is subject to control the active power conducted through the converter. The reactive power can be controlled independently with shift of fundamental harmonic current i_s in respect to voltage v_s.

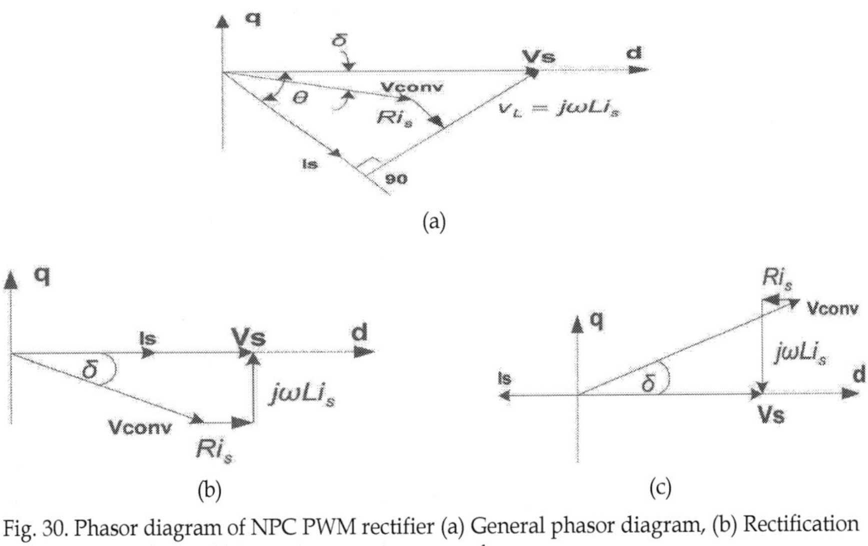

Fig. 30. Phasor diagram of NPC PWM rectifier (a) General phasor diagram, (b) Rectification at unity power factor, (c) Inversion at unity power factor

Fig. 30 presents the general phasor diagram and both rectification and regeneration phasor diagrams when unity power factor operation is required. The figure clearly shows that the voltage vector v_{conv} is higher during the regeneration than the rectification mode. It means that these two modes are not symmetrical.
Now,

$$v_s = v_{conv} + v_L \tag{22}$$

where

$$v_L = L.\frac{di_s}{dt} \tag{23}$$

Assuming v_s to be sinusoidal, the fundamental-frequency component of v_{conv} and supply current i_s can be expressed as phasors v_{conv1} and i_{s1}, respectively (not shown in the phasor diagram). Choosing v_s arbitrarily as the reference phasor $v_s = v_s e^{jo^\circ}$, at line frequency $\omega = 2\pi f$, we can write

$$v_s = v_{conv1} + v_{L1} \tag{24}$$

where

$$v_{L1} = j\omega L i_{s1} \tag{25}$$

Here the resistance R is assumed to be small and hence neglected.
The real power 'P' supplied by the phase A of the three-phase AC source to the converter is given as,

$$P = v_s i_{s1} \cos\theta = \frac{v_{s^2}}{\omega L}\left(\frac{v_{conv1}}{v_s}\sin\delta\right) \tag{26}$$

since in Fig. 30, $v_{L1}\cos\theta = \omega L i_{s1}\cos\theta = v_{conv1}\sin\delta$ (with v_L, v_{conv} and i_s replaced by their fundamental components as v_{L1}, v_{conv1} and i_{s1}, respectively and R=0).
In this phasor diagram, the reactive power 'Q' supplied by phase A of the AC source is positive and can be expressed as,

$$Q = v_s i_{s1} \sin\theta = \frac{v_{s^2}}{\omega L}\left(1 - \frac{v_{conv1}}{v_s}\cos\delta\right) \tag{27}$$

since in this figure, $v_s - \omega L i_{s1}\sin\theta = v_{conv1}\cos\delta$.
It is worth noting that 'Q' is the sum of the reactive power absorbed by the converter and the reactive power consumed by the inductance L. However, at very high switching frequencies, L can be made to be quite small; thus, Q can be approximated as the reactive power absorbed by the converter. The current i_{s1} can be written as,

$$i_{s1} = \frac{v_s - v_{conv1}}{j\omega L_s} \tag{28}$$

From equations (26) to (28), it is clear that for a given line-voltage, v_s and the chosen inductance L, the desired values of P and Q can be obtained by controlling the magnitude and the phase of v_{conv1}.
In the above general analysis, two cases are of special interest: rectification and inversion at a unity power factor. These two cases are depicted in Fig. 30(b) and Fig. 30(c).
In both cases,

$$v_{conv1} = [v_{s^2} + (\omega L i_{s1})^2]^{1/2} \quad \text{(with } R=0\text{)} \tag{29}$$

Thus, for the desirable magnitude and the direction of power flow as well as Q, the magnitude of v_{conv1} and the phase angle δ with respect to the line voltage v_s must be controlled.

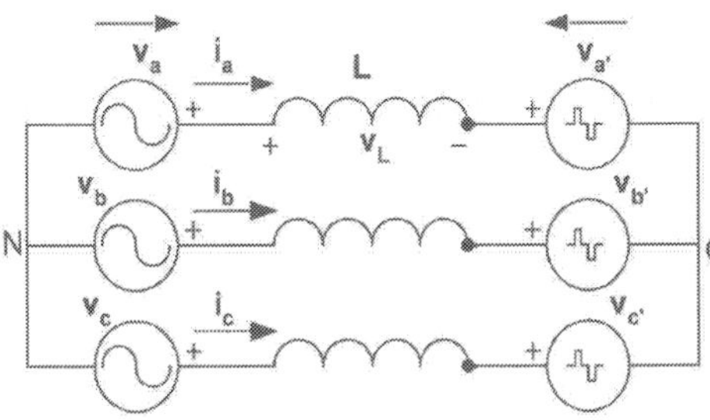

Fig. 31. Equivalent circuit of three-phase NPC rectifier

Now, the equivalent circuit of three-phase NPC rectifier of Fig. 20, as shown in Fig. 31, is considered for following analysis.
The space vectors are defined as below:

$$v = \tfrac{2}{3}\left(v_a + a.v_b + a^2.v_c\right) \tag{30}$$

$$v' = \tfrac{2}{3}\left(v_a' + a.v_b' + a^2.v_c'\right) \tag{31}$$

$$i = \tfrac{2}{3}\left(i_a + a.i_b + a^2.i_C\right) \tag{32}$$

The voltage vector equation can be written as below:

$$v = L.\tfrac{di}{dt} + v' \tag{33}$$

This equation can be expressed in a rotating reference frame *(d-q)*, with the d-axis oriented in the direction of source voltage vector v as depicted in Fig. 32. Thus the equation (33) can be written as,

$$v_d = L.\frac{di_d}{dt} - \omega.L.i_q + v_{d'} \tag{34}$$

$$v_q = 0 = L.\frac{di_q}{dt} + \omega.L.i_d + v_{q'} \tag{35}$$

where ω is the angular frequency of three-phase voltage and v_d, v_d', i_d and v_q, v_q', i_q are the components of v, v' and i_s in the d and q-axis respectively.

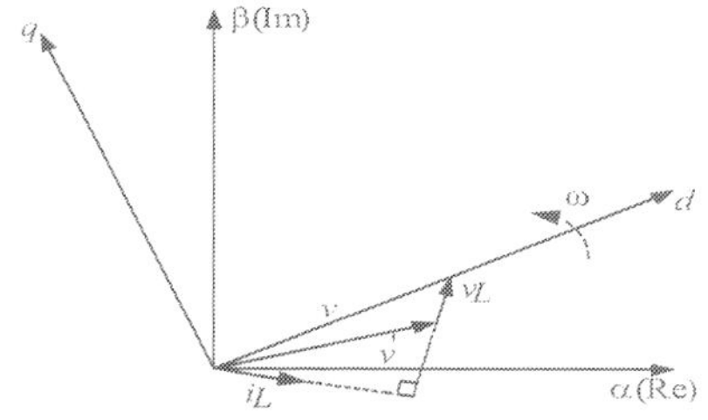

Fig. 32. Vector diagram of SVM-based NPC rectifier

Equations (34) and (35) show that the behaviour of currents i_d and i_q can be controlled by using the voltages v_d' and v_q' generated by the rectifier. In this way, the active as well as the reactive powers delivered by the mains to rectifier can be controlled.

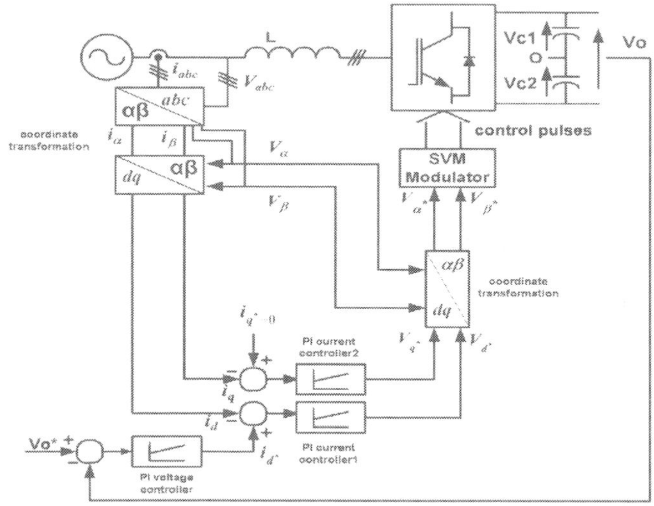

Fig. 33. Block diagram of voltage-oriented control scheme

The control strategy of this rectifier is the same as employed in two-level PWM rectifiers using SVPWM [4]. It is shown in Fig. 33. A PI controller is used to control the converter output voltage V_o. The output of this controller, i_d^* is used as reference for an inner closed-loop used to control the direct current i_d. The current in the q-axis, i_q is controlled by a similar loop with reference, $i_q^*=0$ to obtain operation with unity power factor. It is to be emphasized that this method only controls the total DC-bus voltage V_o and does not ensure the balance of capacitor voltages V_{c1} and V_{c2}. For proper converter operation, $V_{c1} = V_{c2}$. The current-controllers deliver the reference values for the voltages in the d and q-axis, V_d^* and V_q^* respectively. By using coordinates transformation, we obtain V_α^* and V_β^* in the stationary reference frame (α, β). Voltages V_α^* and V_β^* are used to derive the reference command input voltage V^* and its angle θ. These are delivered as inputs to the space vector modulator which generates the control pulses for converter switches using the 'Nearest Three Vector' (NTV) approach of SVM.

Applying the definition of equation (31) to all the 27 possible conduction states of power semiconductors, the converter generates 19 different space vectors as shown in Fig. 34(a). This figure also depicts the commutation states used to generate each space vector. It can be proved that the neutral current will flow through point o in all the states except the zero switching states located at the origin and large voltage vectors located at the outer hexagon corner states as shown in this figure. The complex plane is divided into six sectors and 24 triangles with four triangles (also called regions) in each sector. Fig. 34(b) shows the 'sector 1' triangle formed by voltage vectors V_o, V_7 and V_9. Assuming the command voltage vector V^* to be in region R_3, the following equations should be satisfied for SVPWM:

$$V_1T_a + V_8T_b + V_2T_c = V^*T_s / 2 \tag{36}$$

$$T_a + T_b + T_c = T_s / 2 \tag{37}$$

where V_1, V_8 and V_2 are the space vectors at the corners of region R_3, T_a, T_b, and T_c are the respective vector time intervals of these vectors and T_s is the sampling time.

Taking *sector 1* as the reference, the analytical time expressions for T_a, T_b, *and* T_c for all the regions can be derived and are written as:

$$\text{Region-1:}\quad \begin{aligned} T_a &= 2kT_s\sin\left(\pi/3-\theta\right)\\ T_b &= T_s\left[1-2k\sin\left(\theta+\pi/3\right)\right]\\ T_c &= 2kT_s\sin\theta \end{aligned}$$

$$\text{Region-2:}\quad \begin{aligned} T_a &= 2T_s\left[1-k\sin\left(\theta+\pi/3\right)\right]\\ T_b &= 2kT_s\sin\theta\\ T_c &= T_s\left[2k\sin\left(\pi/3-\theta\right)-1\right] \end{aligned}$$

$$\text{Region-3:}\quad \begin{aligned} T_a &= T_s\left[1-2k\sin\theta\right]\\ T_b &= T_s\left[2k\sin\left(\theta+\pi/3\right)-1\right]\\ T_c &= T\left[1+2k\sin\left(\theta-\pi/3\right)\right] \end{aligned}$$

$$\text{Region-4:}\quad \begin{aligned} T_a &= T_s\left[2k\sin\theta-1\right]\\ T_b &= 2kT_s\sin\left(\pi/3-\theta\right)\\ T_c &= 2T_s\left[1-k\sin\left(\theta+\pi/3\right)\right] \end{aligned} \tag{38}$$

where θ is the command voltage vector angle and $k = 2/\sqrt{3}\,(V^*/V_0)$, V^* = command voltage vector and V_o = DC-bus voltage of the rectifier. The above time intervals are distributed appropriately so as to generate symmetrical PWM pulses for the rectifier

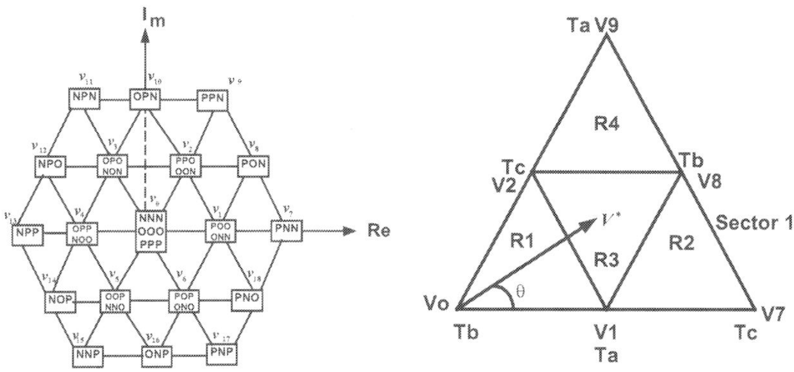

(a) (b)

Fig. 34. (a) Space vector diagram showing all the 27 switching states (b) Sector *1* space vectors with switching times

10. Performance evaluation

The performance of the SVPWM-based NPC converter is evaluated under ideal mains conditions and various performance indices are obtained after exhaustive simulations. The same system parameters as used in the SPWM-based converter (given in Table 4) are employed for the simulation of rectifier.

The steady-state performance of rectifier is evaluated with a constant source voltage and fixed load. Before the evaluation of performance of the rectifier, it is operated on three-phase supply without any control algorithm being implemented for its control. Fig. 35 shows the phase *A* source voltage and line-current waveforms of the converter when operated on an R-L load without the SVPWM algorithm. The line current drawn by the rectifier is highly distorted and rich in harmonics. As shown in Fig. 36, the line-current THD is 20.21% which is unacceptably large. Now, the performance of the rectifier is evaluated by applying the control algorithm for control of load voltage and wave shaping of source current drawn by rectifier from the mains. Fig. 37 shows phase *A* source voltage and line current waveforms in the rectification mode of operation. The frequency spectrum of line-current is shown in Fig. 38. It shows a line-current THD of only 1.45%, which is less than half the THD of source current obtained in SPWM technique. Fig. 39 shows the voltage generated at the rectifier input terminals after the SVPWM algorithm is applied. The source voltage and line-current waveforms of the converter in the inversion mode of operation are shown in Fig. 40. The line current is still free from harmonic pollution and has a THD of just 1.60% as shown in the frequency spectrum of current in Fig. 41. The transition from rectification mode to inversion mode of operation is shown in Fig. 42. It takes about two cycles for the converter to completely transition from rectification to inversion mode of operation. The load voltage waveform is shown in Fig. 43. It is observed that the load voltage is regulated with a negligible ripple factor of only 0.55 volt, which is a highly desirable feature in most of the applications. Fig. 44 shows the DC-bus capacitor voltages, V_{c1} and V_{c2}. Since the control scheme does not consider the capacitor voltage balancing issue, a somewhat higher capacitor voltage unbalance is produced. It is found that the SVPWM-based rectifier gives a better performance than that of SPWM-based rectifier at the same switching frequency. In other words, for the same performance, the SVPWM-based rectifier can be operated at a reduced switching frequency which reduces the switching stress of the devices and also the power losses of the converter. Moreover, there is no need of synchronizing the modulating and carrier waves and a better utilization of DC-link voltage is possible. The space vector PWM has a large number of redundant switching states which can be utilized for the balancing of DC-bus capacitor voltages. In view of these facts, the space vector PWM is preferred over sinusoidal PWM modulation strategy.

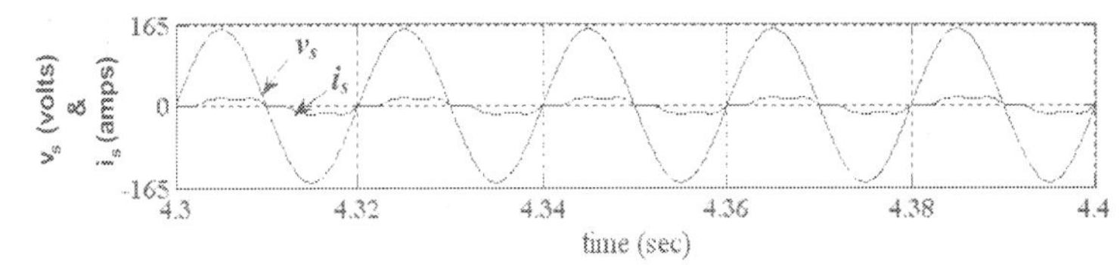

Fig. 35. Phase *A* source voltage and source current waveforms without control

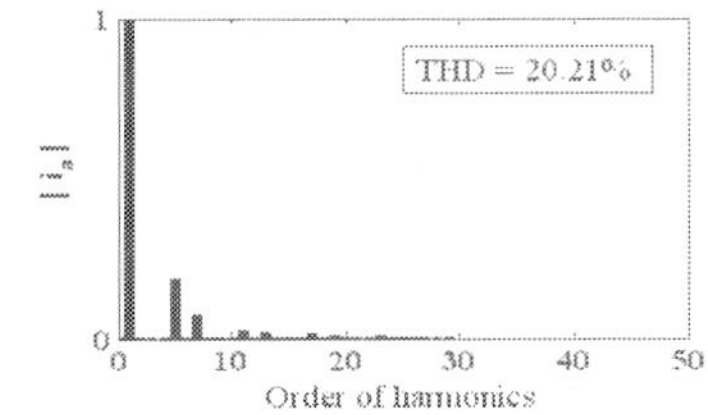

Fig. 36. Frequency spectrum of source current

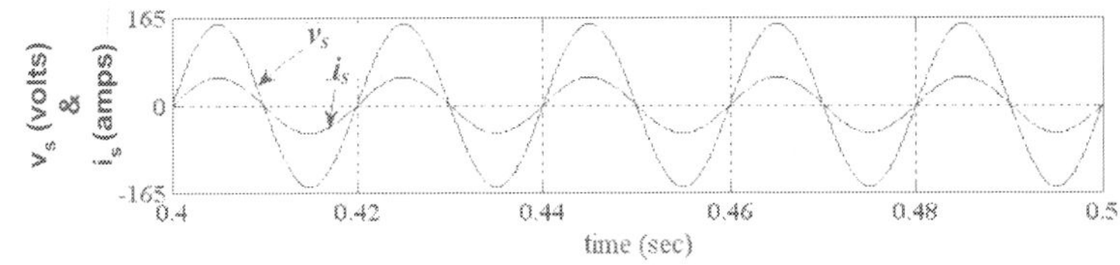

Fig. 37. Phase *A* source voltage and source current waveforms for rectification mode of operation

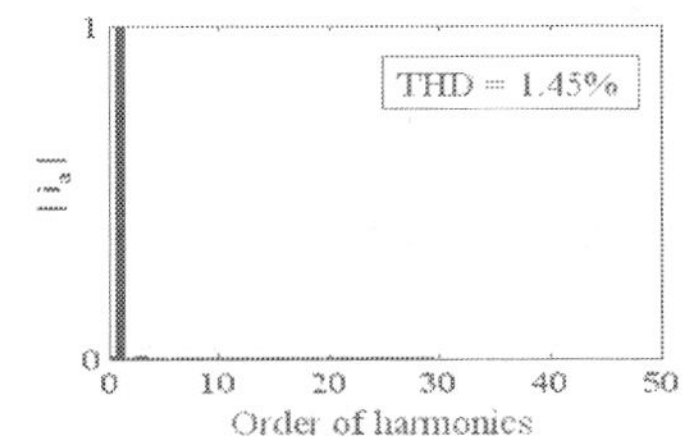

Fig. 38. Frequency spectrum of source current

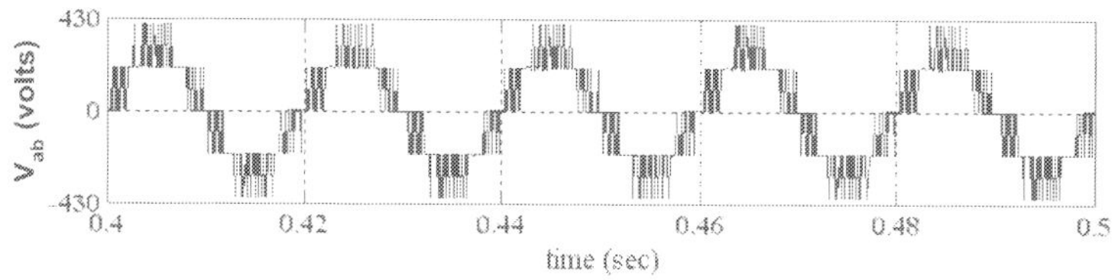

Fig. 39. Voltage reflected at the rectifier input terminals

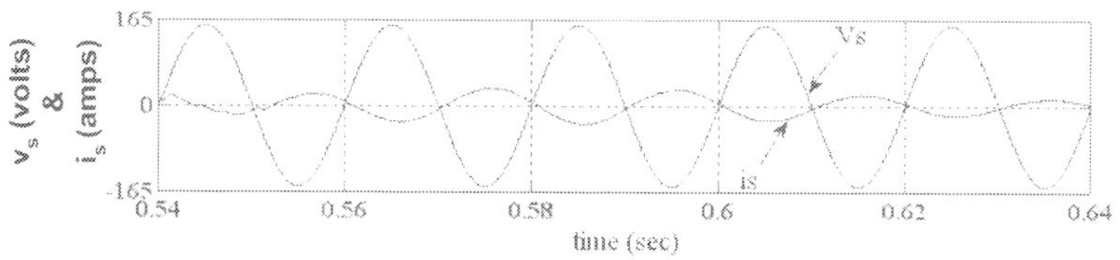

Fig. 40. Phase *A* source voltage and source current waveforms in inversion mode of operation

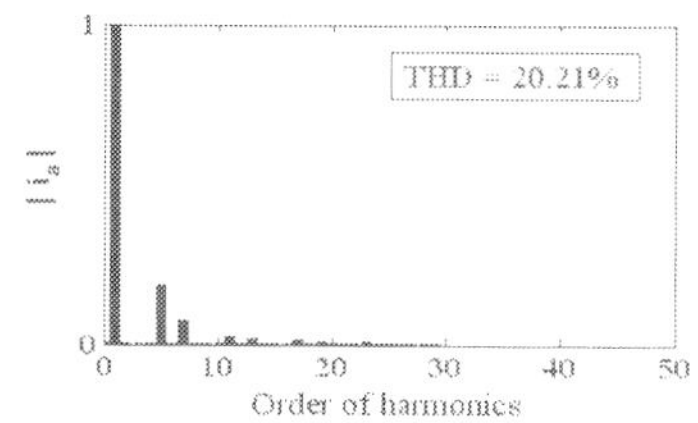

Fig. 41. Frequency spectrum of source current in inversion mode

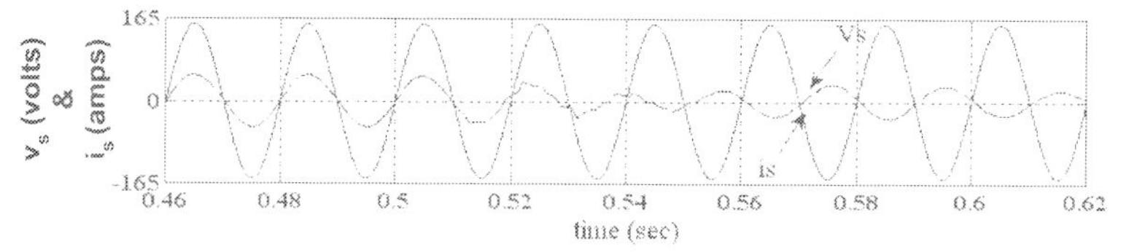

Fig. 42. Phase *A* voltage and line-current waveforms for transition from rectification to inversion mode (Rectification-to-Inversion transition starts at t = 0.51 sec.)

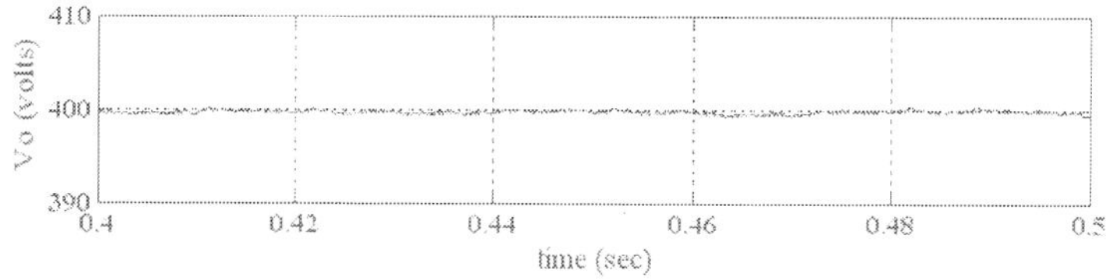

Fig. 43. Load voltage waveform

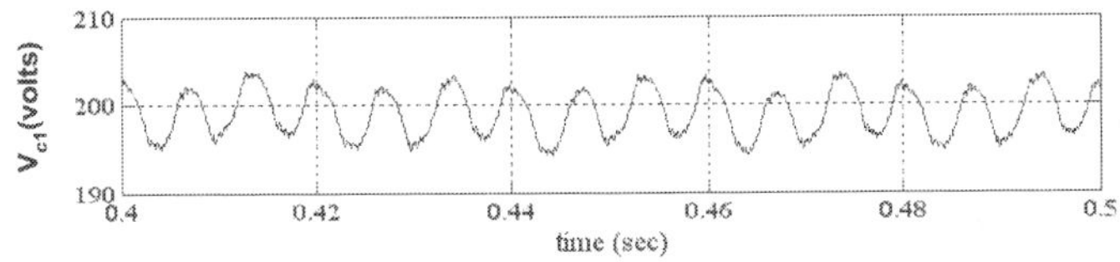

(a)

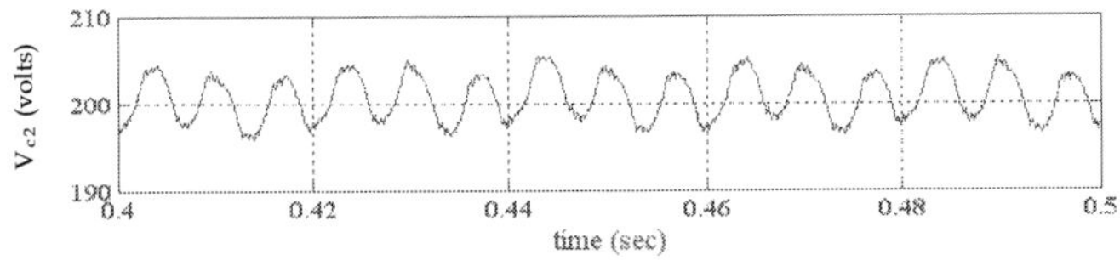

(b)

Fig. 44. DC-Bus capacitor voltages (a) Voltage across C_1 (b) Voltage across C_2

11. Experimental validation

A laboratory prototype of the three-phase, improved power quality neutral-point clamped rectifier is developed by integrating the power circuit, the control hardware and the DSP and tested in the laboratory to experimentally validate the simulation results. The performance of the system is investigated experimentally under steady-state. The system parameters selected for the experimental verification are given in Table - 5.

System Parameters	
Supply-side parameters	**Load-side parameters**
Supply voltage: 40 volts (peak), 50 Hz	Load: R_o = 15 Ω, L_o = 5 mH
Boost Inductor: L = 4.5 mH, R = 0.4 Ω	DC-Bus capacitors: $C_1 = C_2$ = 4700 μF
Reference DC voltage: V_o^* = 100 volts	
Sampling frequency, f_s = 5 kHz	

Table 5. System parameters for Experimentation

For real-time implementation of SVPWM control algorithm, DSP *DS1104* of dSPACE has been used. Fig. 45 shows the block diagram of laboratory prototype of DSP-based real-time implementation of a three-level improved power quality converter.

The DSP *DS1104* R&D Controller Board of dSPACE is a standard board that can be plugged into a PCI slot of a PC. The DS1104 is specifically designed for the development of high-speed multivariable digital controllers and real-time simulations in various fields. It is a complete real-time control system based on a 603 PowerPC floating-point processor running at 250 MHz. For advanced I/O purposes, the board includes a slave-DSP subsystem based on the TMS320F240 DSP microcontroller. For the purposes of rapid control prototyping (RCP), specific interface connectors and connector panels provide easy access to all input and output signals of the board. Thus, DS1104 R&D Controller Board is the ideal hardware for the dSPACE prototyper development system for cost-sensitive RCP applications. It is used for the real-time simulation and implementation of the control algorithm in real-time. In real-time simulation, the real plant (converter) is controlled by the controller (SVPWM modulator) that is simulated in real time. This technique is called rapid control prototyping (RCP).

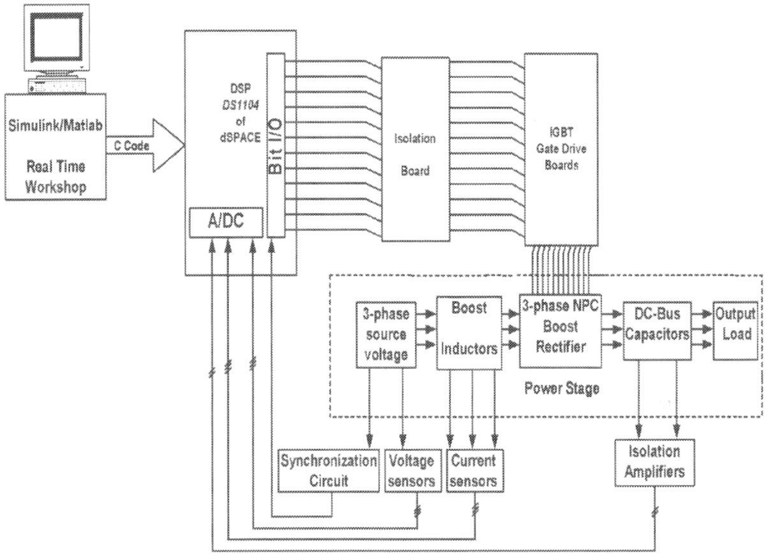

Fig. 45. Block diagram of DSP-Based NPC Rectifier Implementation

The major feature of real-time simulation is that the simulation has to be carried out as quickly as the real system would actually run, thus allowing to combine the simulation and the converter (real plant). The sensed AC and DC voltages and source currents are fed to the dSPACE board via the available ADC channels on its connector panel. In order to add an I/O block (like ADCs and master bit I/Os in this case) to the simulink model, the required block is dragged from the dSPACE I/O library and dropped into the simulink model of the SVPWM modulator. In fact, adding a dSPACE I/O block to a simulink model is almost like adding any simulink block to the model. In this case, twelve master bit I/Os, configured in the output mode, are connected to the model for outputting the twelve gating signals to the IGBTs of NPC rectifier bridge. In addition, eight ADCs are connected to the model for inputting the three sensed AC voltage signals, three source current signals and two DC bus capacitor voltages to the DSP hardware. These sensed signals are used for processing in the space vector PWM modulation algorithm. Because real-time simulation is such a vital aspect for control engineering, the same is true for the automatic generation of real-time code, which can be implemented on the hardware. For dSPACE systems, Real-Time Interface (RTI) carries out this linking function. Together with Real-Time Workshop from the MathWorks, it automatically generates the real-time code from simulink models and implements this code on dSPACE real-time hardware. This saves the time and effort twice as there is no need to manually convert the simulink model into another language such as C and we do not need to be concerned about a real-time program frame and I/O function calls, or about implementing and downloading the code onto the dSPACE hardware. RTI carries out these steps and we just need to add the required dSPACE blocks (I/O interfaces, etc.) to our simulink model. In other words, RTI is the interface between Simulink and various dSPACE platforms. It is basically the implementation software for single-board hardware and connects the simulink control model to the I/O of the board. In the present case, the optimized C-code of the simulink model of control algorithm is automatically generated by the Real-Time Workshop of MATLAB in conjunction with dSPACE's Real-Time Interface (RTI). The generated code is then automatically downloaded into the dSPACE hardware where it is implemented in real-time and the gating pulses are generated. The gating pulses for the power switches of converter are outputted via the master-bit I/Os available on the dSPACE board. The CLP1104 Connector/LED Combi panel provides easy-to-use connections between DS1104 board and the devices to be connected to it. The panel also provides an array of LEDs indicating the states of digital signals (gating pulses). The gating pulses are fed to various IGBT driver circuits via the opto-isolation circuit boards.

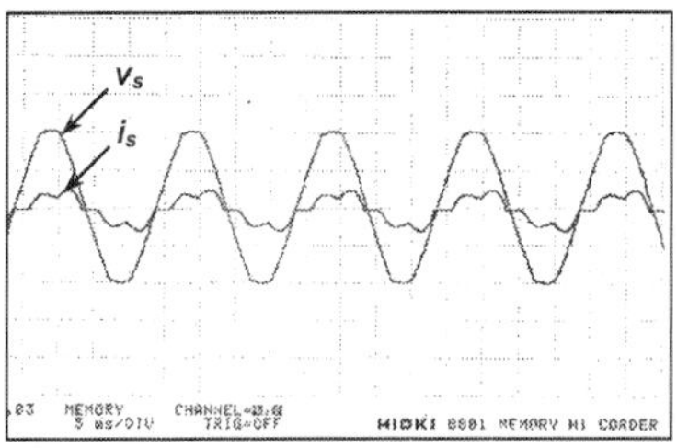

(a)

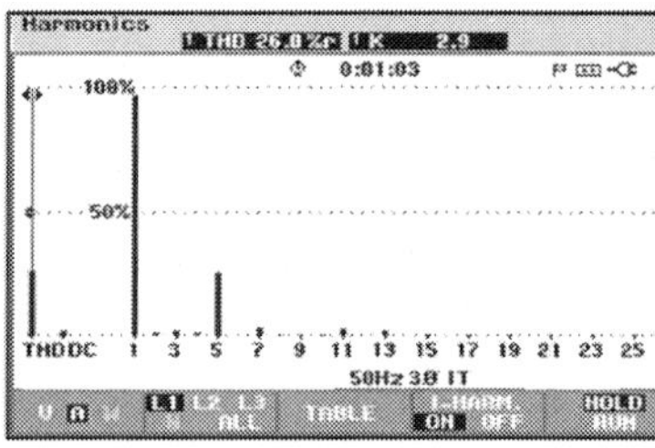

(b)

Fig. 46. (a) Phase A voltage and line-current waveforms before implementing SVPWM algorithm X-axis: time - 5 mS/div. Y-axis: v_a - 20 volts/div, i_a - 10A/div. (b) Frequency spectrum of phase A line-current

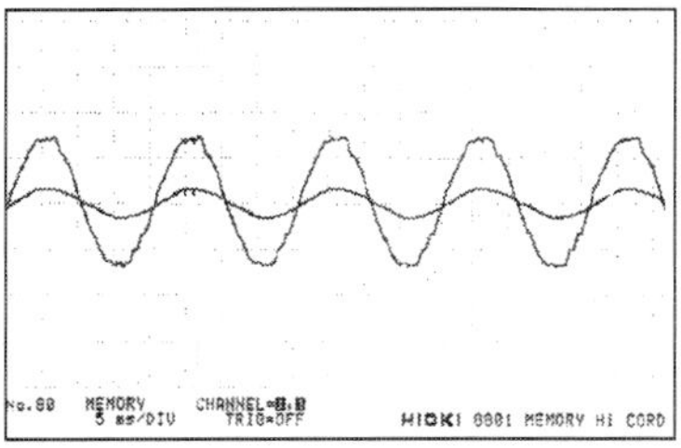

(a)

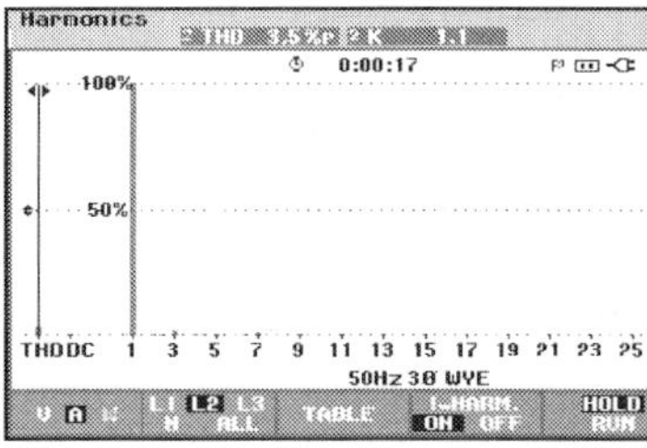

(b)

Fig. 47. (a) Phase A voltage and line-current waveforms after implementing SVPWM algorithm X-axis: time - 5 mS/div. Y-axis: v_a - 25 volts/div, i_a - 10A/div. (b) Frequency spectrum of phase A line-current

Fig. 46(a) shows the phase-A source voltage and line current drawn by the rectifier with an R-L load before the implementation of the space vector PWM algorithm. It is observed that the current drawn by the rectifier is highly distorted in nature and rich in harmonics. This is shown in the frequency spectrum of phase-A line current, as depicted in Fig. 46(b). This figure shows a line-current THD of 26%. After the implementation of the SVPWM control algorithm using dSPACE, the line current drawn by the rectifier becomes nearly sinusoidal in nature, as shown in Fig. 47(a). The current is drawn by the rectifier at unity power factor. This makes the rectifier work as a linear load in the system, causing almost no distortion of line-currents and no reactive power flow. The line-current THD is reduced drastically from an alarmingly large value of 26% to a mere 3.5% as shown in the frequency spectrum of line-current in Fig. 47(b). The THD is below the 5% limit prescribed by IEEE-519 standards.

Fig. 48 shows the voltage generated between the input terminals of the rectifier. Fig. 49(a) shows the phase-*A* voltage and current waveforms when the converter operates in inversion mode, thus regenerating power back to source. Thus the bidirectional neutral-point clamped converter is capable of operating in both the rectification as well as inversion modes. It is worth noting that the space vector control algorithm keeps the line currents sinusoidal. It is clear from the frequency spectrum of source current as shown in Fig. 49(b). The source current THD is 4.8% which is below the imposed limit of 5% by IEEE 519.

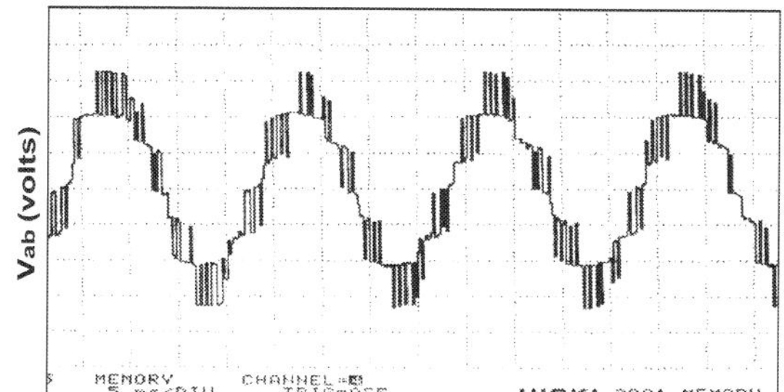

Fig. 48. Voltage at the rectifier input terminals

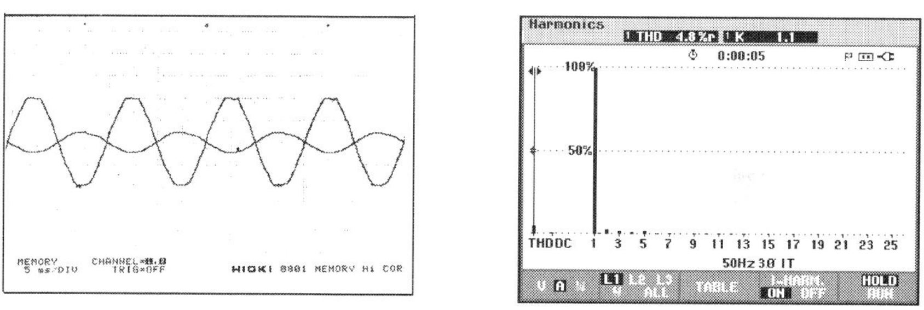

(a) (b)

Fig. 49. (a) Phase *A* voltage and current waveforms in inversion mode X-axis: time - 5 mS/div. Y-axis: v_a - 20 volts/div & i_a - 10 A/div. (b) Frequency spectrum of source current

Fig. 50 shows the DC-bus capacitor voltages of the converter. Fig. 51 shows the load voltage (DC-bus voltage) impressed across the load. It is observed that the load voltage is regulated at the desired reference value of 100 volts. Thus the experimental waveforms of source current and load voltage validate the simulation results obtained. Various harmonic components present in the source current before and after the implementation of SVPWM control algorithm are shown in Tables 6 and 7. Corresponding harmonic components obtained by simulation for almost similar conditions are also presented for comparison.

Thus summarising, the performance of space-vector PWM based neutral-point clamped rectifier model has been thoroughly investigated and the simulation results have been experimentally validated. It is found that the converter operation gives good power quality both at the line-side and load-side of the converter like sinusoidal source currents at nearly unity power factor, reduced line-current THD and regulated and reduced rippled DC bus voltage. The given converter finds tremendous applications in various industrial applications as it elegantly addresses the burning power quality issues created by the use of conventional diode bridge rectifiers and line-commutated converters.

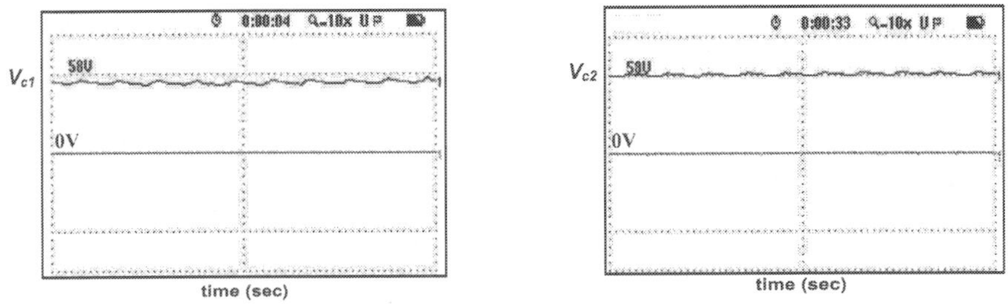

(a) (b)

Fig. 50. DC-Bus Capacitor Voltages (a) Voltage across capacitor C_1 (b) Voltage across capacitor C_2

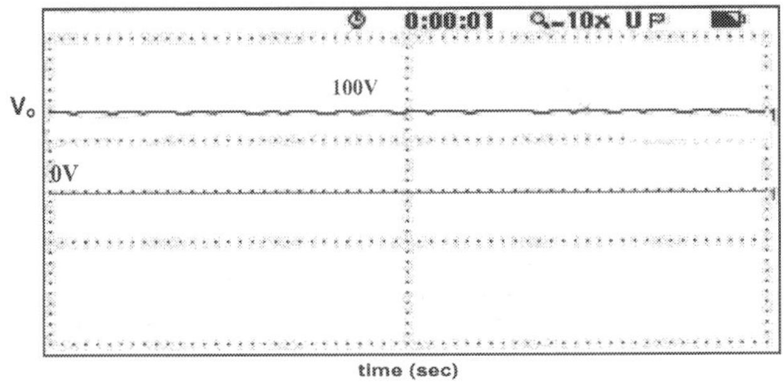

Fig. 51. DC-Bus (Load) Voltage

Order of harmonics	Harmonic components (% of fundamental) (Source current)	
	Simulation	Experimental
3	0.01	1.9
5	23.47	25.3
7	7.27	3.2
9	0.0	0.2
11	3.78	2.8
13	2.90	2.3
15	0.0	0.4
17	1.30	2.1
19	1.07	2.07
% THD	**25.13%**	**26%**

Table 6. Harmonic components present in source current without applying SVPWM algorithm

Order of harmonics	Harmonic components (% of fundamental) (source current)	
	Simulation	Experimental
3	0.88	1.09
5	0.28	0.25
7	0.09	0.85
9	0.3	0.62
11	0.11	0.12
13	0.16	0.38
15	0.23	0.20
17	0.07	0.15
19	0.08	0.09
% THD	**1.45%**	**3.5%**

Table 7. Harmonic components present in source current after applying SVPWM algorithm

12. References

[1] Akagi H., "New Trends in active filters for power conditioning," IEEE Trans. on Industry Applications, vol. 32, Nov./Dec. 1996, pp. 1312-1322.

[2] Bendre A., Krstic S., Meer J. V., and Venkataramanan G., "Comparative Evaluation of Modulation Algorithms for Neutral-Point-Clamped Converters", IEEE Trans. on Industry Applications, vol. 41, no. 2, March/April 2005, pp. 634-643.

[3] Carlton D. and Dunford W. G., "Multilevel, unidirectional AC-DC converters, a cost effective alternative to bi-directional converters," in Proc. IEEE PESC'01, 2001, pp. 1911-1917.

[4] Dixon J. W., "Boost type PWM rectifiers for high power applications," Ph.D dissertation, Dept. Elect. Comput. Eng., McGill Univ., Montreal, QC, Canada, Jun. 1988.

[5] El-Habrouk M., Darwish M. K., and Mehta P., "Active power filters: a review", Proc. IEE-Electric Power Applications, vol. 147, pp. 493-513, Sept. 2000.

[6] Fukuda S. and Hasegawa H., "Current source rectifier/inverter system with sinusoidal currents," in Conf. Rec. IEEE-IAS Annual Meeting, 1988, pp. 909-914.

[7] Fuld B., Kern S., and Ridley R., "A combined buck and boost power-factor-controller for three-phase input," in Power Electronics and Applications, European Conference, Sep. 1993.

[8] Groben T., Menzel E., and Enslin J. H. R., "Three-phase buck active rectifier with power factor correction and low EMI," Proc. IEE-Electric Power Applications., vol. 146, no. 6, pp. 591-596, Nov. 1999.

[9] Hengchun Mao, Fred C. Y. Lee, Boroyevich Dushan, and Silva Hiti, "Review of high-performance three-phase power-factor correction circuits," IEEE Trans. on Industrial Electronics, vol. 44, no. 4, August 1997, pp.437-446.

[10] IEEE Recommended Practices and Requirements for Harmonics Control in Electric Power Systems, IEEE std. 519, 1992.

[11] Jain S., Agarwal P., and Gupta H. O., "Design, simulation and experimental investigations on a shunt active power filter for harmonics and reactive power compensation", Electric Power Components and Systems, vol. 32, no. 7, July 2003.

[12] Kolar J. W., Ertl. H., And Zach F. C., "A novel single-switch three-phase ac-dc buck-boost converter with high-quality input current waveforms and isolated dc

output," in Proc. of 15th International Telecommunications Energy Conference, Paris, vol. 2, 1993, pp. 407-414.

[13] Kolar J. W., Ertl. H., and Zach F. C., "A novel three-phase single-switch discontinuous-mode AC-DC buck-boost converter with high quality input current waveforms and isolated output," IEEE Trans. on Power Electronics, vol. 9, Mar. 1994, pp. 160-172.

[14] Konishi Y., Arai N., Kousaka K., and Kumagai S., "A large capacity current source PWM converter with sinusoidal inputs and high power factor," in Proc. IEEE PESC'92 1992, pp. 1361-1367.

[15] Lai J. S. and Peng F. Z., "Multilevel converters: A New Breed of Power Converters", IEEE Trans. on Industry Application, vol. IA-32, No. 3, May/June 1996, pp. 509-517.

[16] Mahfouz A., Holtz J., and El-Tobshy A., "Development of an integrated high voltage 3-level converter-inverter system with sinusoidal input-output for feeding 3-phase induction motors," Power Electronics and Applications, Fifth European Conference, vol. 4, Sep. 1993, pp. 134-139.

[17] Marchesoni M. and Tenca P., "Diode-clamped multilevel converters: A practicable way to balance dc-link voltages," IEEE Trans. on Industrial Electronics, vol. 49, no. 4, August 2002.

[18] Midavaine H., Moigne P. L., and Bartholomeus P., "Multilevel three-phase rectifier with sinusoidal input currents," in Proc. IEEE PESC'96, 1996, pp. 1595-1599.

[19] Omedi T. J. and Barlik R., "Three-phase AC-DC unidirectional PWM rectifier topologies - selected properties and critical evaluation", in IEEE ISIE'96, 1996, pp. 784-789.

[20] Ridriguez J. Rodriguez D., Silva C., and Wiechmann E., "A simple Neutral Point Control for Three-Level PWM Rectifiers", European Power Electronics Conference, EPE'99, 1999, pp. 1-8.

[21] Sinha G. and Lipo T. A., "A four-level rectifier-inverter system for drive applications," IEEE Industry Applications Magazine, vol. 4, Jan./Feb. 1998, pp. 66-74.

[22] Singh B., Al. Haddad K., and Chandra A., "A review of active filters for power quality improvement," IEEE Trans. on Industrial Electronics, vol. 46, Oct. 1999, pp. 960-971.

[23] Singh B., Singh B. N., Chandra A., Al-Haddad K., Pandey A., and Kothari D. P., "A Review of Three-Phase Improved Power Quality AC-DC Converters", IEEE Trans. on Industrial Electronics, vol. 51, No. 3, June 2004, pp. 641-660.

[24] Tamai S., Koyama M., Fujii T., Mizoguchi S. and Kawabata T., "3 level GTO converter-inverter pair system for large capacity induction motor drive," in Conf. Power Electronics and Applications, 1993, pp. 45-50.

[25] Xu L. and Agelidis V. G., "A flying capacitor multilevel PWM converter based UPFC," in Proc. IEEE PESC'01, 2001, pp. 1905-1910.

[26] Yacoubi L., Al-Haddad K., Fnaiech F., and Dessaint L. A., "A DSP-Based Implementation of a New Nonlinear Control for a Three-Phase Neutral Point Clamped Boost Rectifier Prototype," IEEE Trans. on Industrial Electronics, vol. 52, No. 1, Feb. 2005, pp. 197-205.

[27] Zargari N. R. and Joos G., "A current controlled current-source type unity power factor PWM rectifier," in Conf. Rec. IEEE-IAS Annual Meeting, 1993, pp. 793-799.

[28] Zhang R. and Lee F. C., "Optimum PWM pattern for a three-phase boost DCM PFC rectifier", in Proc. IEEE APEC'97, 1997, pp. 895-901.

Part 6

Statistics and Analysysys

15

Time-Frequency Transforms for Classification of Power Quality Disturbances

Alejandro Rodríguez, Jose A. Aguado, Jose J. López,
Francisco Martín, Francisco Muñoz and Jose E. Ruiz
University of Malaga
Spain

1. Introduction

The usual operations on the distribution network such as switching loads and circuits, the proliferation of power electronic equipment and non-linear loads and the distributed generation with renewable energy are several of the most common causes that are leading to an increasing polluted power system in terms of voltage signal distortion.

One way of improving the power quality (PQ) parameters consists of analyzing these disturbances efficiently and understanding them deeply (Dugan, 2000) and PQ monitoring is one major task in order to achieve it. PQ monitoring is not an easy task usually involving sophisticated hardware instrumentation and software packages. Many recent approaches in PQ monitoring try to achieve it through the automated classification of different disturbances.

The different approaches in this field lead their efforts in two directions, the main parts that form an automated classification as depicted in Fig. 1. The first make focus to obtain a

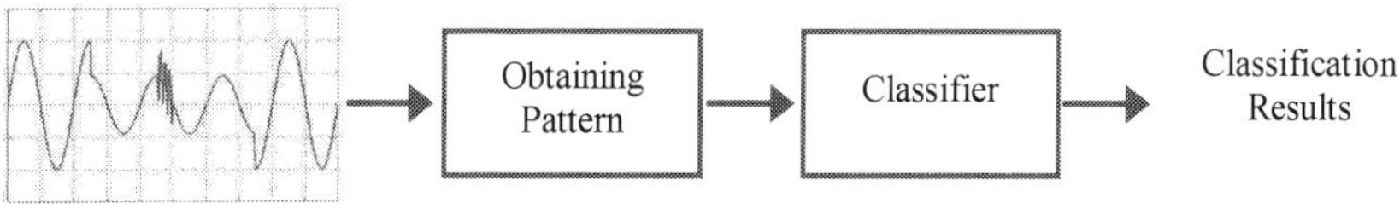

Fig. 1. Automated classification scheme

suitable pattern that allow distinguish clearly each disturbance, by the use of signal processing tool. Among existing signal processing tools the Fourier Transform (FT) results inadequate for analysis of non-stationary events and Short-Time Fourier Transform (STFT) although improves this drawback, it does not achieve good resolution in both time and frequency. Nowadays time-frequency transforms are used to get feature extraction, such as Wavelet transform (WT) (Santoso, 1994) and S-transform (ST) (Dash, 2003). WT extracts information from the signal in time and frequency domains, simultaneously, and provide greater resolution in time for high frequency components of a signal and greater resolution in frequency for the low frequency components of a signal. The ST can conceptually be interpreted as a hybrid of STFT and WT. It uses variable window length and, by using the FT kernel, it can preserve phase information during the decomposition (Stockwell et al., 1996). The frequency-dependent window produces higher frequency resolution at lower frequencies, while at higher frequencies, sharper time localization can be achieved.

The second approach is oriented to use a classifier able to assign each disturbance correctly in its class, so the most of the artificial intelligent techniques have been combined with WT or ST, as Artificial Neural Networks (ANN), Decision Tree Fuzzy Logic, Hidden Markov Model, Support Vector Machines, etc.

In this chapter two different classification systems have been developed, using the WT and ST for pattern extraction, and an ANN as classifier algorithm.

The features obtained from WT are not completely distinctive and it is necessary to add features that give clear information about the signal magnitude. The real mean squared (RMS) value of the voltage signal have been obtained to achieve it.

On the other hand, the features obtained from ST analysis are sufficient to achieve a pattern that can properly classify the disturbances. In order to increase accuracy, simplicity and reliability, this chapter proposes a reduced and simple set of features extracted from the ST. Even in the presence of complex disturbances with different levels of noise, these features characterize the signals in a suitable way.

This chapter is organized as follows. In the second section the time-frequency transforms used in power quality are presented, particularly the WT and ST, and the obtaining pattern using these two transforms. A brief description of ANN used as classifier algorithm used in this chapter is given in section three. In section fourth the classification results obtained using the resulting classification system are presented. These systems are checked by signals obtained from electric power simulation in section fifth. Finally, conclusions are presented.

2. Time-frequency transforms and obtaining pattern

2.1 Fourier transform

The most used classical signal processing is the FT. This transform represents a signal as a sum of sinusoidal terms of different frequencies, named the frequency spectrum. This technique is suitable for stationary signals, but it is not efficient when the signal contents short-term transient disturbances.

In order to solve this drawback, a technique based on the FT is applied to short time intervals. This method is known as "Short Time Fourier Transform" (STFT), and consists of analyzing by the FT a sliding window of the signal. It is not possible reaching a good resolution in time and frequency simultaneously, and therefore it is necessary to adopt a compromise solution between the frequency and time resolutions.

The STFT is obtained by choosing a sliding window (short time interval) where the FT is applied. For narrow windows (or short time intervals) a good time resolution is obtained, suitable for short-term transients; on the other side a relative wide window enables a good frequency resolution but gives inaccurate time resolution.

Therefore, the problem lies in the fact that the window width is a parameter that must be fixed before analyzing the signal, before knowing what resolution is more suitable. At last, the total number of operations for computing STFT is $N \cdot logN$.

2.2 Wavelet transform

Wavelet analysis is a powerful tool widely used in many scientific applications, especially in transient, non stationary, or time-varying situations.

A wavelet can be considered as a small wave which has its energy concentrated in time, and fits certain mathematic properties (Burrus et al., 1998). Fig. 2 shows three examples of wavelet.

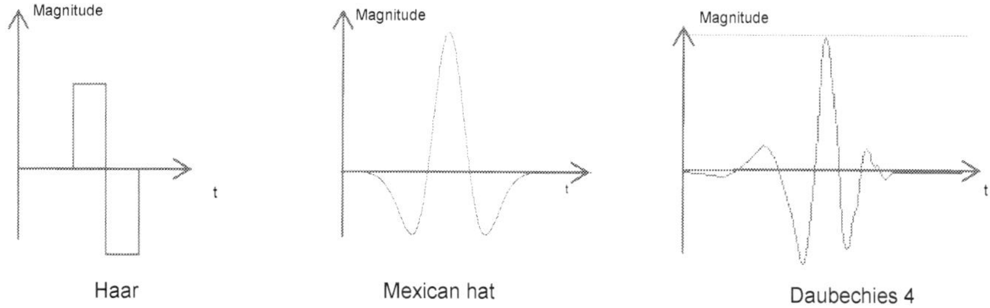

Fig. 3. Three examples of wavelet

The mother wavelet $\psi(t)$ can be scaled and translated in the time, generating a family of functions named the wavelet expansion set (Wavelet system):

$$\psi_{j,k}(t) = 2^{j/2}\psi(2^j t - k) \tag{1}$$

A function $f(t)$ can be expressed as a linear decomposition of this wavelet system as follows:

$$f(t) = \sum c_{j,k}\psi_{j,k}(t) \quad , \; j,k \in Z \tag{2}$$

where j and k are integer indices, $c_{j,k}$ are real coefficients, and $\psi_{j,k}(t)$ is the expansion set. The set of expansion coefficients $c_{j,k}$ is called the *wavelet spectrum* of the function $f(t)$.
If the wavelet expansion system is orthogonal, then:

$$j \neq l \text{ or } k \neq m \;\Rightarrow\; \left\langle \psi_{j,k}(t), \psi_{l,m}(t) \right\rangle = 0 \tag{3}$$

where < > denotes the inner product defined as:

$$\left\langle \psi_{j,k}(t), \psi_{l,m}(t) \right\rangle = \int \psi_{j,k}(t) \cdot \psi_{l,m}(t) dt \tag{4}$$

When the expansion wavelet system is orthogonal the Wavelet spectrum can be computed as follows:

$$c_{j,k} = \int \psi_{j,k}(t) \cdot f(t) \cdot dt \tag{5}$$

The index j is related with the frequency and the index k with time.
In practical applications a minimum frequency have to be established.
For a mother wavelet a basic scaling function $\varphi(t)$ is defined. A set of scaling functions is defined in terms of integer translates of $\varphi(t)$ by

$$\varphi_k(t) = \varphi(t - k) \tag{6}$$

A two-dimensional family of scaling functions can be defined by scaling and translation by

$$\varphi_{j,k}(t) = 2^{2/j}\varphi(2^j t - k) \tag{7}$$

Let consider the minimum frequency is the corresponding to the value $j=J_0$.
The equation (2) becomes to

$$f(t) = \sum_k \sum_j c_{j,k}\Psi_{j,k}(t) = \sum_k a_{J_0,k}\varphi_{J_0,k}(t) + \sum_{j=J_0}^{\infty} \sum_k d_{j,k}\Psi_{j,k}(t), \qquad t \in R \tag{8}$$

where J_0 is an integer. Equation (4) is a linear combination of wavelet coefficients, $(a_{Jo,k}, d_{j,k})$, a set of functions $\varphi_{Jo,k}(t)$, called *scaling functions* and $\Psi_{j,k}(t)$, called *wavelet functions*. Coefficients $a_{Jo,k}$ and $d_{j,k}$ are the *Discrete Wavelet Transform* (*DWT*) of $f(t)$, and can be calculated as:

$$a_{J_0,k} = \langle f(t), \varphi_{J_0,k}(t)\rangle \quad d_{j,k} = \langle f(t), \Psi_{j,k}(t)\rangle \tag{9}$$

Equation (4) can be truncated for $j=J-1$, obtaining:

$$f(t) = \sum_{k=0}^{2^{J_0}-1} a_{J_0,k}\varphi_{J_0,k}(t) + \sum_{j=J_0}^{J-1} \sum_{k=0}^{2^j-1} d_{j,k}\Psi_{j,k}(t), \qquad t \in R \tag{10}$$

The first summation in (6) is a broad representation of $f(t)$ that has been expressed as a linear combination of 2^{J_0} translations of the *scaling function*, $\varphi_{Jo,0}$. The second summation contains the *details* of f(*t*). For each level *j*, a linear combination of 2^j translations of the *wavelet function*, ψ_{j0}, are added to obtain a more accurate approximation of $f(t)$.

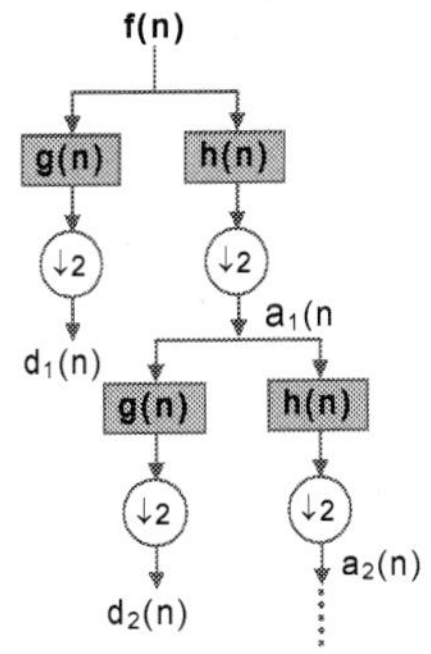

Fig. 2. Mallat algorithm schematics

The Mallat algorithm (Mallat, 1999) has been used in the practical implementation of DWT. The DWT acts as two FIR (Finite Impulse Response) quadrature filters defined by two sequences *h(n)* and *g(n)*. *h(n)* is a high frequency filter and *g(n)* is a low frequency filter. Both filter have the same cut frequency $f_N/2$, where f_N is the Nyquist frequency. Therefore the function *f(n)* is split in two parts, the high frequency part d_1 that contains the higher octave and is called *detail function*, and the low frequency part a_1, that contains the frequencies lower than $f_N/2$, and is called *smoothed function*. Decimation by 2 is done for eliminating redundant information. The algorithm is iterated for a_1, obtaining a second level detail function d_2 and a second level smoothed function a_2, that is again splitted, obtaining a series of detail and broad functions. The total number of operations for computing DWT is N.

The original function *f(n)* is split into a series of detail functions $d_1, d_2, \ldots, d_k$, and a smoothed function a_k, that correspond to the frequencies:
$d_1 : f_N - f_N/2;\ d_2: f_N/2 - f_N/4;\ \ldots\ d_k: f_N/2^n - f_N/2^{k+1}$; and a_n contains the frequencies lower than $f_N/2^{k+1}$.
Fig. 3 and Fig. 4 show an example of the application of the Mallat multi-resolution algorithm to the wavelet spectrum computation of a signal with a sag to 40%.

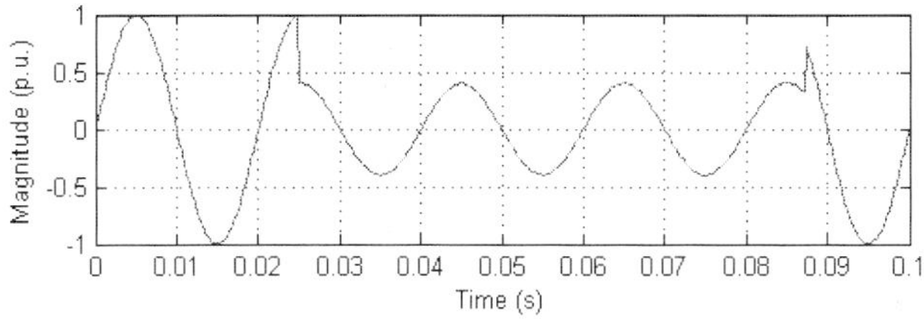

Fig. 3. Voltage sag to 40% and 400 samples length

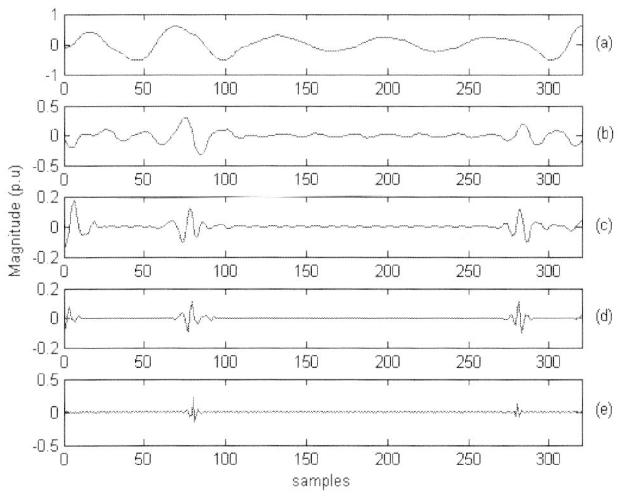

Fig. 4. Voltage sag detail levels wavelet analysis, from d1 to d5, (a) to (e), respectively

It is important to note that the wavelet is not a single specified function but a framework within which can design different wavelets. In this work, Daubechies 3 (db3) has been used as wavelet mother, performing 5 levels of decomposition. A voltage sag signal is shown in Fig. 3, and its wavelet transform decomposition, using db3, can be seen in Fig. 4.

2.3 Wavelet based pattern

The pattern used in this work is based on Parseval's Theorem. This Theorem states that the energy of a signal *f(t)* remains the same whether it is computed in a signal domain (time) or in a transform domain (frequency) as follows:

$$E_{signal} = \frac{1}{T}\int_0^T |f(t)|^2\, dt = \sum_{n=0}^{N} |F[n]|^2 \qquad (11)$$

where *T* and *N* are the time period and the length of the signal, respectively, and *F(n)* is the Fourier transform of the signal. In the case of the DWT, the signal is decomposed in terms of bands of frequencies, thus the energy of a signal can be given as:

$$E_{DWT} = \frac{1}{T}\int_0^T |f(t)|^2\, dt = \sum_{l=-\infty}^{\infty} |a(l)|^2 + \sum_{j=0}^{\infty}\sum_{k=-\infty}^{\infty} \left|d_j(k)\right|^2 \qquad (12)$$

with the energy in the expansion domain partitioned in time by k and scaled by j.
The sampling frequency of the signals is 3.2 kHz, which is equal to $64f_1$ where f_1 is the power system frequency (50 Hz in Europe). The wavelet spectrum contains the total information of the original waveform as shown in Table 1.

DWT coefficients	Frequency band
d_1	$32f_1 \div 16f_1$
d_2	$16f_1 \div 8f_1$
d_3	$8f_1 \div 4f_1$
d_4	$4f_1 \div 2f_1$
d_5	$2f_1 \div f_1$

Table 1. Frequency band information contained in the wavelet spectrum.

In certain signals an energy-based pattern is not completely discriminatory because the energy of the magnitude disturbance depends on the depth of the disturbance and its duration in time.
The spectra of normal signals, sags and swells do not contain energy in the bands up to the fundamental frequency, as these signals only differ in the magnitude of the fundamental frequency component, and present very few energy in the high frequencies.
Therefore, it is necessary to provide a feature based on the magnitude of the signal. The RMS value is a widely accepted tool that provides information about how much the magnitude of the voltage changes. It is a fast and simple algorithm (13) that requires very few computational resources.

$$V_{RMS} = \sqrt{\frac{1}{N}\sum_{n=1}^{N} (f(n))^2} \qquad (13)$$

where *f(n)* is the signal of length *N*.
The digital measurement instruments perform the computation of this parameter from instantaneous values of the samples, choosing a temporary window depending on the frequency of the steady state signal. If the RMS values are updated when a new sample is acquired, the method is called RMS continuum. If the RMS values are updated at a certain interval, usually half cycle, then it is called RMS (1/2). In this work the RMS (1/2) has beencomputed, selecting the maximum and minimum values. These values provide a feature based on the signal magnitude, as shown in Table 3.

	RMS voltage values	
Signal	Max	Min
Perfect signal	1.001	0.999
Normal to 0.91 in 4.5 cycles	1.001	0.909
Sag to 0.8 in 1 cycle	1.001	0.799

Table 3. Maximum and minimum RMS voltage value for different signals

Therefore the pattern is made up in two stages. The first part has five values, the energy of the signal in each frequency band of wavelet decomposition (feature extraction). These values have been normalized with reference to the values obtained from an ideal sinusoidal signal. The second part of the pattern is formed for two values the maximum and minimum RMS value calculated directly (feature selection) from each signal.

2.4 S-transform

The S-Transform (Stockwell et al., 1996) is a time-frequency transform generated by the combination of WT and STFT. The S-transform $s(\tau,f)$ of the signal $x(t)$ is defined as follows:

$$S(\tau,f)=\int_{-\infty}^{\infty}X(t)g_f(\tau-t)\exp(-j2\pi ft)dt \tag{14}$$

where

$$g_f(\tau-t)=\frac{|f|}{\sqrt{2\pi}}\exp\left(\frac{-f^2(\tau-t)^2}{2}\right) \tag{15}$$

$g_f(\tau\text{-}t)$ is the Gaussian window function, τ is a shift parameter for adjusting the position in the time axis and f is the scale parameter.
$X(f)$ is defined as the Fourier transform of $x(t)$. The relationship between the S-transform and the Fourier transform is:

$$X(f)=\int_{-\infty}^{\infty}S(\tau,f)dx \tag{16}$$

The discrete ST is defined by:

$$S\left(\frac{n}{NT},jT\right)=\sum_{m=0}^{N-1}X\left(\frac{m+n}{NT}\right)\cdot\exp\left(-\frac{2\pi^2m^2k^2}{n^2}\right)\exp\left(\frac{i2\pi nj}{N}\right) \tag{17}$$

In order to obtain the ST, the FFT of the original signal is computed [17]. The total number of operations for computing ST is $N\cdot(N+\log N)$.
The multi-resolution ST is a complex matrix whose rows and columns values are frequency and are time values, respectively. Each column represents the local spectrum in time. Frequency-time contours having the same amplitude spectrum are also obtained. This information is used to detect and characterize power disturbance events.
A mesh-dimensional output of the ST yields frequency-time, amplitude-time and frequency-amplitude plots. Examples of multi-resolution ST analysis for signals containing harmonics

and oscillatory transients are presented in Fig. 2 and 3, respectively. Each figure plots the disturbance signal (a), the time-frequency contours (b), and 3-D mesh giving amplitude, frequency and time plots (c).

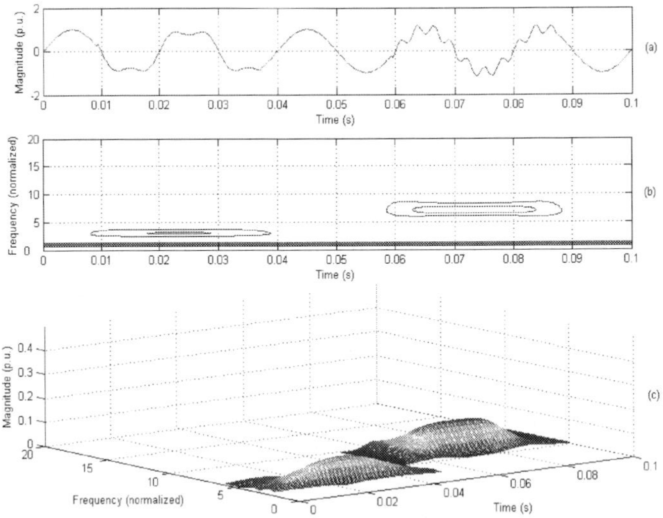

Fig. 2. Signal with 3rd and 7th harmonic content (a). S-transform Time-Frequency contours (b). 3D mesh Time-Frequency-Amplitude (c).

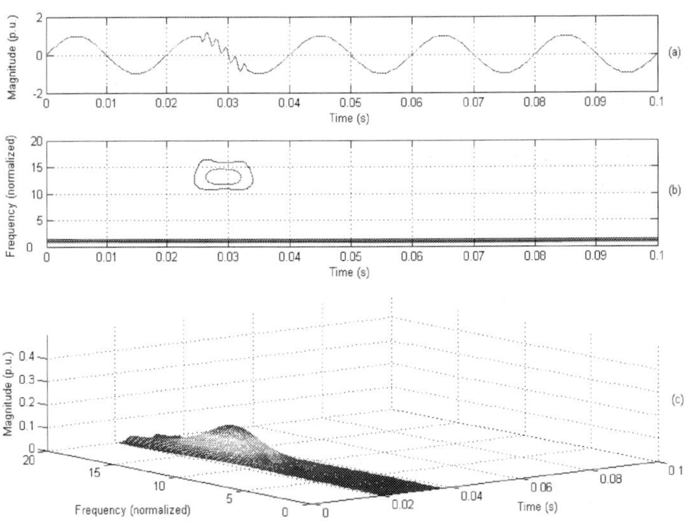

Fig. 3. Oscillatory transient (a). S-transform Time-Frequency contours (b). 3D mesh Time-Frequency-Amplitude (c).

2.4 S-transform based pattern

An efficient pattern can be defined from observation of ST contours. Below some examples of signals with different disturbances are analyzed in order to illustrate the pattern proposed in this approach.

A sinusoidal signal without any disturbance is depicted in Figure 4 (a). The fundamental frequency contour (b) shows a horizontal line. Other frequency contours (c), (d) and (e), corresponding to 150, 250 and 700 Hz show the non-existence of these frequencies in the signal.

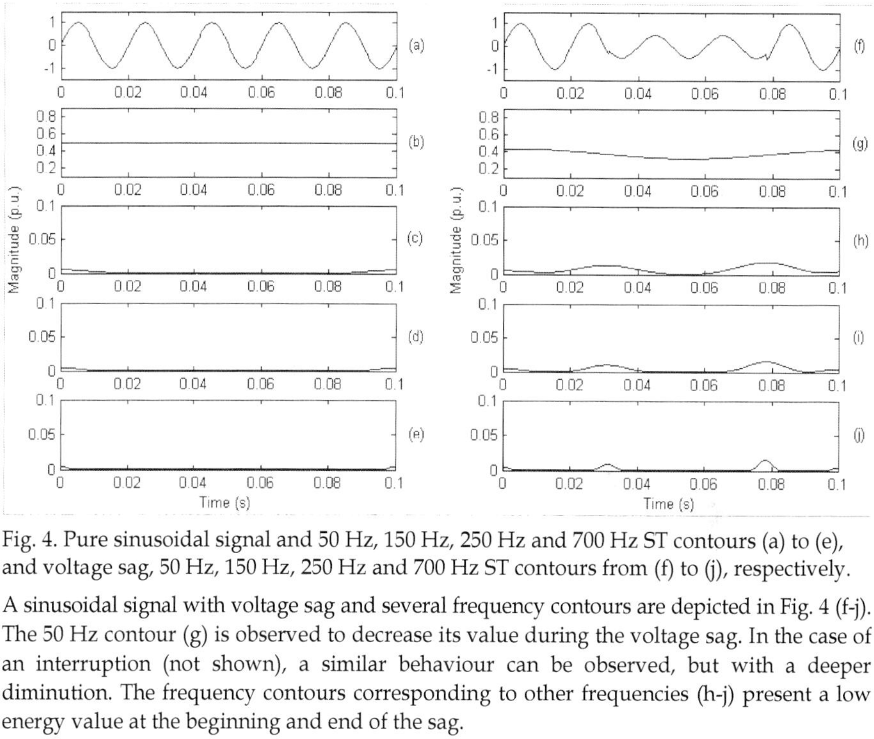

Fig. 4. Pure sinusoidal signal and 50 Hz, 150 Hz, 250 Hz and 700 Hz ST contours (a) to (e), and voltage sag, 50 Hz, 150 Hz, 250 Hz and 700 Hz ST contours from (f) to (j), respectively.

A sinusoidal signal with voltage sag and several frequency contours are depicted in Fig. 4 (f-j). The 50 Hz contour (g) is observed to decrease its value during the voltage sag. In the case of an interruption (not shown), a similar behaviour can be observed, but with a deeper diminution. The frequency contours corresponding to other frequencies (h-j) present a low energy value at the beginning and end of the sag.

An oscillatory transient disturbance is depicted in Figure 6 (a) and the 150 Hz, 250 Hz, 350 Hz and 700 Hz contours in (b) to (e), respectively. A big amount of energy can be noted in the contour corresponding to the frequency present in the transient signal, 700 Hz in this particular case.

Figure 6 (f) depicts a sinusoidal signal with a third and a fifth harmonic simultaneously, and several frequency contours. It clearly shows a big amount of energy in the contours corresponding to the frequencies present in the signal, 150 and 250 Hz, Fig. 6 (h-i).

The examples shown above illustrate the way of taking advantage of this particularity of the ST in order to search for specific frequency disturbances such as any order harmonics or particular oscillatory transients. The results shown above have been taken into account in the election of the characteristic features.

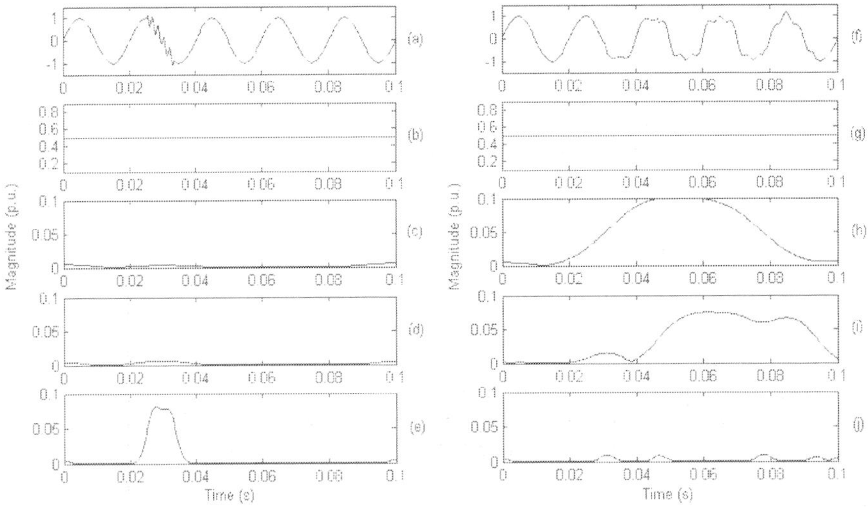

Fig. 4. Oscillatory transient, 50 Hz, 150 Hz, 250 Hz and 700 Hz ST contours (a) to (e), respectively, and signal with third and fifth harmonic component, 50 Hz, 150 Hz, 250 Hz and 700 Hz ST contours from (f) to (j), respectively.

The fundamental frequency contour has proven to contain valuable information about sags, swells and interruptions. Hence, the mean value of the 50 Hz contour has been taken as a distinctive feature. But this value does not clearly discriminate among sags and interruptions, and therefore the minimum value of the 50 Hz contour, which gives an idea of the severity of the disturbance, has been taken as the second feature.

In order to discriminate disturbances with presence of harmonics, the energy of the third, fifth and seventh harmonic (150 Hz, 250 Hz and 350 Hz) contours are used as distinctive features. This approach has been restricted to these frequencies although it could be extended to other harmonics within the Nyquist condition.

The sum of the energies from 600 Hz to 1600 Hz (Nyquist frequency) has also been taken as another characteristic feature. A high value of this energy gives information related to high frequency transient events.

A summary of the distinctive features used in ST based pattern is listed below:

- F1: Mean of the fundamental frequency contour (50 Hz)
- F2: Minimum of the fundamental frequency contour
- F3: Energy of the 3rd harmonic contour (150 Hz)
- F4: Energy of the 5th harmonic contour (250 Hz)
- F5: Energy of the 7th harmonic contour (350 Hz)
- F6: Sum of energy from 600 to 1600 Hz contours

3. Artificial neural network

As algorithm in the automated classification scheme has been selected ANN. Neural networks have emerged as an important tool for classification, and have been successfully applied to a variety of real world classification tasks in industry, business and science

(Widrow et al., 1994). Applications include classification of power quality disturbances (Borras et al. 2001).

An ANN is composed of very simple elements operating in parallel. These elements are inspired by biological nervous systems. As in nature, the network function is determined largely by the connections between elements. An ANN can be trained to perform a particular function by adjusting the values of the connections between elements. The ANNs can approximate any function with arbitrary accuracy, and they are nonlinear models, which makes them flexible in modeling real world complex relationships. Moreover ANNs are data driven self-adaptive methods in that they can adjust themselves to the data without any explicit specification of functional or distributional form for the underlying model (Zhang, 2000). In this chapter two different ANNs have been used, backpropagation and probabilistic.

3.1 Backpropagation

Feedforward backpropagation (BP) is a gradient descent algorithm, in which the network weights are moved along the negative of the gradient of the performance function. The term BP refers to the manner in which the performance function is propagated from the output to backward, and feedforward to the direction of the different connection between elements.

Once the network weights and biases have been initialized, the network is ready for training. The training process requires a set of examples of proper network behaviour, network inputs and target outputs. During training the weights and biases of the network are iteratively adjusted to minimize the network performance function. The default performance function for feedforward networks is mean square error, (i.e. the average squared error between the network outputs and the target outputs).

The BP has been set with two layers, one of them is a hidden layer. This structure can uniformly approximate any continuous function (Cybenko, 1989). The number of nodes in the hidden layer has been chosen as a function of number of inputs (Hecht-Nielsen, 1989), i.e. the number of pattern features, according to the expression $(2n+1)$, where n is the number of inputs. So for each pattern used, the hidden layer has different number of nodes, and the output layer only has one. The transfer functions for the hidden and output layer are tansigmoidal and lineal, respectively. The learning ratio has been 0.1, the epoch 3500 and the training algorithm Levenberg-Marquadt.

3.2 Probabilistic

The basic principle of probabilistic neural network (PNN) is implemented using the probabilistic model, such as Bayesian classifiers (Specht, 1990). The training examples are classified according to their values of probabilistic density function. When an input is presented, the first layer computes distances from the input vector to the training input vectors, and produces a vector whose elements indicate how close the input is to a training input. The second layer sums these contributions for each class of inputs to produce a vector of probabilities as output. Finally, a complete transfer function on the output of the second layer picks the maximum of these probabilities (Mishra et al., 2008). The most obvious advantage of this network is that training is trivial and instantaneous.

4. Classification results

Once the algorithm classifier is selected it has to be trained. At this stage, the classification algorithm is adjusted so that the function that assigns data to its corresponding class has the lowest possible error based on a particular set of training data. In this way the system learns to match the output of the function with the tag or class for each data.

Once the classifier has been adjusted, the results are validated with a different data set, in order to check the effectiveness of the whole system.

The data set consists 6000 signals, where 10% are used to verify the effectiveness of the network, so 900 and 100 for each class, have been used for training and verification stage, respectively.

In this section the most usual power quality disturbances have been considered, sags, interruptions (int), swells, oscillatory transients (OT), harmonics (harm) and signals considered as normal voltage (normal). These signals are generated following mathematical models presented in (Gargoom et al., 2008), simulated in Matlab [Matlab, 2000] with a fundamental frequency of 50 Hz and a virtual voltage of 1V. The signals have a five cycles length (100 ms) and the sampling frequency is 3.2 kHz.

4.1 Classification using WT based pattern and ANN

The resulting classification system using WT based pattern is presented in Fig. 8. It can be observed that the obtained pattern is composed of a feature selection (RMS value) and feature extraction. Selection chooses distinctive features from a set of candidates, while extraction utilizes some transformations, WT in this case, to generate useful and novel features from the original ones (Xu & Wunsch, 2005).

The classification results obtained are presented in Table 4 and 5, using BP and PNN as algorithm classifier, respectively.

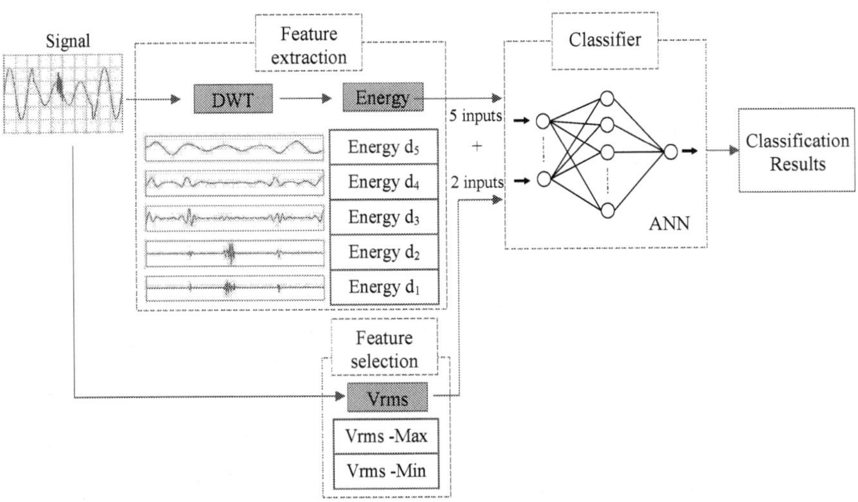

Fig. 8. Resulting classification system using wavelet based pattern and ANN

As can be observed the classification results are slightly better with BP than PNN except with high level of noise, 20dB. For disturbances, the worst results are obtained for sag and interruptions.

	No noise	40 dB	30 dB	20 dB
Normal	0.97	0.99	0.97	0.8
Sag	1	1	0.95	0.91
Int	0.99	0.97	1	0.94
Swell	1	1	0.99	0.98
OT	0.9	0.96	1	1
Harm	1	1	1	0.86
Total	**0.977**	**0.987**	**0.985**	**0.915**

Table 4. Classification results using WT based pattern and BP

	No noise	40 dB	30 dB	20 dB
Normal	1	0.99	0.99	0.99
Sag	0.91	0.93	0.96	0.96
Int	0.74	0.82	0.92	0.92
Swell	0.96	1	1	1
OT	1	0.96	0.99	0.99
Harm	1	1	0.99	0.99
Total	**0.935**	**0.950**	**0.975**	**0.943**

Table 5. Classification results using WT based pattern and PNN

4.2 Classification using ST based pattern and ANN

The resulting classification system using ST based pattern is presented in Fig. 9. The classification results are presented in Table 6 and 7, using BP and PNN as algorithm classifier, respectively.

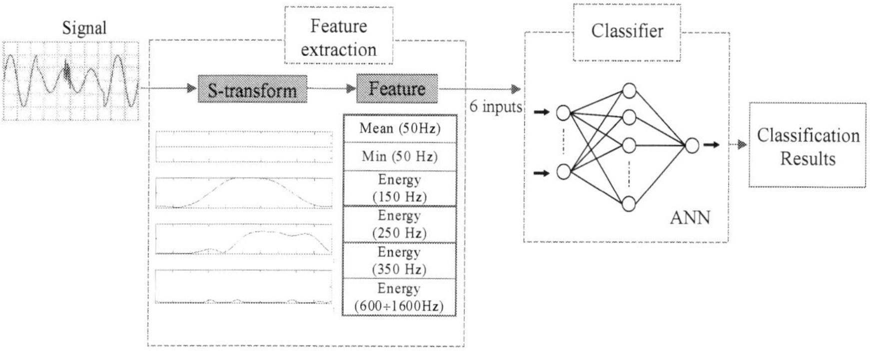

Fig. 9. Resulting classification system using ST based pattern and ANN

	No noise	40 dB	30 dB	20 dB
Normal	1	0.98	0.99	0.94
Sag	0.99	0.96	0.99	0.93
Int	0.79	0.98	1	0.99
Swell	0.97	0.95	0.96	0.99
OT	1	1	1	0.99
Harm	1	1	1	1
Total	**0.958**	**0.978**	**0.990**	**0.973**

Table 6. Classification results using ST based pattern and BP

	No noise	40 dB	30 dB	20 dB
Normal	1	1	1	0.99
Sag	0.79	0.65	0.57	0.62
Int	0.61	0.73	0.69	0.76
Swell	0.89	0.95	0.7	0.94
OT	0.62	0.97	1	1
Harm	0	0	0.64	1
Total	**0.652**	**0.717**	**0.795**	**0.885**

Table 7. Classification results using ST based pattern and PNN

The obtained results using ST based pattern are hardly better with BP than PNN. As can be observed the classification results are quite robust against the noise using BP. For disturbances the worst results are obtained for sag and interruptions.

Anyway, the results using ST based pattern and BP are better than WT based pattern and any ANN used, moreover the ST based pattern no needs additional features, as occurs with WT based pattern. On the other hand, the total number of operations to compute ST is greater than WT.

5. Checking the resulting classification systems

In this section, a data set of power quality disturbances has been generated using the power network simulation environment PSCAD/EMTDC (PSCAD, 2005). This application is an industry standard simulation tool for studying the transient behaviour of electrical networks.

The aim is to obtain different power quality disturbances of data set used for training and validation, in order to check the implemented system.

Several electrical systems with different events, typical and simple, as faults, switching big loads, non linear loads, etc., have been simulated to obtain different power quality disturbances. Moreover different signals can be obtained from a single system by using a facility of software simulation.

One example consists of a low voltage load fed by two parallel lines through a power transformer as shown in Fig. 6. A three phase fault occurs in the middle of one of the two parallel transmission lines at a given time. Approximately two cycles after (t=0.045 s) the fault is cleared. In Fig. 7 (a) can be seen the signal generated, where it can be observed an oscillatory transient produced by the fault, voltage sag during two cycles, and other oscillatory transient produced when the fault is cleared.

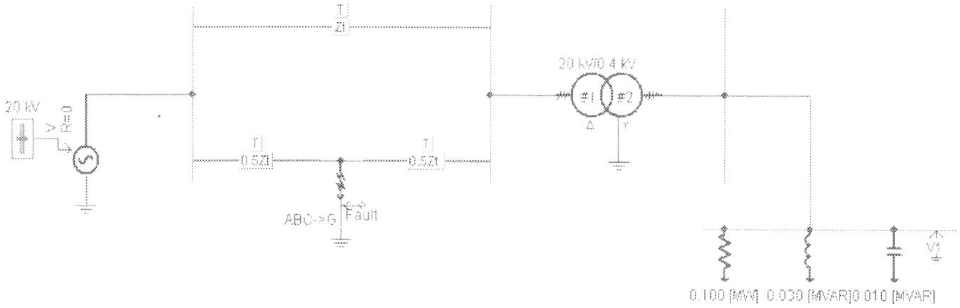

Fig. 6. 3 bus test system scheme in PSCAD

The signals shown in Fig. 7 (b)-(c) belong to initial data set used for training stage of classifier systems. In Fig. 7 (b) it is shown a normal voltage to 0.9, just in the boundary with voltage sag. Fig. 7 (c) depicted a very light oscillatory transient and, just after, a slightly magnitude diminution.

The values of the features in WT based pattern for signals of Fig. 7 (a)-(c) are shown in Table 8, in rows A to C, respectively. As can be observed the signal A(simulation) present a higher value of energy in relative high frequencies D1 ($32f_1 \div 16f_1$) than C, therefore A has an oscillatory transient. The minimum of RMS value for A, lower than B, gives information about the diminution of the magnitude, voltage sag.

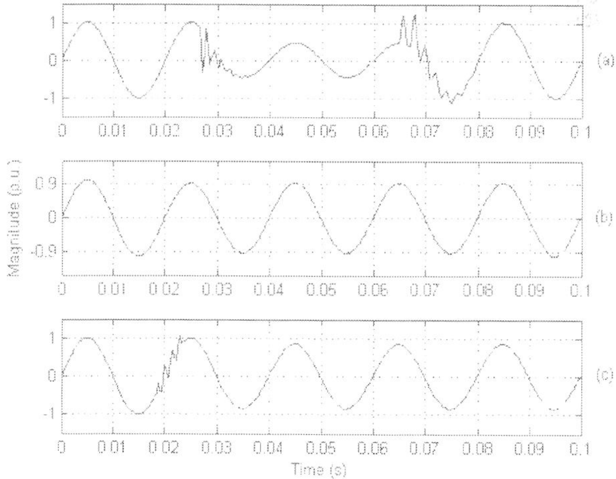

Fig. 7. Signal generated by model simulation (a). Normal voltage to 0.9 (b). Oscillatory transient and a slightly magnitude diminution.

From Table 8 the obtained values of the features for these signals, any efficient classifier can properly assign the disturbance to each corresponding class.

	Energy					RMS	
Signals	D1	D2	D3	D4	D5	Max	Min
A	407.9	336.6	3.164	0.920	0.675	1.043	0.457
B	2.426	1.281	0.959	0.885	0.897	1	0.9
C	351.4	27.22	0.722	0.794	0.801	1.012	0.849

Table 8. Feature values of WT based pattern for different signals

In order to compare the ST based pattern for these same signals are shown several ST contours. In Fig. 8 (a) it is shown the signal from simulation. The fundamental frequency ST contour for this signal and the normal one to 0.9 is showed in (b). The 700 Hz ST contour of the simulation signal and the oscillatory transient is depicted in (c). It can be observed that the simulation signal contours advise the presence of voltage sag, the 50 Hz ST contour for simulation signal (A) is below normal voltage to 0.9 (B), moreover an oscillatory transient because the energy in 700 Hz ST contour for simulation signals (A) is higher than oscillatory transient (C) used as reference. These graphs show that any efficient classifier can properly allocate the disorder to each corresponding class.

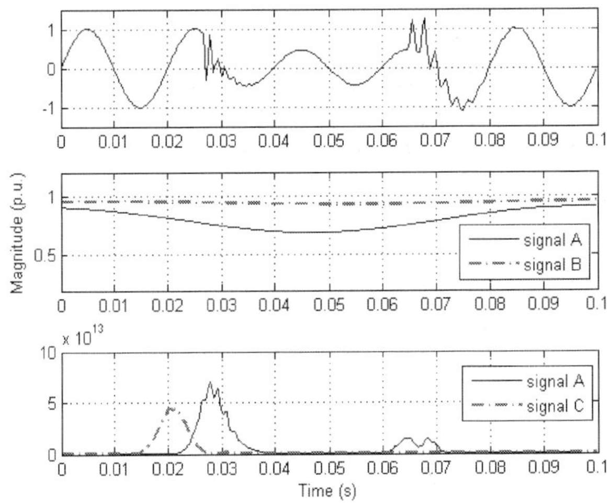

Fig. 8 . Signal generated from simulation (a), comparison of 50 Hz ST contours (b) and comparison of 700 Hz ST contours (c)

Others systems have been simulated using PSCAD and the power quality disturbances obtained are showed in Table 9, which also indicates the classification results using resulting classification systems.

These signals verify the behaviour of the implemented system with power quality disturbances based on electrical models, and therefore close to real ones.

		pattern-ANN		
Disturbance	Number	WT-BP	WT-PNN	ST-BP
Normal	10	0.90	0.90	1
Sag	40	0.875	0.80	0.9
Int	10	0.80	0.70	0.9
Swell	30	0.933	0.866	0.933
OT	30	0.866	0.933	0.933
Harm	30	0.933	0.933	0.933
Total	150	0.893	0.866	0.926

Table 9. Classification results for signals obtained by simulation software

As can be observed the results with signals obtained by electrical networks simulation are worse than those obtained with generated following mathematical models. The BP results do not worsen than PNN. At last the classification system using ST based pattern are better than WT based pattern.

7. Conclusion

Two different automated classification systems has been implemented, based on time-frequency transforms, WT and ST, in combination with Artificial Neural Network as algorithm classifier.

The classification results are very similar using two different algorithm classifiers, a BP and PNN, so it can be affirmed that is more important designing a properly and suitable pattern that the election of the algorithm. Any efficient classifier algorithm can do its expected job.

Therefore the extraction and selection features both are very crucial to the effectiveness of classification applications. Suitable features can greatly decrease the workload and simplify the subsequent process. Generally, ideal features should be of use in distinguishing patterns belonging to different class, immune to noise, easy to extract and interpret.

It can be appreciate that the detail signals of Wavelet transform analysis are very similar in the different disturbances, and in some cases almost identical, therefore becomes necessary to add other features, RMS value, which can distinguish magnitude disturbances. The S-transform generates several contours that put in evidence the nature of the disturbance, and these contours have certain characteristics that are suitable for automated pattern recognition.

On the other hand, the computational cost for Wavelet transform is less that for S-transform, N and $N \cdot (N+logN)$ respectively.

The implemented systems have been tested with signals generated with simulation program obtaining similar results.

A further limitation of many of the studies is that both training and testing are based in signals obtained following mathematical models. In this paper signals generated by power network simulation have been used to verify the classification system. In any case it becomes necessary to have power quality disturbances available to all researchers in this field that can be reliably used to compare classification results of each approach and not every researcher uses their own signals, whether synthetic, simulated or acquired.

The implemented system has been tested with signals generated with simulation program obtaining very good results.

8. References

Borras, D.; Castilla, M. Moreno, N. & Montano, J.C. (2001) Wavelet and neural structure: a new tool for diagnostic of power system disturbances. *IEEE on Industry Applications*, Vol. 37 (Jan. 2001) pp. 184-190

Burrus, C. S.; Gopinath, R. A.; Guo, H. (1998). *Introduction to wavelets and wavelet transforms*, Prentice Hall, ISBN: 0134896009, New Jersey, USA

Cybenko, G. (1989). Approximation by superpositions of a sigmoidal function. *Mathematics of Control, Signals and Systems*, No. 2 (1989) pp. 303-314

Dash, P.K.; Panigrahi, B.K. & Panda; G. (2003). Power quality analysis using S-transform. *IEEE Trans. on Power Delivery*, Vol. 18, (Apr. 2003) pp. 406-411

Dugan, R.S.; McGranaghan, M.F.; Santoso, S. & Beaty, H.W. (2000). *Electrical Power System Quality*, Wiley-Intersciencie, ISBN 007138622X, New York, USA

Gargoom, A.M.; Ertugrul, N. & Soong, W.L. Automatic classification and characterization of power quality events. *IEEE Trans. on Power Delivery*, Vol. 23, (Oct. 2008) pp. 2417-2425

Hecht-Nielsen, R. Counterpropagation networks. *Applied Optics*, No. 26 (Oct. 1989) pp. 4979-4984

Mallat, S. (1999). *A Wavelet Tour of Signals Processing*, Academic Press, ISBN 0123743702

MATLAB (2000). Natick, MA: Math Works, Inc., 2000.

Mishra, S.; Bhende, C.N. & Panigrahi, B.K. (2008). Detection and classification of power quality disturbances using S-transform and probabilistic neural network. *IEEE Trans. on Power Delivery*, (Jan. 2008)Vol. 23.

PSCAD (2005). Manitoba HVDC Research Centre, Inc., 2005.

Santoso, S.; Powers, E.J. & Grady, W.M. (1994). Electric power quality disturbance detection using wavelet transform analysis, *Proceedings of IEEE - SP International Symposium on Time- Frequency and Time Scale Analysis*, pp. 166-169

Specht, D.F. Probabilistic neural networks. *Neural Network*, (1990) Vol. 3, pp. 109-118

Stockwell, R.G.; Mansinha, L. & Lowe, R.P. (1996). Localization of the complex Spectrum: the S-transform. *IEEE Trans. on Signal Processing*, No. 44, (1996) pp. 998-1001

Widrow, B.; Rumelhard, D.E. & Lehr, M.A. (1994) Neural networks: Applications in industry, business and science. *Commun. ACM*, Vol. 37, pp. 93-105

Xu, R. & Wunsch D. (2005). Survey of Clustering Algorithms. IEEE Trans. on Neural Networks. Vol. 16, No. 3 (May. 2005) pp. 645-678

Zhang, G.P. (2000).Neural Networks for Classification: A Survey. *IEEE Trans. on Systems, Man, and Cybernetics – part c: applications and reviews*. Vol. 30, No. 4, (Nov. 2000) pp. 451-462

16

High Performance of An Unified Power Quality Conditioner Based on a Fuzzy Logic

Mekri Fatiha[1], Machmoum Mohamed[2] and Ait Ahmed Nadia[2]
[1]*IRENav : Institut de Recherche de l'Ecole Navale EA3634 - BREST,*
[2]*IREENA : Institut de Recherche en Electrotechnique et Electronique de Nantes Atlantique, Saint Nazaire Cedex, France*

1. Introduction

The quality of energy depends on the loads and their sensitivity to supply voltages variations. The loads being more sophisticated, the voltage disturbances become very expensive for the industrialists in term of loss of production and damage of materials. A universal solution for improving power quality in the network and protecting sensitive loads can be used by combination of the series - parallel active power filters called Unified Power Quality Conditioner (Han et al., 2006 - Khadkikar et al., 2005 - Reza et al., 2009). It is a versatile device that can compensate almost all types of perturbations such as voltage harmonics, voltage unbalance, voltage flicker, voltage sag and swell, current harmonics, current unbalance, reactive current, etc. A typical configuration of UPQC is shown in fig. 1. This paper is focused on voltage sag and along with current and voltage harmonics compensation based on fuzzy hysteresis band control. Initially, the equivalent circuit of the UPQC is presented. The control algorithm for parallel active power filter (PAPF) is discussed in section 3 and the control algorithm for series active power filter (SAPF) is discussed in section 4. The UPQC performances will depend on the design of power semiconductor devices, on the modulation technique used to control the switches, on the design of coupling elements (the decoupling inductance Lf for shunt part, the filter parameters L_{fs} -C_{fs} for series part and the DC link capacitor value C_{dc}), on the method used to determine active filters current and voltage references and on the dynamics and robustness of current and voltage control loops. For the PAPF, the standard instantaneous p-q algorithm is used to determine current references (Fujita & Akagi, 1998 - Mekri et al., 2006). Among various PWM techniques, hysteresis fixed band current control is popularly used because of its simplicity of implementation. This known technique does not need any information about system parameters and has the disadvantage of uncontrolled frequency. As a result, the switching losses are increased and current sources contain excess ripples. The current controller performances can be improved by using adaptive control system theory (Rahman et al., 1997- Tzou & Shiu, 1998). An adaptive hysteresis band current control PWM technique can be programmed as a function of supply and APF parameters in order to maintain a fixed modulation frequency. Unfortunately, adaptive control is very sensitive to parameters system and its global stability is hard to be proved. A new technique, based on the same concept, but where the hysteresis band is implemented with fuzzy logic is proposed to optimize the PWM performances. For the SAPF, a robust PLL system is used for supply voltage disturbances identification. The PLL is developed to achieve good results under unbalanced, interruption or distorted voltage conditions based on a fuzzy logic regulator. Appropriate fuzzy hysteresis band voltage controller is also synthesized for SAPF output voltage control. Simulation results will be shown and discussed in the last section to verify the performances of the proposed UPQC in different conditions such as voltage and current harmonics.

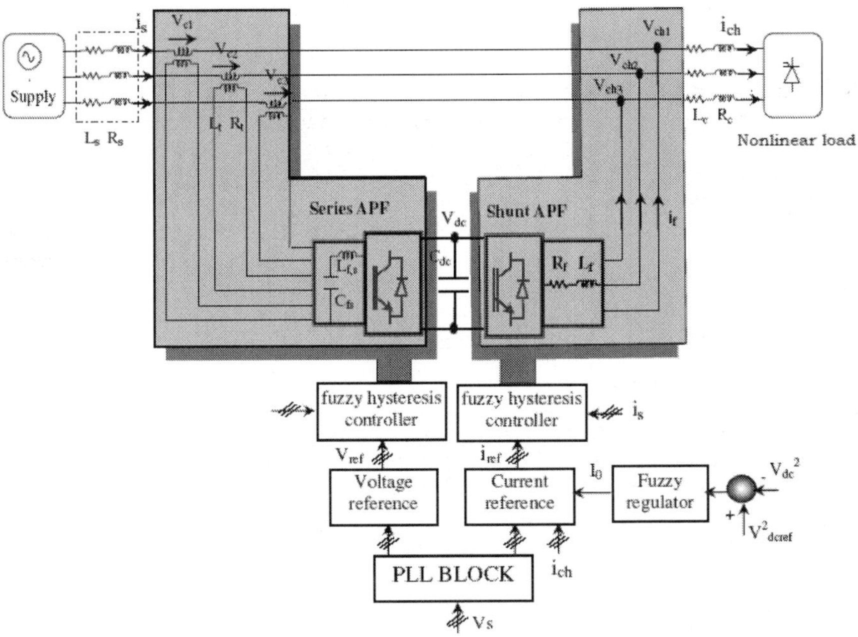

Fig. 1. Global diagram of Unified Power Quality Conditioner

2. Equivalent scheme of an UPQC

The UPQC is controlled in such a way that the voltage at load bus is always sinusoidal and at desired magnitude. The function of the PAPF is to compensate current harmonics, to maintain the dc link voltage at constant level and to provide the var required by the load. The SAPF serves us to mitigate the mains voltages perturbations. The single phase equivalent circuit for a UPQC is shown in fig. 2. The source voltage; terminal voltage at PCC and load voltage are e_s, v_s and v_{ch} respectively. The source and load currents are i_s and i_{ch} respectively. The voltage injected by SAPF is v_c *and* i_f is the current injected by PAPF.

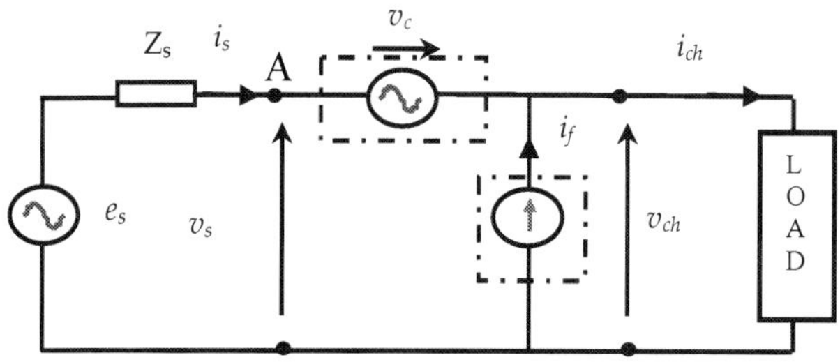

Fig. 2. Equivalent Circuit of a UPQC

3. Parallel active power filter control

3.1 Current reference extraction

The objective is to get sinusoidal line currents in phase with the supply voltages at the common coupling point. The well known instantaneous p-q algorithm is used to determine the current references (Akagi, 1996). The distorted supply voltages conditions may result in partial compensated source current. To overcome this problem, a robust PLL system is first used to extract the fundamental positive sequence voltage components. The generated sinusoidal voltage signals are then used as new inputs of the p-q algorithm for instantaneous current references extraction fig. 3. The PLL is a common block between PAPF and SAPF and will be detailed in section IV.

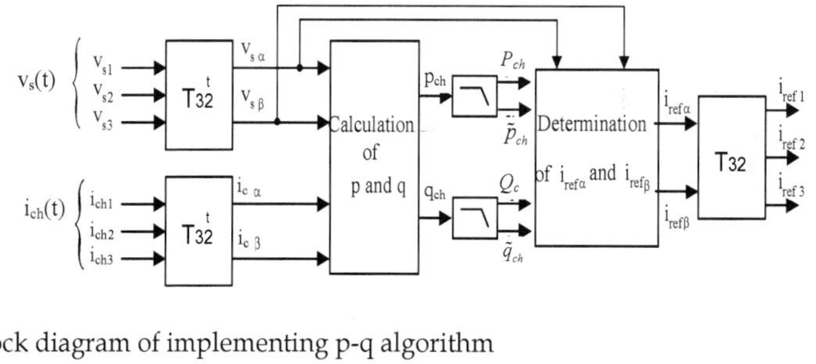

Fig. 3. Block diagram of implementing p-q algorithm

3.2 DC voltage control

In the UPQC the management of DC bus concerns the role of the PAPF. This one determines the active power (current) necessary to keep constant the DC voltage in steady state or transient conditions (Mekri et al., 2006). There are three principal factors that affect the voltage fluctuations of the DC capacitor. The first is the alternating power of the load to be compensated, the second is the active power imbalance during transients and the third is the active power absorbed by the SAPF part for compensating network voltage sag.

If a power imbalance occurs; because of load changing or voltage dips; the PAPF should consume or supply real power. This power is given by:

$$P_f = \bar{p}_{ch} - P_s = 3.V_s.(I_{ch1}.\cos\varphi_1 - I_s) = \pm 3V_sI_0 \quad (1)$$

I_{ch1} is the fundamental current of the load and φ_1 it phase, $\overline{p}_{ch}$ is the DC power consumed by the non-linear load, P_S is the active power provided by the supply and I_0 component is given by:

$$I_0 = \left| I_{ch1} \cos \varphi_1 - I_s \right|$$

The active fundamental current necessary to ensure the balance of the active power called $i_0(t)$, must be sinusoidal and in phase with the supply voltage (fig. 4). It combines of two fundamental components:

$$i_o(t) = I_{ms} \sin(\omega t) + I_{mo} \sin(\omega t) \tag{2}$$

The first fundamental component corresponds to the active power consumed by the series active filter during a long term voltage dip for the network, and the second term refers to the power losses consumed by the power transistors in both inverters.

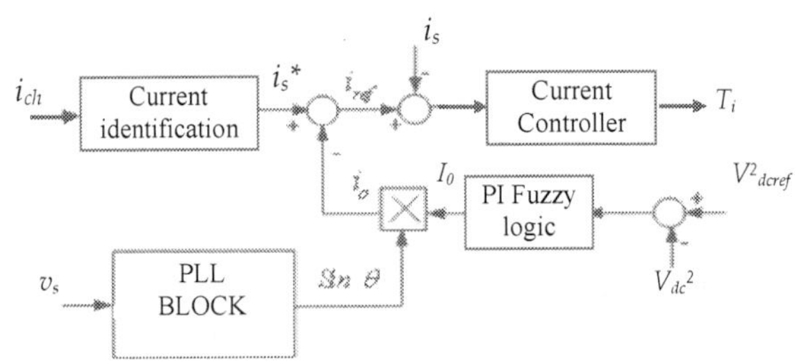

Fig. 4. Block diagram of shunt APF control

If switching losses are neglected, the real power absorbed by the PAPF can be expressed as:

$$P_f = -\frac{1}{2} C_{dc} \frac{dv_{dc}^2(t)}{dt} \tag{3}$$

The aim of the synthesized voltage regulator is to adjust $v_{dc}^2(t)$ to its reference, to reject the internal disturbance dI of the system due to the variation of real load power and to assure a good filtering of the external one dE relating to DC voltage ripple due to the alternating power (Mekri et al., 2006). To realize these objectives, a fuzzy logic controller is considered. It consists of fourth stages: fuzzification, knowledge base, inference mechanisms and defuzzification. The knowledge bases are designed in order to obtain a good dynamic response under uncertainty in process parameters and external disturbances. In our application, the fuzzy controller is based on processing the voltage error and its derivation $\dot{e}$. The input variable 'e' is:

$$e = V_{dc}^{*2} - V_{dc}^2 \tag{4}$$

Triangle shaped membership function has the advantages of simplicity and easier implementation and is chosen in this application (Tzou & Shiu, 1998 - Xiong & Gatland, 1996). Fig. 5 shows the membership functions of the input and the output linguistic variables.

In the design of a fuzzy control system, the formulation of its rule set plays a key role in improvement of system performances. The rule table is constructed to contain the 49 rules as shown in Table 1, where PL, PM, etc, are linguistic codes (PL: positive large, PM: positive medium, PS: positive small, ZE: zero, NL: negative large, NM: negative medium, NS: negative small). In this paper, we applied max-min inference method to get implied fuzzy set of the turning rules.

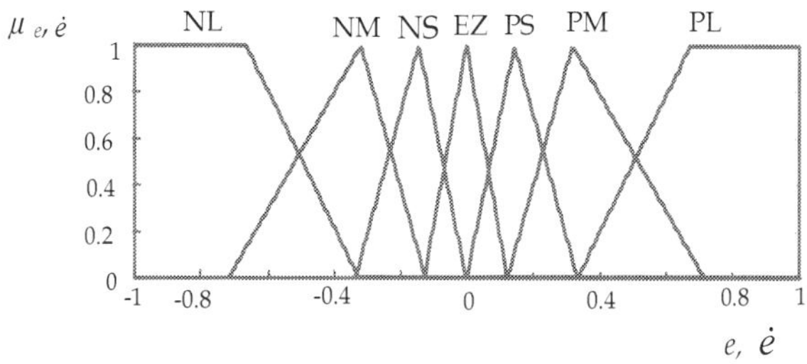

Fig. 5. Membership functions for input variables (e, $\dot{e}$) and output variables.

$\dot{e}$ /e	NL	NM	NS	EZ	PS	PM	PL
NL	NL	NL	NL	NL	NM	NS	EZ
NM	NL	NL	NL	NM	NS	EZ	PS
NS	NL	NL	NM	NS	EZ	PS	PM
EZ	NL	NM	NS	EZ	PS	PM	PL
PS	NM	NS	EZ	PS	PM	PL	PL
PM	NS	EZ	PS	PM	PL	PL	PL
PL	EZ	PS	PM	PL	PL	PL	PL

Table 1. Rules of inference for the DC voltage

3.3 Fuzzy hysteresis band current control

The shunt active filter is realized with a current controlled PWM inverter. The core of active filter is the control section that must be able to derive the reference current waveform matching the harmonic content of the line current and to drive the inverter producing a filtering current faithfully tracking the reference one. So there are two sections: the reference current generation, that, the standard instantaneous p, q algorithm is used to determine the current references (Mekri et al., 2006 - Akagi, 1996), and the current control.

To compensate harmonics and fundamental reactive power, two current control methods can be considered, according that the regulated variables are respectively the line currents or the active power filter currents. Both methods are equivalent. In fact, the error signal at the input of the current regulator has the same dynamic and only its sign is modified Our algorithm will be focused on the following on calculating and controlling directly the mains currents. The objective is to get sinusoidal line currents in phase with the supply voltages at the common coupling point.

The current control strategies, often used, can be classified in the hysteresis current control, the ramp comparison control methods (natural, asymmetrical or optimal PWM) associated with linear controller and the predicted current control. The first method is very simple and easy to implement, but has the disadvantage of an uncontrollable high switching frequency. This high frequency produces a great stress for the power transistors and induces important switching losses. The second and third methods allow operating at a fixed switching frequency and are usually performed by software using the system parameters. In this case, the operating conditions must be known to meet sufficient and accurate control (Xiong & Gatland, 1996). Consequently, a fuzzy hysteresis band circuit control for a sinusoidal input current is involved for our application.

Fixed hysteresis band method is very sample and easy to implement, but has several known disadvantages such as uncontrollable high switching frequency and induced important switching losses. To improve this control, an adaptive and fuzzy hysteresis band current control technique is studied (Mekri et al., 2006).

An adaptive hysteresis band current control PWM technique can be programmed as a function of the active filter and supply parameters to minimize the influence of current distortions on modulated waveform (Mekri et al., 2006 - Jiang & Chang, 2004).

The band (HB) can be modulated at different points of fundamental frequency of the cycle to control the PWM switching pattern of the inverter.

The hysteresis band is given by:

$$HB_j = \frac{V_{dc}}{6.f_m.L_f}\left(1 - \frac{9L_f^2}{4V_{dc}}\left(\frac{v_{sj}(t)}{L_f} + \frac{di_{sj}^*}{dt}\right)^2\right) \quad j = 1,2,3 \tag{5}$$

Where f_m is the modulation frequency, i_s^* is the source reference current and di_s^*/dt represents its slope.

Fig. 6 shows the block diagram of the adaptive hysteresis band current control using (5). Current control with an adaptive hysteresis band current control needs a precise knowledge of the PAPF parameters (L_f and V_{dc}). To improve active filter performances, equation (5) is implemented in our case with fuzzy logic. In order to establish a fuzzy logic controller, input and output variables must be treated. The supply voltage wave $v_s(t)$ and mains current reference slope di_s^*/dt, are selected as input variables and HB as output variable.

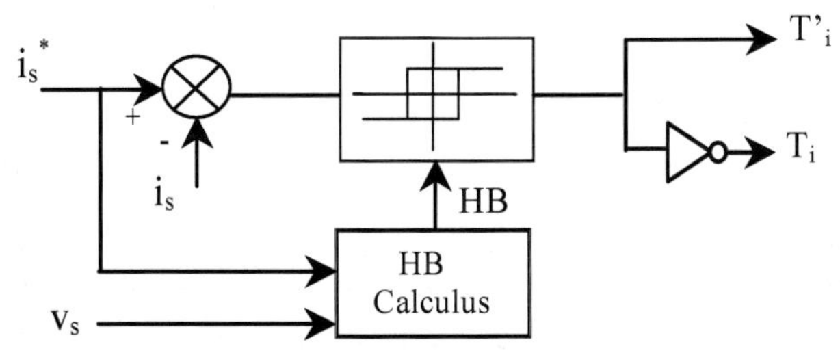

Fig. 6. Simplified model for fuzzy hysteresis-band current control

The following step is to determine the set of linguistic values associated to each variable. Triangle memberships are chosen. Each input variables is transformed into linguistic size with five fuzzy subsets, PL: positive large, PM: positive medium, PS: positive small, EZ: zero, NL: negative large, NM: negative medium, NS: negative small and for the output variables are: PVS: positive very small, PS: positive small, PM: medium positive, PL: positive large, PVL: positive very large. Then, fig. 7 shows the membership functions of the input and the output variables. The resulting rule is presented in Table2.

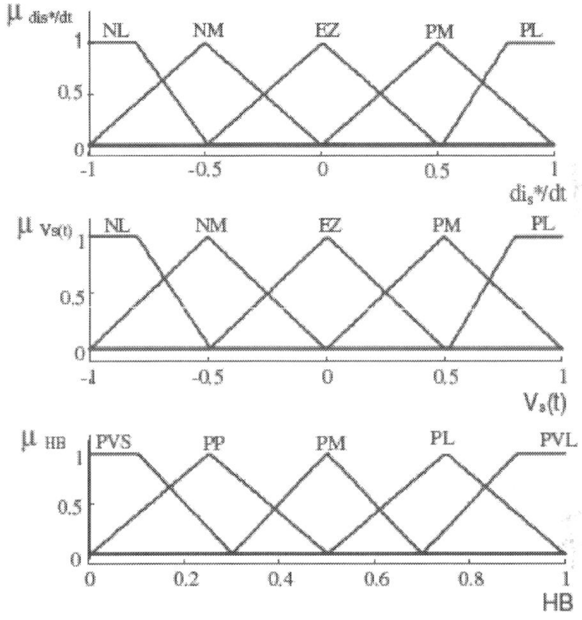

Fig. 7. Membership functions for input variables $\frac{di_s^*}{dt}$, $v_s(t)$ and output variable HB

di_{s1}*/dt $v_s(t)$	NL	NM	EZ	PM	PL
NL	PS	PS	PM	PS	PS
NM	PS	PM	PL	PM	PS
EZ	PVS	PM	PVL	PM	PVS
PM	PS	PM	PL	PM	PS
PL	PS	PS	PM	PS	PS

Table 2. Rules of inference for Fuzzy hysteresis current control

4. Series active power filter control

The main function of a series active power filter is the protection of sensitive loads from supply voltage perturbations such as sags or voltage harmonics.

4.1 Voltage reference calculation

There are various methods of the identification of the network voltage perturbations (Khadkikar et al., 2005 - Awad et al., 2003 - Svensson & Sannio, 2002). The proposed method is based on a robust PLL system and is able to detect quickly any voltage drop due to dips or flickers besides voltage harmonics in the network (figure 8) (Mekri et al., 2010). The PLL block allows to detect the amplitude and phase (V_s and $\hat{\theta}_d$) of fundamental positive sequence components of the utility voltages. The direct component voltages ($V_{sd1,2,3}$) will be subtracted from the instantaneous network voltages ($V_{s1,2,3}$) to determine disturbing homopolar and inverse components ($V_{dif1,2,3}$). The difference between the desired voltage amplitude ($V_{nom\text{-}desired}$) and the direct voltage amplitude provided by the PLL, gives, through Park inverse transformation, the regulated load voltages ($V_{r1,2,3}$). Finally, the reference voltages ($V_{ref\text{-}1,2,3}$), which will be used for control of the series active filter, are obtained by the sum of disturbing components ($V_{dif1,2,3}$) and regulated load voltages ($V_{r1,2,3}$).
The detailed block diagram of the PLL, which can be completely implemented in software, is presented in fig. 9. The PLL block allows to control of an estimated phase angle $\hat{\theta}_d$ with respect to the angle θ_s of mains voltage.
Assuming that the measured voltages at the common coupling point are sinusoidal and balanced, the d-q supply voltage components are given by:

$$\begin{bmatrix} V_{sd} \\ V_{sq} \end{bmatrix} = \sqrt{3}\,V_s \begin{bmatrix} \sin(\theta_s - \hat{\theta}_d) \\ -\cos(\theta_s - \hat{\theta}_d) \end{bmatrix} = \sqrt{3}\,V_s \begin{bmatrix} \sin(\Delta\theta) \\ -\cos(\Delta\theta) \end{bmatrix} \tag{6}$$

According to (6), the PLL will be locked out of the supply voltages when the error between the phase of the supply voltage and the exit of PLL is null $\left(\Delta\theta = \theta_s - \hat{\theta}_d\right)$. In this case $V_{sd} = 0$ and V_{sq} gives the amplitude of the fundamental component of the utility voltage.

$$V_{sd} = 0 \text{ and } V_{sq} = -\sqrt{3}.V_s \tag{7}$$

Thus, it is possible to control θ_s by regulating V_{sd} to zero.
The synthesized regulator must answer the following objectives: the control of V_{sd} must be carried out without bias and with a fast dynamics, the predominant undesirable frequencies present on voltage component V_{sd} under unbalanced mode (100Hz) or distorted conditions (300 Hz and 600 Hz) must be filtered by the regulator. For this purpose, several regulators are studied and compared to satisfy these constraints (Mekri et al., 2010). The functional diagram of the PLL based on a fuzzy PI regulator is shown in fig. 10 where E is the error $E = V_{sd}^* - V_{sd}$ and $\dot{E}$ is the derivation of the error.
We choose the triangular and trapezoidal functions for the variables of entries and exit in the same manner than presented in fig. 5. They allow an easy establishment and the stage of fuzzification requires a little computing time during its evaluation in real time. The knowledge obtained on the behavior of the system is put in the form of rules, which are summarized in Table 1 of inference.

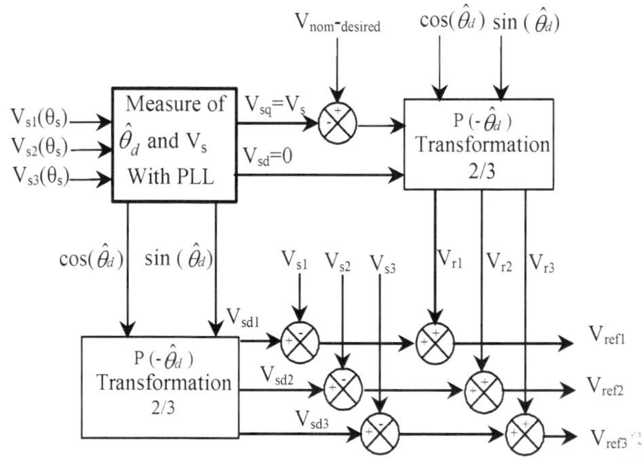

Fig. 8. Determination of the series active filter voltage references

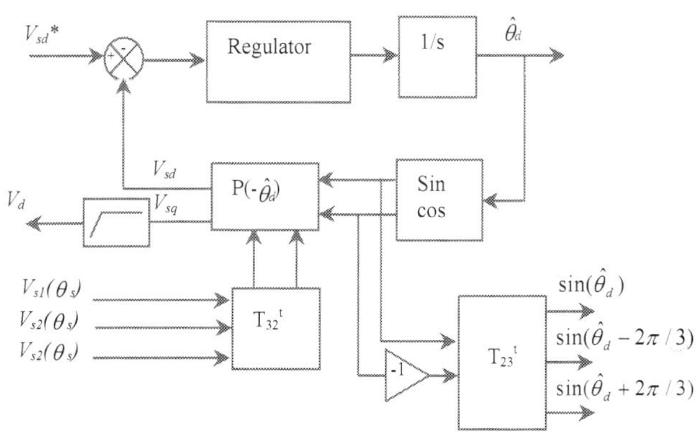

Fig. 9. Detailed block Diagram of the PLL

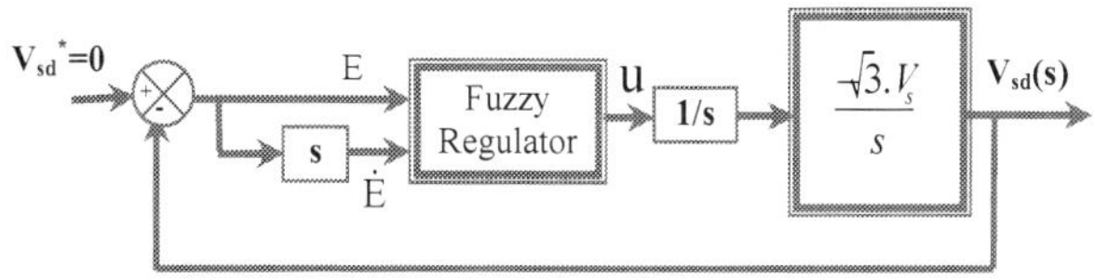

Fig. 10. Functional Diagram of the PLL with the fuzzy regulator

4.2 Series active filter output voltage regulation

The control part of the SAPF must be able to derive the reference voltage waveform with matching the harmonic content of the line voltage. A fuzzy hysteresis band control is adopted allowing operating at nearly fixed frequency which is detailed in (Mekri et al., 2010). The adaptive hysteresis band is given by:

$$HB = \frac{V_{dc}}{6f_c R_{fs} C_{fs}}\left(1 - \frac{9(R_{fs}C_{fs})^2}{4V_{dc}^2}\left(\frac{V_{sl}(t)}{R_{fs}C_{fs}} + \frac{dV_{ref}(t)}{dt}\right)^2\right) \quad (8)$$

with: $V_{sl} = V_{ref} + L_{fs}C_{fs}\frac{dV_{ref}^2}{dt^2} - (L_{fs}\frac{dI_{ch}}{dt} + R_{fs}I_{ch})$

This equation shows that the hysteresis band can vary while keeping the switching frequency nearly constant.

In order to establish a fuzzy logic controller, the input and the output variables are again treated. The voltage reference slope dV_{ref}^*/dt and it derivation $(V_x(t)=d(dV_{ref}^*/dt)/dt)$ are selected as input variables and HB as output variable. The following step is to determine the set of linguistic values associated to each variable. Fig. 11 shows the membership functions of the input and the output variables. The resulting rules are the same that presented in Table 2.

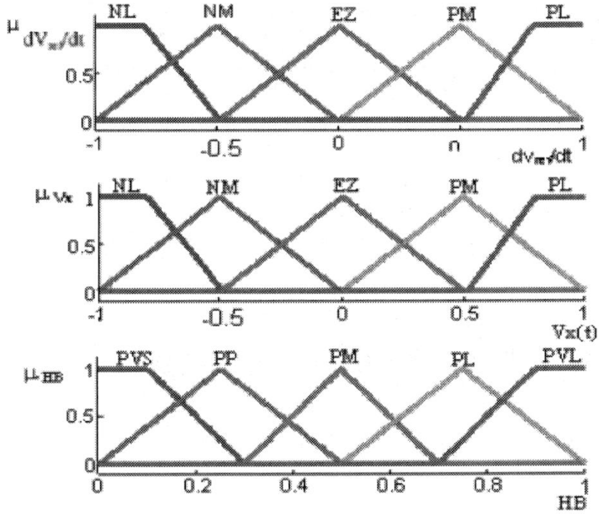

Fig. 11. Membership functions for input variables dv_{ref}/dt, $v_x(t)$ and output variable HB

5. Simulation results

The system (UPQC) shown in fig.1 is simulated using Matlab/ Simulink software. The parameters of simulation are: $V_{dc}=700V$, $V_s=220V$-$50Hz$, $L_{fs}=4*10^{-4}H$, $C_{fs}=73*10^{-4}F$, $f_c=8kHz$, $R_s=0.5\ m\Omega$, $L_s=11\ \mu H$ and $L_f=150\mu H$. The non-linear load is composed by a diode rectifier feeding an R-L load $(Rd=0.66\Omega\ L_d=4mH)$.

Fig. 12 illustrates the behavior of the PLL block (see fig. 9) under excessive distorted and unbalanced supply voltages conditions. The rms values are V_{s1}=230V, V_{s2}=180V and V_{s3}=300V and the main voltages are affected by 5th and 7th harmonic components with individual harmonics distortion rate respectively 20% and 24,5%. These results of simulations show us that the application of fuzzy logic in the control loops makes it possible to fulfil the desired requirements concerning the locking of PLL, even under the most unfavorable conditions. The estimated pulsation is constant and the extraction of unity sinusoids synchronized with the direct supply voltages components at the point of common coupling (PCC) is quasi instantaneous.

Fig. 13 shows the behavior of the UPQC under distorted supply voltages conditions. Before t=0.06s, the PAPF is operational. It shows that the source current is sinusoidal and in phase with the supply voltage. The THD of the source current decreases from 29,5% before filtering to 2%. At t=0.06s a sag (50% during 40 ms) is introduced on the system, both SAPF and PAPF are operating together as UPQC. Results show that UPQC is maintaining the load voltage sinusoidal and at desired constant level even during the sag. While SAPF is providing the required real power to the load, the PAPF is maintaining the DC link voltage at constant level and the source delivered more current. This extra power flows from source to PAPF, PAPF to SAPF via DC link and from SAPF to the load. The THD of the load voltage decreases from 27.5 % before compensation to less than 0.5%.

The waveforms of the nonlinear load current i_{ch}, the line current and its reference (i_s, i_s*), injected shunt active filter current i_f and current error δ given by hysteresis fuzzy band current control for the PAPF are shown in fig. 14. The current error is reduced. The performance of the proposed control algorithm of the active power filter is found to be excellent and the source current is practically sinusoidal. The THD decreases from 29.5% before filtering to 2% for fuzzy hysteresis band control after filtering.

Fig. 15 shows the response of the UPQC with 50% voltage sag. As it can be seen the voltage sag is very fast detected by the voltage perturbations identification method and compensated by the series part of the UPQC. The voltage of the DC bus goes down to about 674 V when sag occurs, but this value is enough for the series filter to compensate the perturbations.

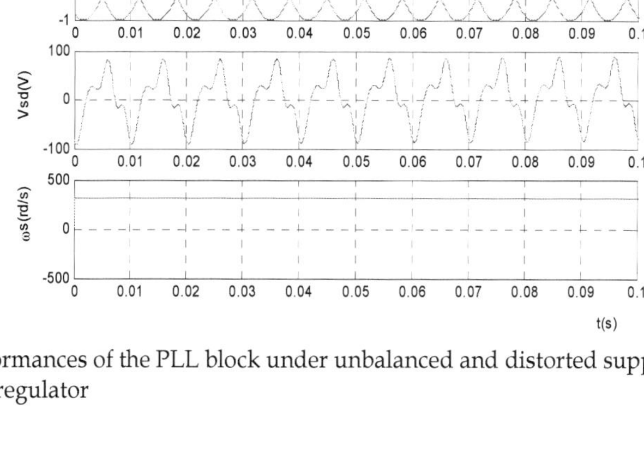

Fig. 12. Performances of the PLL block under unbalanced and distorted supply voltages using fuzzy regulator

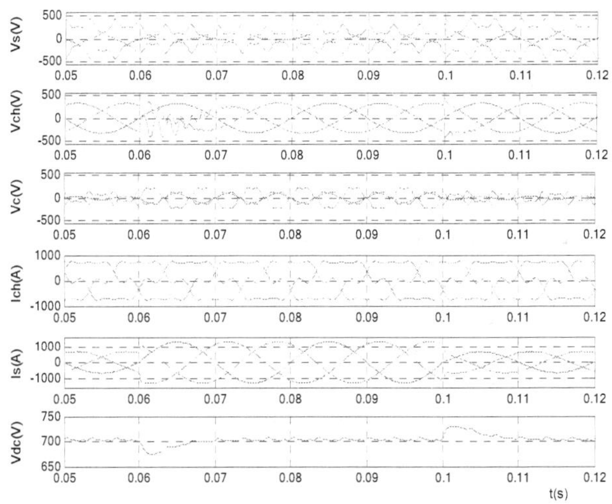

Fig. 13. Performances of the UPQC for voltage sag and harmonics compensation and a 3-phase non linear load

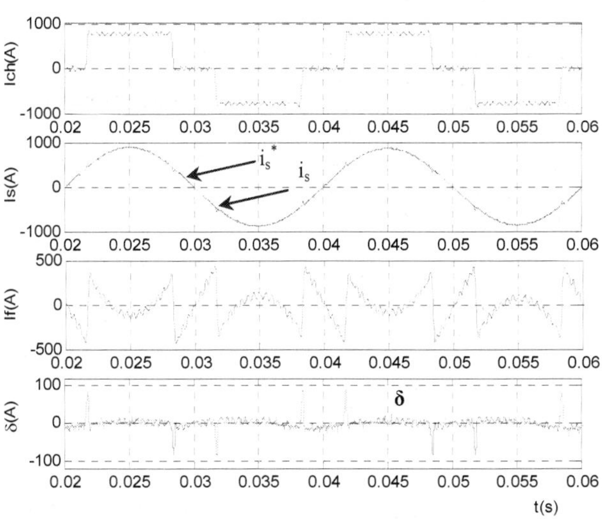

Fig. 14. Compensation of harmonic currents with fuzzy hysteresis band control

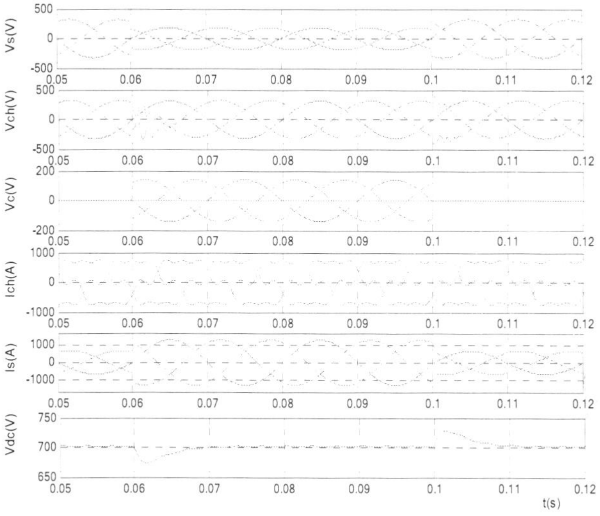

Fig. 15. Response of the UPQC for a three phase 50% voltage sag and non linear load

7. Conclusion

This paper demonstrates the validation of simpler control approach for the Unified Power Quality Conditioner based on the fuzzy logic. In the first, for the parallel active power filter while basing on the method of the instantaneous real and imaginary powers to identify the currents of references, and for determination of voltage references for series active power filter we based on a robust three-phase digital locked loop (PLL) using fuzzy logic regulator. The PLL system has a good reliability, a fast tracking performances and assures a good attenuation of undesirable supply voltage frequencies. In other hand, the UPQC has shown the ability to compensate sag, unbalanced voltages and current and/or voltage harmonics based on a new hysteresis fuzzy band voltage and current control which constraints constant switching frequency without being influenced by parameters of active filter. The current and voltage bands can be easily implemented with fuzzy logic to maintain the modulation frequency nearly constant for each control. Simulation results confirms the viability of the proposed approach and proves that the UPQC, thanks to robust voltage and current controllers, allows to improve power quality by maintaining the load voltage at desired level even during unbalanced, distorted or supply voltage sag conditions. Therefore, the proposed control can easily be adapted to others more severe constraints

8. References

B. Han, B. Bae, H. Kim, S. Baek, (2006), Combined operation of unified power quality conditioner with distributed generation, *IEEE Trans on Power Electronics*, 21 (1), (2006), 330-338.

V. Khadkikar, A. Chandra, A.O. Barry, A.O., and T.D. Nguyen, (2005), Steady state power flow analysis of unified power quality conditioner (UPQC) , IEEE *International Conference on Industrial Electronics and Control Applications* (ICIECA), November 29-December 2, pp. 1-6, Quito, Ecuador.

H. Reza. Mohammadi, A.Y. Varjani, H. Mokhtari (2009), Multiconverter Unified Power-Quality Conditioning System: MC-UPQC" *IEEE transactions on power delivery* ISSN 0885-8977, vol. 24, no3, pp. 1679-1686,

H. Fujita, H. Akagi, (1998),The unified power quality conditioner: the integration of series-and shunt-active filter", *IEEE Trans. on Power Electronics*, vol. 13, no. 2, March 1998, pp. 315-322.

F. Mekri, B. Mazari, M. Machmoum, (2006), Control and optimization of shunt active power filter parameters by fuzzy logic, *Canadian Journal of Electrical and Computer Engineering*, 31 (3) (2006) 127-134

K.M. Rahman, M.R. Khan, M.A. Choudhury, (1997),Variable band hysteresis current controllers for PWM voltage source inverters", *IEEE Trans on Power Electronics* , vol. 12, no. 6, November 1997, pp. 964-970.

Ying-Yu Tzou, Shiu-Yung Lin, (1998),Fuzzy -tuning current-vector control of a three phase PWM inverter- performance AC drives", *IEEE Trans. on Industrial Electronics,* October 98, pp. 782-791.

H. Akagi, (1996), New trends in active filters for power conditioning", IEEE Trans. on Ind. Appl., vol. 32, no.6, pp 1312-1322, 1996

H. Xiong. Li, H. B. Gatland, (1996), Conventional fuzzy control and its enhancement, *IEEE Trans. on Systems*, vol. 26, N°5, October 1996.

Jiang Zeng a, Chang Yu b Qingru Qi (2004), A novel hysteresis current control for active power filter with constant frequency, *Electric Power Systems Research* 68 (2004) 75_/82

H. Awad, J. Svensson, M. Bollen, (2003), Phase locked loop for static series compensator", in CD-ROM of the 2003 *European Power Electronics Conference,* France.

J. Svensson, A. Sannio, (2002), Active filtering of supply voltage with series-connected voltage source converter", *EPE Journal,* Vol. 12, n° 1, February 2002, pp. 19-25.

F. Mekri, M. Machmoum, N. Aït-Ahmed, B. Mazari, (2010), A comparative study of voltage controllers for series active power filter, *Electric Power System Research,* Ed. Elsevier, Volume 80, June 2010, Issue 6, pp 615-738.

17

Exploiting Higher-Order Statistics Information for Power Quality Monitoring

Danton D. Ferreira[1], Cristiano A. G. Marques[2], José M. de Seixas[3], Augusto S. Cerqueira[4], Moisés V. Ribeiro[5] and Carlos A. Duque[6]

[1]*Department of Engineering, Federal University of Lavras*
[2,3]*Signal Processing Lab, COPPE - Poli, Federal University of Rio de Janeiro*
[4,5,6]*Electrical Engineering Department, Federal University of Juiz de Fora*
Brazil

1. Introduction

The basic requirement for a Power Quality (PQ) monitoring system is the automatic detection of the PQ disturbances. Furthermore, the monitoring system should be capable of distinguishing different classes of disturbances. Usually, the PQ disturbance detection and classification tasks are performed in the power system voltage signal, because the majority of disturbances are mainly related to changes in the voltage signal. As a result, disturbance detection and classification in voltage signals can, indeed, assist the identification of the underlying source of disturbances. Disturbances that often appear in the power system voltage signals comprise sags, swells, outages, harmonics, oscillatory transients, notching and spikes, causing failure or miss-operation of the electrical equipments (Morsi & El-Hawary, 2008). Several authors use the denomination of waveform distortion for periodic distortion added to the voltage signal, such as harmonic distortion, notchings, etc. In this work we will use the term disturbance indistinctly for both periodic or non periodic distortions.

In order to improve PQ, the underlying sources and causes of such disturbances must be known before appropriate actions be taken. However, for determining the causes and sources of disturbances, one must have, *a priori*, the ability to automatic detect and classify such disturbances. In the last two decades, researchers have attempted to use appropriate signal processing and computational intelligence techniques for that aim. A good review about the main signal processing tools applied to the PQ problem can be found in Bollen et al. (2009).

In general, a detection and/or classification system comprises processing blocks as shown in Figure 1. The preprocessing step can be viewed as the application of signal decomposition techniques (de Aguiar et al., 2009; Ferreira, de Seixas & Cerqueira, 2009; Gaouda et al., 1999), aiming at expliciting the primitive components of the electric signals, and signal segmentation (Bollen et al., 2009; Styvaktakis et al., 2002), which envisage the data partition into frames with a fixed number of samples. The preprocessing can range from a simple signal processing technique up to more complex signal decomposition techniques as it is well discussed in Duda et al. (2000).

Feature extraction is an essential step towards a successful disturbance detection and classification. It is based on subjective knowledge gathered from power system specialists or objective information extracted from the voltage signal. Regarding objective information,

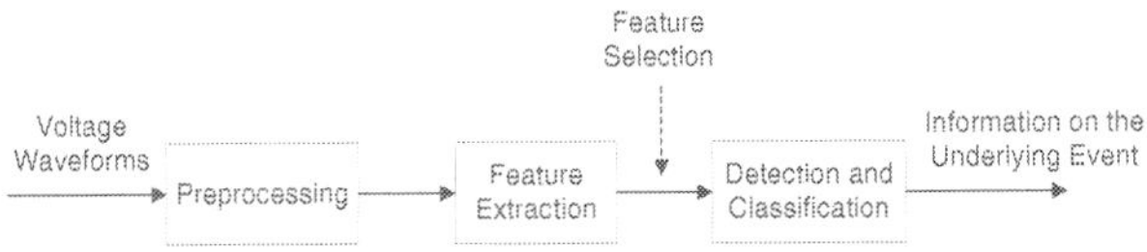

Fig. 1. Basic detection and classification system.

features can sometimes be extracted without considering the specific nature of voltage signals, for instance by using the outputs from signal transformations, which includes the Discrete Fourier Transform (DFT), Short Time Fourier Transform (STFT), wavelets, and other time-frequency signal decomposition methods, and/or using second-order statistics, Root Mean Square (RMS) values, etc (Bollen et al., 2009). Several works found in the literature have pointed out the Discrete Wavelet Transform (DWT) (Mallat, 1999) and S-Transform (ST) (Stockwell et al., 1996) are efficient feature extraction techniques for PQ disturbances, mainly due to their capability to represent the signal frequency components, preserving the time information (Mishra et al., 2008; Samantaray, 2010).

The main goal of the feature extraction, in both detection and classification contexts, is to represent the data set in a new feature space in which the probability to distinguish classes is higher than the one in the original space. Typically, feature selection techniques will result in low computational burden approaches for detection and/or classification of disturbances. Nevertheless, some important aspects must be considered for choosing the feature extraction tool for PQ monitoring. Among them, the following ones must be carefully taken into account:

1. The sensitivity to noise. The monitoring system performance can be strongly corrupted by the presence of noise in the electric signals;
2. The sensitivity to power frequency variations. A variation of 2% can be found in the nominal voltage frequency, and this variation can also reduce the performance of the monitoring system;
3. Computational burden. For real-time applications, low cost and complex techniques should be chosen.

The DWT is a very attractive tool, as a feature extraction technique for PQ disturbance detection and classification. However, difficulties may arise when the signals are corrupted by noise and/or when the number of samples of the signal window is reduced. In order to overcome these limitations, some recent works (Ferreira, Cerqueira, Duque & Ribeiro, 2009; Ribeiro et al., 2006; 2007) have been exploiting the usage of Higher-Order Statistics (HOS) (Mendel, 1991) as features for PQ monitoring. HOS measures are extensions of second-order measures (such as the autocorrelation function and power spectrum) to higher orders (Mendel, 1991). Any Gaussian signal is completely characterized by its mean and variance. Consequently, the HOS information for Gaussian signals is useless. However, many signals encountered in practice have non-zero mean HOS, and many measurement noises are Gaussian, and so, in principle, the HOS is less affected by Gaussian background noise than the second-order measures (Mendel, 1991). This is the main motivation for the usage of the HOS in voltage signals.

This chapter focuses on detection and classification of PQ disturbances based on HOS for feature extraction. Cumulants (Mendel, 1991) are extracted from the power system signals using a simple estimation for finite length vectors, as proposed in Ribeiro et al. (2006). Based on these features, a Bayesian detection system followed by a neural classifier are both

designed. Additionally, a filter bank is used envisaging multiple disturbance decoupling, in order to increase the performance of the neural classifier in presence of more complex disturbance contexts.
This chapter is organized as follows. Next section presents a model to describe the power signal in the presence of PQ disturbances. Section 3 discusses the main features of the HOS information. In Section 4, both preprocessing and feature extraction based on HOS for PQ are presented. Section 5 illustrates the power of HOS features for PQ automatic disturbance detection and classification and, the conclusions are derived in Section 6. Finally, future tends in PQ detection and classification issues are pointed out in Section 7.

2. PQ problem formulation

The discrete version of the monitored voltage signal can be segmented into non-overlapped frames of N samples, which are expressed as an additive contribution of several types of phenomena, as previously formulated in Ribeiro & Pereira (2007):

$$v[n] = v(t)|_{t=\frac{n}{f_s}} = f[n] + h[n] + i[n] + t[n] + r[n] \tag{1}$$

where $n = 0, \cdots, N-1$, f_s is the sampling frequency, the sequences $f[n]$, $h[n]$, $i[n]$, $t[n]$ and $r[n]$ are the fundamental component, harmonics, interharmonics, transient and background noise, respectively. Each of these signals is defined as follows:

$$f[n] = A_0[n]\cos[2\pi\frac{f_0[n]}{f_s}n + \theta_0[n]], \tag{2}$$

$$h[n] = \sum_{m=1}^{M} h_m[n], \tag{3}$$

$$i[n] = \sum_{j=1}^{J} i_j[n], \tag{4}$$

$$t[n] = t_{imp}[n] + t_{not}[n] + t_{osc}[n], \tag{5}$$

and $r[n]$ is independently and identically distributed (i.i.d.) normal noise $\mathcal{N}(0, \sigma_n^2)$ and independent of $f[n]$, $h[n]$, $i[n]$ and $t[n]$. In (2), $A_0[n]$, $f_0[n]$ and $\theta_0[n]$ refer to the magnitude, fundamental frequency, and phase of the fundamental component, respectively. In (3) and (4), $h_m[n]$ and $i_j[n]$ are the mth harmonic and the jth interharmonic, respectively, which are defined as:

$$h_m[n] = A_m[n]\cos[2\pi m\frac{f_0[n]}{f_s}n + \theta_m[n]][u[n - n_{h_{m,i}}] - u[n - n_{h_{m,f}}]], \tag{6}$$

and

$$i_j[n] = A_{I,j}[n]\cos[2\pi\frac{f_{I,j}[n]}{f_s}n + \theta_{I,j}[n]][u[n - n_{i_{j,i}}] - u[n - n_{i_{j,f}}]], \tag{7}$$

in which $u[n]$ denote the unit step sequence, $n_{h_{m,i}}$ and $n_{h_{m,f}}$ refer to the start and end samples of the harmonics, respectively; Similarly, $n_{i_{j,i}}$ and $n_{i_{j,f}}$ refer to the start and end samples of the interharmonics, respectively. In (6), $A_m[n]$ is the magnitude and $\theta_m[n]$ is the phase of the mth harmonic. In (7), $A_{I,j}[n], f_{I,j}[n]$, and $\theta_{I,j}[n]$ are the magnitude, frequency, and phase

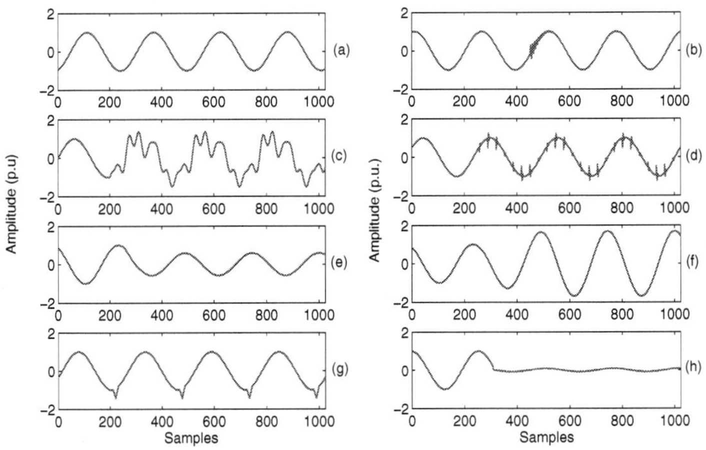

Fig. 2. Examples of voltage signals and disturbances: (a) nominal voltage signal, (b) oscillatory transient, (c) harmonics, (d) notching, (e) sag, (f) swell, (g) spikes and (h) outage.

of the jth interharmonic, respectively. In (5), $t_{imp}[n]$, $t_{not}[n]$ and $t_{osc}[n]$ denote impulsive transients, named spikes, notching and oscillatory transients, which can be expressed by (Ribeiro & Pereira, 2007):

$$t_{imp}[n] = \sum_{i=1}^{N_{imp}} t_{imp,i}[n][u[n - n_{t_{imp,i}}] - u[n - n_{t_{imp,f}}]], \tag{8}$$

$$t_{not}[n] = \sum_{i=1}^{N_{not}} t_{not,i}[n][u[n - n_{t_{not,i}}] - u[n - n_{t_{not,f}}]], \tag{9}$$

$$t_{osc}[n] = \sum_{i=1}^{N_{osc}} A_{osc,i}[n]exp[-\alpha_{osc,i}[n - n_{osc,i}]][u[n - n_{t_{osc,i}}] - u[n - n_{t_{osc,f}}]], \tag{10}$$

where $t_{imp,i}[n]$ and $t_{not,i}[n]$ are the nth samples of the ith spike and the ith notching, respectively. $n_{t_{imp,i}}$, $n_{t_{not,i}}$ and $n_{t_{osc,i}}$ denote the start of each impulsive transient and $n_{t_{imp,f}}$, $n_{t_{not,f}}$ and $n_{t_{osc,f}}$ denote the end of each.

Based in these equations, a nominal voltage waveform is shown in Figure 2(a), and some corrupted voltage waveforms are displayed in Figure 2(b)-(h), which are: (b) oscillatory transient, (c) harmonics, (d) notching, (e) sag, (f) swell, (g) spikes and (h) outage.

3. Higher-order statistics

Higher-Order Statistics (HOS) have been found wide applicability in many fields: sonar, radar, plasma physics, biomedicine, seismic data processing, image reconstruction, harmonic retrieval, time delay estimation, adaptive filtering, array processing and blind equalization (Mendel, 1991). An important example of HOS features is the cumulant, which is defined for various orders. The main characteristic of cumulants is to be blind to Gaussian processes.

Hence, cumulant-based signal processing tools can handle colored Gaussian measurement noise more efficiently.

The expressions of the second-, third- and fourth-order cumulants of a real and random sequence $\{x[n]\}$, such that $E\{x[n]\} = 0$, are expressed by Mendel (1991)

$$c_{2,x}[i] = E\{x[n]x[n+i]\}, \tag{11}$$

$$c_{3,x}[i] = E\{x[n]x^2[n+i]\}, \tag{12}$$

and

$$c_{4,x}[i] = E\{x[n]x^3[n+i]\} - 3c_{2,x}[i]c_{2,x}[0]. \tag{13}$$

For a finite N-length vector, the stochastic approximations (Kushner & Yin, 2003) can offer the following expressions:

$$\hat{c}_{2,x}[i] = \frac{2}{N}\sum_{n=0}^{N/2-1} x[n]x\,[n+i]\,, \tag{14}$$

$$\hat{c}_{3,x}[i] = \frac{2}{N}\sum_{n=0}^{N/2-1} x[n]x^2\,[n+i] \tag{15}$$

and

$$\hat{c}_{4,x}[i] = \frac{2}{N}\sum_{n=0}^{N/2-1} x[n]x^3\,[n+i] - \frac{2}{N^2}\sum_{n=0}^{N/2-1} x[n]x\,[n+i]\sum_{n=0}^{N/2-1} x^2[n], \tag{16}$$

respectively, where $i = 0, 1, \cdots, N/2-1$, in which i is called the sample lag. Note that equations (14)-(16) can not be used if $i > N/2-1$ because $n+i$ surpasses N, thus, alternative approximations were introduced in Ribeiro & Pereira (2007), considering $i = 0, 1, \cdots, N$. These approximations are expressed as:

$$\hat{c}_{2,x}[i] = \frac{1}{N}\sum_{n=0}^{N-1} x[n]x\,[\mathrm{mod}(n+i,N)]\,, \tag{17}$$

$$\hat{c}_{3,x}[i] = \frac{1}{N}\sum_{n=0}^{N-1} x[n]x^2\,[\mathrm{mod}(n+i,N)] \tag{18}$$

and

$$\hat{c}_{4,x}[i] = \frac{1}{N}\sum_{n=0}^{N-1} x[n]x^3\,[\mathrm{mod}(n+i,N)] - \frac{1}{N^2}\sum_{n=0}^{N-1} x[n]x\,[\mathrm{mod}(n+i,N)]\sum_{n=0}^{N-1} x^2[n], \tag{19}$$

respectively, where $\mathrm{mod}(n+i,N)$ is the integer rest of the division of $n+i$ by N. The approximations presented in (17-19) lead to a very appealing approach for problems where one has a finite-length vector from which higher-order-based features are extracted for applications. One has to note that the use of $\mathrm{mod}(\cdot)$ operator means that we are considering that the vector is an N-length periodic vector. The rationale is that by using such very simple assumption, we can evaluate the approximation of HOS with all available N samples. Note that this is a heuristic approach that had emerged from the periodically nature of the voltage signals in power systems.

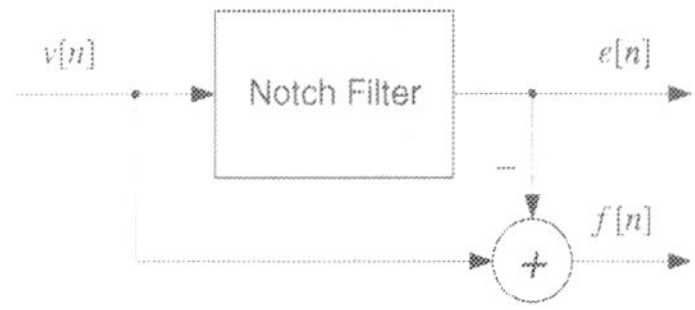

Fig. 3. Notch filter for the voltage decomposition into the fundamental and transient components.

4. HOS-based features for power quality monitoring

In PQ monitoring, the feature extraction based on HOS could be performed directly over voltage signals, as proposed in Ferreira, Cerqueira, Duque & Ribeiro (2009) or after pre-processing as proposed in Ribeiro et al. (2006). The pre-processing step follows the idea of signal decomposition and can be implemented by a notch filter-based methodology, which divides the acquired voltage signal $v[n]$ into two derived signals, $e[n]$ and $f[n]$, as shown in Figure 3, where the signal $e[n]$ is the remaining of $v[n]$ after filtering the fundamental component and $f[n] = v[n] - e[n]$ is an estimation of the fundamental component.
Due to the low computational cost and the reasonable selectivity in the frequency of interest, an IIR filter structure of second order has been used to design the notch filter (Hirano et al., 1974). The transfer function of the notch filter in z-domain is given by:

$$H_0(z) = \frac{1 + a_0 z^{-1} + z^{-2}}{1 + \rho_0 a_0 z^{-1} + \rho_0^2 z^{-2}}; \tag{20}$$

in which $a_0 = -2\cos\omega_0$, ω_0 is the notch frequency, and ρ_0 is the *notch* factor, with $0 \ll \rho_0 < 1$. On the other hand, the usage of a non-adaptive notch filter may generate erroneous results if power frequency deviation occurs. Actually, the power frequency normally varies very slowly over a small frequency range, however for some power systems the frequency variation can be large, about 2% of its nominal value (IEEE, 2008). For resolving this problem, the usage of the *enhanced phase locked-loop* (EPLL) technique (K.-Ghartemani & Iravani, 2004), that controls the notch frequency, is an interesting solution and it is suggested for those scenarios where the power frequency variation is expected.

4.1 Application of HOS for feature extraction

In order to illustrate the efficiency of the HOS feature extraction for PQ monitoring, 200 events of each disturbance class (notching, spike, harmonics, outage, sag, swell and oscillatory transient) and 200 nominal voltage signals were generated following the recommendations of the IEEE standard (IEEE, 1995). A sampling frequency of 15,360 Hz and a signal to noise ratio (SNR) of 30 dB were considered. These signals were applied to the decomposition system shown in Figure 3.
Let us first analyze the signal $e[n]$. The signal was segmented into non-overlap frames with $N = 1,024$ samples (4 cycles of the fundamental component). Hence, the expressions (17) and (19) were applied to these frames and a feature vector $\mathbf{p} = [c_{2,e} \;\; c_{4,e}]$ was obtained. It is important to point out that, for PQ events, the second and forth order cumulants can achieve better results with respect to the third-order cumulant, as it was been shown in Ferreira, Cerqueira, Duque & Ribeiro (2009). Therefore, for the present application, results were obtained considering only second- and forth-order cumulants. As a result of this, a total of $2 \times N$ features were extracted for each event.

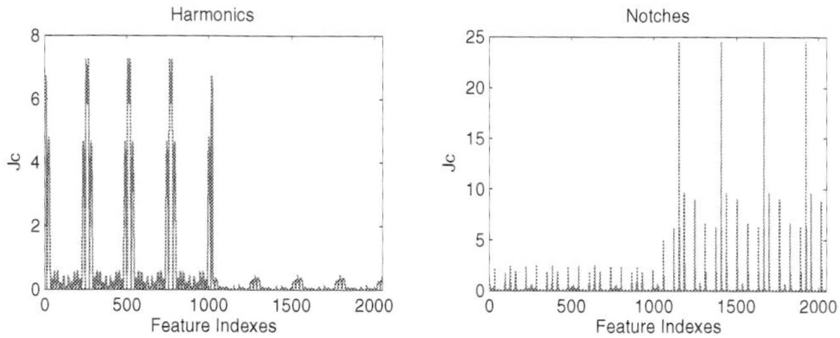

Fig. 4. Values of the FDR criterion attained for the harmonics and notching classes.

The HOS feature extraction leads to a high-dimensional feature space ($2 \times N$). Therefore, a feature selection must be performed in order to maximize the separation border between classes and also to reduce the dimension of the feature space and consequently the computational burden and processing time. In this context, some feature selection techniques (Duda et al., 2000) can be used. Recent works, such as de Aguiar et al. (2009); Ferreira, Cerqueira, Duque & Ribeiro (2009); Ferreira, de Seixas & Cerqueira (2009) have used the Fisher's discriminant ratio (FDR) (Duda et al., 2000) for feature selection, which is given by

$$\mathbf{J}_c = \mathbf{\Lambda}_{\mu_0,\mu_1}\mathbf{\Lambda}^{-1}_{\sigma_0,\sigma_1}, \tag{21}$$

in which

$$\mathbf{\Lambda}_{\mu_o,\mu_1} = diag\{(\mu_{0,1} - \mu_{1,1})^2, (\mu_{0,2} - \mu_{1,2})^2, \ldots, (\mu_{0,N} - \mu_{1,N})^2\} \tag{22}$$

and

$$\mathbf{\Lambda}_{\sigma_o,\sigma_1} = diag\{(\sigma^2_{0,1} + \sigma^2_{1,1}), (\sigma^2_{0,2} + \sigma^2_{1,2}), \ldots, (\sigma^2_{0,N} + \sigma^2_{1,N})\}. \tag{23}$$

Assuming that $\mathbf{x}_{FDR}$ is constituted by the element in the main diagonal of the matrix $\mathbf{J}_c$, such that $x_{FDR}(1) > x_{FDR}(2) > \cdots > x_{FDR}(N)$, then, a set of k features associated with the k highest values in the vector $\mathbf{x}_{FDR}$ can be selected.

Figure 4 illustrates the FDR ($\mathbf{J}_c$) for the harmonics against all other classes and notching against all other classes, obtained using the feature vector $\mathbf{p}$. The first indexes (1...1,024) comprise the second order cumulants ($c_{2,e}$), and the remaining comprise the fourth order cumulants ($c_{4,e}$). These examples point out that for some classes, the second order cumulant is more discriminative than fourth order, as it could be seen for harmonics and the other way round for other classes, as it can be seen for notching disturbances. Therefore, the combination of both second and fourth order cumulants is powerful, since they carry distinct information, as discussed in Mendel (1991).

Figure 5 shows the discrimination capability of the second and fourth order cumulants. Analyzing the events in the feature space, it is possible to notice that notching, nominal voltage waveforms, sags, swells and outages are more homogeneous classes, while harmonics, spikes and oscillatory transients classes are scattered in the feature space. It is also important to notice that there are only interceptions between the nominal voltage, sag and swell signals. Therefore, most classes may be separated using just these two features.

Additionally, the usage of the information related to the fundamental component ($f[n]$) may lead to a better separation between the nominal voltage, sag and swell signals. Figure 6 shows the feature space obtained with cumulants that were extracted from the fundamental

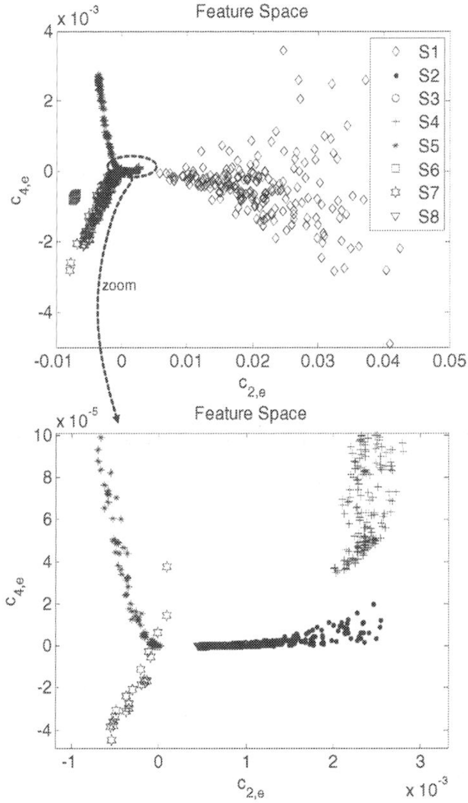

Fig. 5. Feature space for: (S1) harmonics, (S2) sag, (S3) swell, (S4) outage, (S5) spike, (S6) notching, (S7) oscillatory transient and (S8) nominal voltage waveform.

component. In this new feature space it is easy to recognize the nominal voltage, sag and swell classes.

5. Disturbance detection and classification based on HOS

In this section, the HOS based features are used for automatic detection and classification of PQ disturbances. Once the cumulant based features are extracted from the incoming signal, the next step consists in applying the detection and classification techniques. At this point, it is important to consider the computational complexity of the chosen techniques. In general, the techniques with high performance may lead to large computational cost. Then, the challenge is to develop a low-complexity technique that achieves high performance.

5.1 PQ disturbances detection using HOS

The aim of the detection techniques is to provide a real-time and source reliable detection of a variety of disturbances, so that event classification and underlying identification can be both

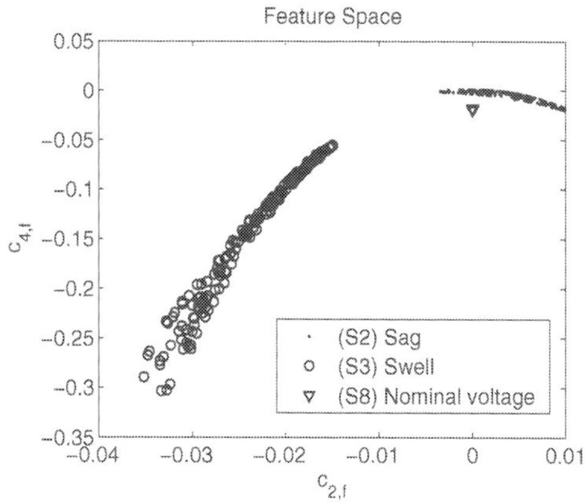

Fig. 6. Feature space for: (S2) sag, (S3) swell and (S8) nominal voltage waveform.

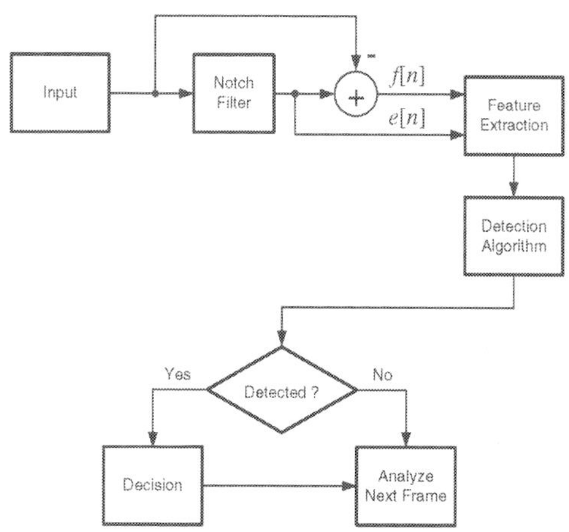

Fig. 7. Detection system.

achieved. Several methods have been proposed in the literature and the most used techniques are based on wavelet transforms (WT) (Chen et al., 2009; Lin et al., 2008; Wang & Wang, 2007; Yang & Liao, 2001). However, the attained results with WT may be seriously affected by the system noise (Yang & Liao, 2001). Other methods that may be mentioned include S-transform (Bhende et al., 2008; Mishra et al., 2008), Hilbert transform (Chun-Ling et al., 2009), fractals (Li et al., 2005) and support vector machines (Moraveja et al., 2010). Each of these techniques have advantages and disadvantages. Disturbance detection based on HOS have the following characteristics: i) it is more insensitive to the presence of background noise; and ii) it is capable of detecting the occurrence of disturbances in frames corresponding to 1/16 of the fundamental component. As a result, the HOS-based techniques can be used in noisy applications and situations where the detection of disturbances in frames whose lengths correspond to submultiples or multiples of one fundamental cycle is needed.

Figure 7 portrays the block diagram of a HOS-based detection technique proposed in Ribeiro et al. (2007). In this diagram, the Input block contains discrete samples of the power line signal and the block NF0 implements a infinite impulse response (IIR) digital notch filter given by Equation (20). Subsequently, two discrete signals are generated which are, the fundamental component $f[n]$ and the error component $e[n]$. Then, 2nd and 4th order's cumulants of N-length vectors constituted by samples of $f[n]$ and $e[n]$ are extracted by the Feature Extraction block. During the design stage, the Fisher's criterion is applied in order to select the best features, as explained in Section 4. Then, in the operational stage, only the previously chosen features are computed. Finally, the Detection Algorithm block performs the decision by using a Bayesian detector (Trees, 2001) based on maximum likelihood criterion (Theodoridis & Koutroumbas, 2006).
Considering the vectors $\mathbf{f} = [f[n] \cdots f[n-N-1]]^T$ and $\mathbf{e} = [e[n] \cdots e[n-N-1]]^T$ built from samples of the signals $f[n]$ and $e[n]$, respectively, the detection problem can be formulated as a hypothesis test problem (Ribeiro et al., 2007).

$$\begin{aligned}
\mathcal{H}_1 &: \mathbf{e} = \mathbf{r}_e \\
\mathcal{H}_2 &: \mathbf{f} = \mathbf{f}_{ss} + \mathbf{r}_f \\
\mathcal{H}_3 &: \mathbf{e} = \mathbf{i} + \mathbf{t} + \mathbf{h} + \mathbf{r}_e \\
\mathcal{H}_4 &: \mathbf{f} = \mathbf{f}_{ss} + \Delta\mathbf{f}_{ss} + \mathbf{r}_f,
\end{aligned} \tag{24}$$

where $\mathbf{i} = [i[n] \cdots i[n-N-1]]^T$, $\mathbf{t} = [t[n] \cdots t[n-N-1]]^T$, $\mathbf{h} = [h[n] \cdots h[n-N-1]]^T$, $\mathbf{r}_e + \mathbf{r}_f = \mathbf{r} = [r[n] \cdots r[n-N-1]]^T$. The vector $\Delta\mathbf{f}_{ss}$ represents a sudden variation in the fundamental component and the vector $\mathbf{f}_{ss}$ denotes the steady-state component of the fundamental component. The hypotheses formulation introduced in (24) emphasizes the need to analyze abnormal events through the so-called primitive components of voltage signals that are represented by the vectors $\mathbf{f}$ and $\mathbf{e}$. While the hypotheses $\mathcal{H}_1$ and $\mathcal{H}_2$ are related to standard operation, both hypotheses $\mathcal{H}_3$ and $\mathcal{H}_4$ are associated with abnormal conditions. Equation (24) means that we are looking for some kind of abnormal behavior in one or two primitive components of the input signal, so that a decision about disturbance occurrences can be accomplished. This concept is very attractive, because the vectors $\mathbf{f}_{ss} + \Delta\mathbf{f}_{ss} + \mathbf{r}_f$ and $\mathbf{i} + \mathbf{t} + \mathbf{h} + \mathbf{r}_e$ can reveal insightful and different information from the voltage signals.
Although four hypotheses are given in (24), for the detection problem we can consider only two hypotheses: the hypothesis $\mathcal{H}_a = \mathcal{H}_1 \cup \mathcal{H}_2$ which comprises standard operational condition of the monitored voltage signal and hypothesis $\mathcal{H}_b = \mathcal{H}_3 \cup \mathcal{H}_4$, which comprises abnormal conditions (disturbances). Based on the Bayes decision theory (Theodoridis & Koutroumbas, 2006), the detection through the vector $\mathbf{p}$, which was selected by the FDR, can be performed as follows:

$$\frac{p(\mathbf{p}|\mathcal{H}_b)}{p(\mathbf{p}|\mathcal{H}_a)} \underset{\mathcal{H}_b}{\overset{\mathcal{H}_a}{\gtrless}} \frac{P(\mathcal{H}_a)}{P(\mathcal{H}_b)}, \tag{25}$$

where $P(\mathcal{H}_i)$, $i = a, b$, represents the *a priori* probability and $p(\mathbf{p}|\mathcal{H}_i)$ represents the conditional probability density function of the class $\mathcal{H}_i$. The conditional probability density function used here is expressed by

$$p(\mathbf{p}|\mathcal{H}_i) = \frac{1}{(2\pi)^{L/2}|\sum_i|^{1/2}} e^{-\frac{1}{2}(\mathbf{p}-\mu_i)^T \sum_i^{-1}(\mathbf{p}-\mu_i)}, \tag{26}$$

N	Detection Rates (%)
256	100
128	100
64	99.8
32	99.8
16	98.6

Table 1. Detection rates for disturbance detection

where $\mu_i = E\{\mathbf{p}\}$ is the average vector of the class $\mathcal{H}_i$, $\sum_i$ is the covariance matrix of the same class defined as

$$\sum_i = E\{(\mathbf{p} - \mu_i)(\mathbf{p} - \mu_i)^T\}, \tag{27}$$

and $|\sum_i|$ denotes the determinant of $\sum_i$. Note that μ_i and $\sum_i$ are obtained in the design stage. Supposing that $P(\mathcal{H}_a)$=$P(\mathcal{H}_b)$=1/2 and assuming the probability density functions referred in (26) the detector described in (25) assumes the following form:

$$\frac{|\sum_a|^{\frac{1}{2}} e^{-\frac{1}{2}(\mathbf{x}-\mu_b)^T \sum_b^{-1}(\mathbf{p}-\mu_b)}}{|\sum_b|^{\frac{1}{2}} e^{-\frac{1}{2}(\mathbf{p}-\mu_a)^T \sum_a^{-1}(\mathbf{p}-\mu_a)}} \underset{\mathcal{H}_b}{\overset{\mathcal{H}_a}{\gtrless}} 1, \tag{28}$$

Thus, the left side of (28) is applied to the feature vector, and if the evaluated value is higher or equal to 1, a disturbance in the voltage signal is detected, otherwise, the voltage signal is considered to be without any disturbance.

5.1.1 Results

To verify the performance of the detection technique, simulations were carried out with several waveforms of voltage signals with signal-to-noise rate (SNR) equals to 30 dB and sampling rate $f_s = 256 \times 60$ Hz. The generated disturbances, with 600 waveforms each, were sag, swell, outages, oscillatory transient, notching, spikes and harmonics. In order to show how detection efficiency deteriorates with a reduced number of samples, the number of samples used to detect the disturbances were N= 256, 128, 64, 32 and 16 samples. The notch factor of the notch filter was 0.997. The achieved results in terms of detection rates are presented in Table 1. It is important to note that detection rates are higher than 98 %, even when only 16 samples are considered.

5.2 PQ disturbance classification using HOS

An important step in designing pattern recognition systems is the feature extraction, which aims to find the best features $\mathbf{p}$ envisaging classes separation on the feature space. The application of cumulant-based features for disturbance classification, proved to be efficient, as from the results obtained by Ferreira, Cerqueira, Duque & Ribeiro (2009). However, this work did not consider power line signals corrupted by various disturbances occurring simultaneously, i.e., multiple disturbances. In this context, Ribeiro & Pereira (2007) proposed the principle of divide and conquer, which was applied to decompose an electric signal into a set of primitive components for classification of single and multiple disturbances in electric network. In the present chapter, the main goal of the pre-processing is to decouple the multiple disturbances into single disturbances before classifying. This procedure is motivated

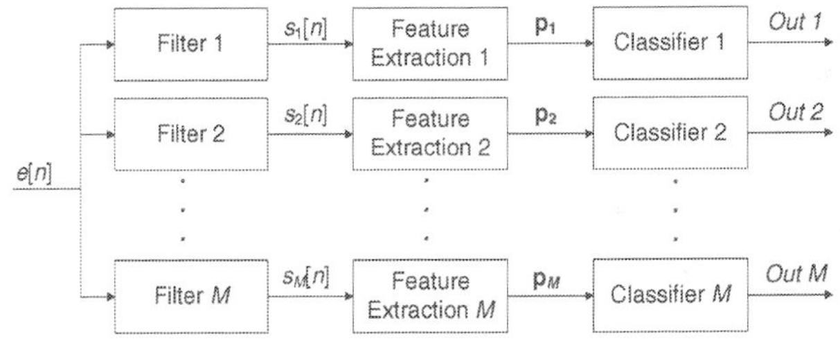

Fig. 8. Filter bank for disturbance decoupling.

by the assumption that the voltage signal is composed by the additive contribution of several types of disturbances, as formulated in (1).

Nevertheless, a digital filter bank may be used to decouple multiple disturbances (Ferreira et al., 2010). According to IEEE (1995), each disturbance class is well defined in terms of specific variables, such as magnitude, frequency range, and others. Hence, a well defined set of simulated disturbances may provide consistent spectral information about each class and, then, a simple and efficient filter bank can be designed. Figure 8 illustrates the filtering approach. The signal $e[n]$ is firstly filtered and the output of each filter is individually analyzed. Each classifier can be designed to be assign to a specific class or a reduced group of classes. Finally, the outputs *Out 1*, *Out 2*, ..., *Out M* feed a final logic which defines the type of disturbance (multiple or single) presented in $e[n]$. The final logic may also incorporate information based on $f[n]$, which is very important to separate standard events from sags and swells, as shown in Section 4.

5.2.1 Classifier design

A block diagram of the automatic classification system proposed in Ferreira et al. (2010) can be seen in Figure 9. The disturbances related to the fundamental component (sags and swells) are handled directly using second and fourth order cumulants. The filter bank was designed using the spectrum content of the disturbances related to the error signal $e[n]$ (harmonics, transients and notching). In such a way, the majority of the energy from the harmonic is presented at the output of Filter 1 ($s_1[n]$), which is a low-pass filter with cut-off frequency f_C=500 Hz. The high-pass filter (Filter 3) with f_c=3 kHz selects the disturbances with high frequencies in its output ($s_3[n]$), which in this case corresponds mainly to notching class of disturbance. Filter 2 is a band-pass filter with f_{Ci}=500 Hz and f_{Cs}=3 kHz, which basically reduces the energy from harmonics and notching from the remaining disturbances (oscillatory transients and spikes) at its output.

Considering the filtering approach, the classification problem can be formulated as the following:

(i) From signal $f[n]$, the hypothesis test for disturbance classification for the fundamental component becomes:

$$\begin{aligned}
\mathcal{H}_{f,1} &: \mathbf{f} = \mathbf{f}_{ss} + \mathbf{r}_f \\
\mathcal{H}_{f,2} &: \mathbf{f} = \mathbf{f}_{under} + \mathbf{r}_f \\
\mathcal{H}_{f,3} &: \mathbf{e} = \mathbf{f}_{over} + \mathbf{r}_f \\
\mathcal{H}_{f,4} &: \mathbf{f} = \mathbf{f}_{inter} + \mathbf{r}_f,
\end{aligned} \tag{29}$$

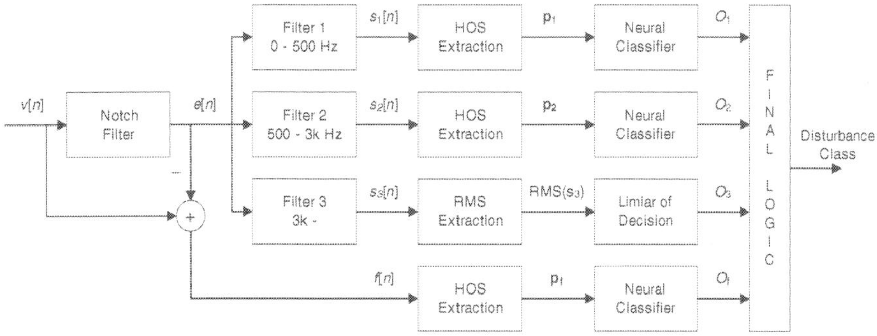

Fig. 9. Classification system.

where the vectors $\mathbf{f}_{under}$, $\mathbf{f}_{over}$, and $\mathbf{f}_{inter}$ denote an undervoltage or sag, a disturbance called overvoltage or swell, and a disturbance named sustained interruption or outage, respectively.

(ii) From signal $s_1[n]$, disturbance classification for the error component is formulated as:

$$\begin{aligned} \mathcal{H}_{s_1,1} &: \mathbf{s}_1 = \mathbf{r}_{s_1} \\ \mathcal{H}_{s_1,2} &: \mathbf{s}_1 = \mathbf{h} + \mathbf{r}_{s_1}, \end{aligned} \tag{30}$$

where $\mathbf{s}_1 = [s_1[n] \cdots s_1[n-N-1]]^T$ and $\mathbf{r}_{s_1}$ is the filtered version of the noise vector $\mathbf{r}$ by Filter 1.

(iii) From signal $s_2[n]$, the classification of disturbances in the error component is formulated as:

$$\begin{aligned} \mathcal{H}_{s_2,1} &: \mathbf{s}_2 = \mathbf{r}_{s_2} \\ \mathcal{H}_{s_2,2} &: \mathbf{s}_2 = \mathbf{t}_{osc} + \mathbf{r}_{s_2} \\ \mathcal{H}_{s_2,3} &: \mathbf{s}_2 = \mathbf{t}_{imp} + \mathbf{r}_{s_2}, \end{aligned} \tag{31}$$

where $\mathbf{t}_{osc} = [t_{osc}[n] \cdots t_{osc}[n-N-1]]^T$, $\mathbf{t}_{imp} = [t_{imp}[n] \cdots t_{imp}[n-N-1]]^T$ and $\mathbf{r}_{s_2}$ is the filtered version of the noise vector $\mathbf{r}$ by Filter 2.

(iv) From signal $s_3[n]$, disturbance classification for the error component is formulated as:

$$\begin{aligned} \mathcal{H}_{s_3,1} &: \mathbf{s}_3 = \mathbf{r}_{s_3} \\ \mathcal{H}_{s_3,2} &: \mathbf{s}_3 = \mathbf{t}_{not} + \mathbf{r}_{s_3}, \end{aligned} \tag{32}$$

where $\mathbf{t}_{not} = [t_{osc}[n] \cdots t_{not}[n-N-1]]^T$ and $\mathbf{r}_{s_3}$ is the filtered version of the noise vector $\mathbf{r}$ by Filter 3.

The three filters were designed as IIR (Infinite Impulse Response) (Mitra, 2005) of fourth order (see Equation (33)). The elliptic approximation was used for designing the filters. Elliptic filters have an equiripple pass-band and an equiripple stop-band. Because the ripples are distributed uniformly across both bands, these filters are optimum in the sense of having the smallest *transition* width for a given filter order, cut-off frequency and pass-band and stop-band ripples (Mitra, 2005).

$$H(z) = \frac{b_0 + b_1 z^{-1} + \cdots + b_4 z^{-4}}{a_0 + a_1 z^{-1} + \cdots + a_4 z^{-4}} \tag{33}$$

The block diagram (Figure 9) shows that the HOS features are extracted for $s_1[n]$, $s_2[n]$ and $f[n]$. As signal $s_3[n]$ is mainly composed by notching, a simple feature extraction was used (the root mean squared value (RMS)). As for the detection system presented in Section 5.1, the 2nd and 4th order's HOS features of N-length vectors constituted by samples of $s_1[n]$, $s_2[n]$ and $f[n]$ were extracted and the Fisher's criterion was applied in order to select the best features.

The Bayesian classifier minimizes the error probability, however, not all problems are well suited to such approaches as the involved probability density functions are complicated and their estimation is not an easy task. In such cases, it may be preferable to compute decision surfaces directly by means of alternative costs, as is the case of the neural networks, support vector machines, etc. Therefore, for the classification of power quality disturbances, several works use neural networks, support vector machines, fuzzy classifiers, decision threes, among others.

In Ferreira et al. (2010), for each vector of extracted features given by $\mathbf{p}_1$, $\mathbf{p}_2$ and $\mathbf{p}_3$, an expert pattern recognition system was be used. Due to its good ability to distinguish disturbances, a reduced number of cumulants is enough, as discussed in Section 4. Consequently, simple neural classifiers may be used. This is an important advantage in classification systems, mainly for real-time applications. The three pattern recognition systems used were from multilayer feedforward artificial neural networks (Haykin, 2009). The neural networks comprise a single hidden layer. The RPROP algorithm (Riedmiller & Braun, 1993) was used to train the neural classifiers. The hyperbolic tangent was used as activation function. For $RMS(s_3)$, a simple threshold was used.

The Final Logic block combines the classifier outputs O_1, O_2, O_3 and O_f, using a logical operation based on the logical gate AND. Thus, a large group of multiple disturbance classes can be easily incorporated by the Final Logic block.

5.2.2 Results

The following disturbance classes were considered in this application: outages; harmonics; sags; swells; oscillatory transients; notching; spikes; sag with harmonics; swell with harmonics; sag with oscillatory transient; swell with oscillatory transient; sag with notching; swell with notching; notching with harmonics; oscillatory transient with harmonics; sag with oscillatory transient and harmonics; sag with notching and harmonics; swell with notching and harmonics; and swell with oscillatory transient and harmonics. Figure 2 (b)-(h) illustrates examples of single disturbances. Examples of multiple disturbances are illustrated in Figure 10.

Five hundred events from each class were simulated. Three hundred of each class were used to design the classifier and the remaining data were used for testing. The classification design comprises the design of filters, the feature selection and the neural training. The achieved results can be seen in Table 2.

The main advantage of this system is its capability of classifying multiple disturbances with reasonable efficiency. Eight classes comprising two simultaneous disturbances and four classes formed by three simultaneous disturbances were correctly classified with efficiencies above 97.2 %, as shown in Table 2. Additionally, others classes of multiple disturbances can be addressed by combining the classifier outputs O_1, O_2, O_3 and O_f through the Final Logic.

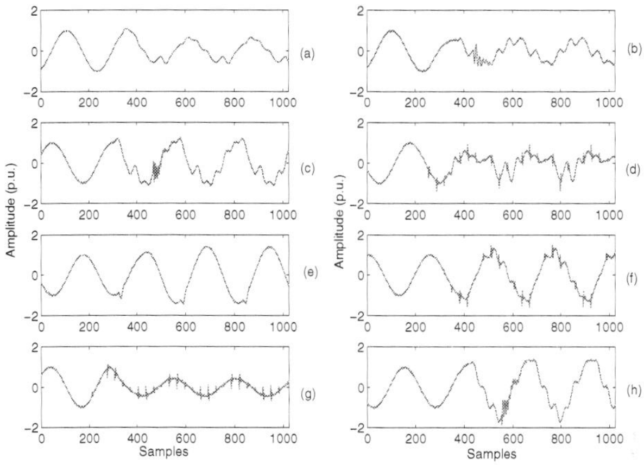

Fig. 10. Examples of multiple disturbances: (a) sag with harmonics, (b) sag with oscillatory transient and harmonics, (c) oscillatory transient with harmonics, (d) sag with notching and harmonics, (e) swell with spikes, (f) swell with notching and harmonics, (g) sag with notching, and (h) swell with oscillatory transient and harmonics.

Disturbance Classes	Efficiency in %
Outage	100
Harmonics	99.0
Sag	100
Swell	100
Oscillatory Transient	99.0
Notching	100
Spike	100
Sag + harmonics	99.0
Swell + harmonics	97.2
Sag + oscillatory transient	100
Swell + oscillatory transient	99.4
Sag + notching	100
Swell + notching	99.0
Notching + harmonics	99.4
Oscillatory transient + harmonics	98.2
Sag + oscillatory transient + harmonics	98.2
Sag + notching + harmonics	98.4
Swell + notching + harmonics	98.8
Swell + oscillatory transient + harmonics	97.8

Table 2. Classification results based on a filter bank.

6. Conclusions

The performance of the PQ monitoring system is directly related to the pre-processing and feature extraction techniques used. Therefore, the identification of efficient pre-processing and feature extraction techniques is very important.
The usage of HOS as a feature extraction technique for PQ monitoring systems is very promising and several recent works presented good results with respect to both detection and classification tasks.
The main advantages of the HOS as a feature extraction technique is its immunity to Gaussian noise and also the capability to reveal non-linear characteristics from the data, which is important for pattern recognition applications.
Some results shown that HOS is able to detect disturbances even using short acquisition time windows, which represents an important characteristic for several power systems applications such as protection, signal segmentation and disturbance localization. Being specific, the results shown that the detection of disturbances can be accomplished in less than a quarter of cycle, which is excellent for protection application, where speed and accuracy need to be combined to guarantee selectivity and reliability during the occurrence, for example, of a fault in a system.
Regarding the usage of HOS in PQ classification, the results shown that combining techniques allows efficient classification of single and simultaneous disturbances, and more, the usage of the second and fourth orders HOS features, for a specific lag chosen from the FDR criterion, has been enough to deal with the majority of the disturbances considered.

7. Future trends

Several problems in PQ are still open. Among them, load identification and source localization, both related to each other. Given that the PQ Analyzer has detected and classified a disturbance, what kind of event and load have caused that problem?
For example, if a sag is detected and classified, the next step is answering if that sag was generated by a fault in the system or by a start of a big motor. If a transient is detected, what kind of event has caused it, a fault or a switching capacitor bank? Can HOS be used to overcome this problem? These questions are under investigation at the moment.
Other promising application of HOS is in protection issues. HOS can be used in fault detection, classification and localization as shown by recent works, but there are a few works in this area and several questions to be answered. The challenge is to guarantee simultaneously speed, reliability and selectivity.
Another area where HOS can bring good results is in diagnose of electrical equipments, such as transformers, motors and generators. Can cumulants, of voltage and current signals, carry useful information about the status of the equipment? All these questions surely make PQ research attractive and full of challenges.

8. References

Bhende, C., Mishra, S. & Panigrahi, B. (2008). Detection and classification of power quality disturbances using s-transform and modular neural network, *Electric Power Systems Research* Vol. 78(No. 1): 122–128.

Bollen, M. H. J., Gu, I. Y. H., Santoso, S., Mcgranaghan, M. F., Crossley, P. A., Ribeiro, M. V. & Ribeiro, P. F. (2009). Bridging the gap between signal and power, *IEEE Signal Processing Magazine* Vol. 23(No. 4): 12–31.

Chen, C., Liang, W. & Xu, T. (2009). Transient disturbance detection based on hilbert phase-shifting and wavelet de-noising, *9th International Conference on Electronic Measurement and Instruments. ICEMI'09*, 10.1109/ICEMI.2009.5274129, Beijing, pp. 4–125–4–130.

Chun-Ling, C., Yong, Y., Tongyu, X., Yingli, C. & Xiaofeng, W. (2009). Power quality disturbances detection based on hilbert phase-shifting, *Asia-Pacific Power and Energy Engineering Conference, APPEEC 2009*, 10.1109/APPEEC.2009.4918613, Wuhan, China, pp. 1–4.

de Aguiar, E. P., Marques, C. A. G., Duque, C. A. & Ribeiro, M. V. (2009). Signal decomposition with reduced complexity for classification of isolated and multiple disturbances in electric signals, *IEEE Transactions on Power Delivery* Vol. 24(No. 4): 2459–2460.

Duda, R. O., Hart, P. E. & Stork, D. G. (2000). *Pattern Classification*, Wiley-Interscience.

Ferreira, D. D., Cerqueira, A. S., Duque, C. A., de Seixas, J. M. & Ribeiro, M. V. (2010). Automatic system for classification of isolated and multiple disturbances in electric signals, *SBA Controle & Automação* (Accepted for publication).

Ferreira, D. D., Cerqueira, A. S., Duque, C. A. & Ribeiro, M. V. (2009). HOS-based method for classification of power quality disturbances, *Electronics Letters* Vol. 23(No. 3): 183–185.

Ferreira, D. D., de Seixas, J. M. & Cerqueira, A. S. (2009). ICA-based method for power quality disturbance analysis, *15th International Conference on Intelligent System Applications to Power Systems. ISAP '09*, 0.1109/ISAP.2009.5352915, Curitiba, Brazil, pp. 1–6.

Gaouda, A. M., Salama, M. M. A., Sultan, M. R. & Chikhani, A. Y. (1999). Power quality detection and classification using wavelet-multiresolution signal decomposition, *IEEE Transactions on Power Delivery* Vol. 14(No. 4): 2459–2460.

Haykin, S. (2009). *Neural networks and Learning Machines*, NJ: Englewood Cliffs, Prentice Hall.

Hirano, K., Nishimura, S. & Mitra, S. K. (1974). Design of digital notch filters, *IEEE Transactions on Communications* Vol. 22(No. 7): 964–970.

IEEE (1995). IEEE recommended practice for monitoring electric power quality, *Technical report*, IEEE Standards Coordinating Committee 22 on Power Quality.

IEEE (2008). Ieee guide for the design and application of power electronics in electrical power systems on ships, *Technical report*, IEEE Industry Applications Society.

K.-Ghartemani, M. & Iravani, M. R. (2004). Robust and frequency-adaptive measurement of peak value, *IEEE Transaction on Power Delivery* Vol. 19(No. 2): 481–489.

Kushner, H. J. & Yin, G. G. (2003). *Stochastic Approximation and Recursive Algorithms and Applications*, New York: Springer-Verlag.

Li, G., Zhou, M., Luo, Y. & Ni, Y. (2005). Power quality disturbance detection based on mathematical morphology and fractal technique, *2005 IEEE/PES Transmission and Distribution Conference and Exhibition: Asia and Pacific*, 10.1109/TDC.2005.1547030, Dalian, pp. 1–6.

Lin, W.-M., Wu, C.-H., Lin, C.-H. & Cheng, F.-S. (2008). Detection and classification of multiple power-quality disturbances with wavelet multiclass svm, *IEEE Transaction on Power Delivery* Vol. 23(No. 4): 2575–2582.

Mallat, S. (1999). *A Wavelet Tour of Signal Processing*, New York: Academic.

Mendel, J. M. (1991). Tutorial on higher-order statistics (spectra) in signal processing and system theory: theoretical results and some applications, *Proceedings of the IEEE* Vol. 79(No. 3): 278–305.

Mishra, S., Bhende, C. N. & Panigrahi, B. K. (2008). Detection and classification of power quality disturbances using s-transform and probabilistic neural network, *IEEE Transactions on Power Delivery* 23(1): 280–287.

Mitra, S. K. (2005). *Digital Signal Processing*, McGraw-Hill Science.

Moraveja, Z., Abdoosa, A. A. & Pazokia, M. (2010). Detection and classification of power quality disturbances using wavelet transform and support vector machines, *Electric Power Components and Systems* Vol. 38(No. 2): 182–196.

Morsi, W. G. & El-Hawary, M. E. (2008). A new perspective for the ieee standard 1459-2000 via stationary wavelet transform in the presence of nonstationary power quality disturbance, *IEEE Transactions on Power Delivery* Vol. 23(No. 4): 2356–2365.

Ribeiro, M. V., Marques, C. A. G., Duque, C. A., Cerqueira, A. S. & Pereira, J. L. R. (2006). Power quality disturbances detection using hos, *IEEE Power Engineering Society General Meeting*, 10.1109/PES.2006.1709131, Montreal, Que, pp. 1–6.

Ribeiro, M. V., Marques, C. A. G., Duque, C. A., Cerqueira, A. S. & Pereira, J. L. R. (2007). Detection of disturbances in voltage signals for power quality analysis using hos, *EURASIP Journal on Advances in Signal Processing* Vol. 2007(Article ID 59786): 13 pages.

Ribeiro, M. V. & Pereira, J. L. R. (2007). Classification of single and multiple disturbances in electric signals, *EURASIP Journal on Advances in Signal Processing* Vol. 2007(Article ID 56918): 18 pages.

Riedmiller, M. & Braun, H. (1993). A direct adaptive method for faster backpropagation learning: the rprop algorithm, *IEEE International Conference on Neural Networks*, 10.1109/ICNN.1993.298623, San Francisco, CA, pp. 586–591.

Samantaray, S. R. (2010). Phase-Space-Based Fault Detection in Distance Relaying, *IEEE Transaction On Power Delivery*, Vol. PP (No. 99): 1-1.

Stockwell, R. G., Mansinha, L. & Lowe, R. P. (1996). Localization of the complex spectrum: the s transform, *IEEE Transaction on Signal Processing* Vol. 44(No. 4): 998–1001.

Styvaktakis, E., Bollen, M. H. J. & Gu, I. Y. H. (2002). Expert system for classification and analysis of power system events, *IEEE Transactions on Power Delivery* Vol. 17(No. 2): 423–428.

Theodoridis, S. & Koutroumbas, K. (2006). *Pattern Recognition*, Academic Press.

Trees, H. L. V. (2001). *Detection, Estimation, and Modulation Theory*, John Wiley & Sons.

Wang, J. & Wang, C. (2007). Detection of power quality disturbance based on binary wavelet transform, *TENCON 2007 - 2007 IEEE Region 10 Conference*, 10.1109/TENCON.2007.4428928, Taipei, pp. 1–3.

Yang, H.-T. & Liao, C.-C. (2001). A de-noising scheme for enhancing wavelet-based power quality monitoring system, *IEEE Transaction on Power Delivery* Vol. 16(No. 3): 353–360.